THE OXFORD HANDBOOK OF

# NEW SCIENCE FICTION CINEMAS

# THE OXFORD HANDBOOK OF
# NEW SCIENCE FICTION CINEMAS

*Edited by*
J. P. TELOTTE

# OXFORD
UNIVERSITY PRESS

Oxford University Press is a department of the University of Oxford. It furthers
the University's objective of excellence in research, scholarship, and education
by publishing worldwide. Oxford is a registered trade mark of Oxford University
Press in the UK and certain other countries.

Published in the United States of America by Oxford University Press
198 Madison Avenue, New York, NY 10016, United States of America.

© Oxford University Press 2023

All rights reserved. No part of this publication may be reproduced, stored in
a retrieval system, or transmitted, in any form or by any means, without the
prior permission in writing of Oxford University Press, or as expressly permitted
by law, by license, or under terms agreed with the appropriate reproduction
rights organization. Inquiries concerning reproduction outside the scope of the
above should be sent to the Rights Department, Oxford University Press, at the
address above.

You must not circulate this work in any other form
and you must impose this same condition on any acquirer.

Library of Congress Cataloging-in-Publication Data
Names: Telotte, J. P., 1949– editor.
Title: The Oxford handbook of new science fiction cinemas / edited by J. P. Telotte.
Description: New York, NY : Oxford University Press, [2023] |
Series: Oxford handbooks series | Includes bibliographical references and index.
Identifiers: LCCN 2022040751 (print) | LCCN 2022040752 (ebook) |
ISBN 9780197557723 (hardback) | ISBN 9780197557754 | ISBN 9780197557747 (epub)
Subjects: LCSH: Science fiction films—History and criticism. | LCGFT: Film criticism.
Classification: LCC PN1995.9.S26 O987 2023 (print) | LCC PN1995.9.S26 (ebook) |
DDC 791.43/615—dc23/eng/20220923
LC record available at https://lccn.loc.gov/2022040751
LC ebook record available at https://lccn.loc.gov/2022040752

DOI: 10.1093/oxfordhb/9780197557723.001.0001

Printed by Integrated Books International, United States of America

# CONTENTS

| | |
|---|---|
| *Slant Screens Image Captions* | vii |
| *Acknowledgments* | xiii |
| *Contributors* | xv |

## I. INTRODUCTION

1. Introduction: Slant Screens/Slant Screenings
   J. P. TELOTTE ......... 3

## II. THE SLANT SCREENS OF NEW SCIENCE FICTION CINEMA

2. Afrofuturist Cinema
   DE WITT DOUGLAS KILGORE ......... 23

3. Biopunk Film
   LARS SCHMEINK ......... 38

4. Cli-Fi Cinema
   MARK BOULD ......... 52

5. Ethnogothic Film
   SUSANA M. MORRIS ......... 71

6. Femspec Science Fiction
   SUSAN A. GEORGE ......... 88

7. Heterotopias
   JOAN GORDON ......... 101

8. Kaiju Film
   BRADLEY SCHAUER ......... 113

9. Magical Realism Science Fiction
   GERALD DUCHOVNAY ......... 129

vi CONTENTS

10. Steampunk Cinema 141
THOMAS LAMARRE

11. Superhero Science Fiction 162
ANGELA NDALIANIS

## III. NEW SLANTS ON SCIENCE FICTION FILMS

12. The Anthropocene and Ecosophy 181
GERRY CANAVAN

13. Biopolitics and Bioethics 193
SHERRYL VINT

14. Cult Behaviors 206
JEFFREY ANDREW WEINSTOCK

15. Digital Science Fictions 220
CHUCK TRYON

16. Feminist Materialism 236
FRANCES MCDONALD

17. Object-Oriented Ontology and Science Fiction Cinema 249
LEVI R. BRYANT

18. Posthumanism 265
J. P. TELOTTE

19. Queer Cinema 278
CATHERINE CONSTABLE AND MATT DENNY

20. Utopianism 295
CAROLINE EDWARDS

Filmography 317
Index 325

# Slant Screens Image Captions

## Chapter 1

1.1: The time-bound nature of genres: television as sf icon in 1936's *Things to Come*. United Artists. — 5

1.2: SF as improvisational form: sf merges with film noir and a punk aesthetic in *Blade Runner* (1982). Warner Bros. — 11

1.3: Digital effects visualize a world that might be: the world-train of *Snowpiercer* (2013) careens through a climate catastrophe. CJ Entertainment. — 18

## Chapter 2

2.1: "Thin" Afrofuturism offers a Black hero (Will Smith) in a conventional space adventure: *After Earth* (2013). Sony Pictures. — 26

2.2 "Strong" Afrofuturism and a new image of Africa: entering Wakanda in *Black Panther* (2018). Walt Disney Studios. — 35

## Chapter 3

3.1: Trapped in the patterns of DNA, Vincent (Ethan Hawke) of *Gattaca* (1997). Sony Pictures. — 41

3.2: Children as "freakish abominations" in the postapocalyptic *The Girl with All the Gifts* (2016). Warner Bros. — 48

## Chapter 4

4.1: A frozen New York harbor in *The Day After Tomorrow* (2004). 20th Century-Fox. — 54

4.2: A "strategic realism": the controlled space habitat of *Elysium* (2013). Sony Pictures. — 60

# Chapter 5

5.1: The developing Ethnogothic tradition seen in *Def by Temptation* (1990). Shapiro-Glickenhaus Entertainment.     80

5.2: Trapped in racial terror: Chris (Daniel Kaluuya) in *Get Out* (2017). Universal Pictures.     85

# Chapter 6

6.1: The femspec vantage: woman (Sigourney Weaver) as "last human standing" in *Alien* (1979). 20th Century-Fox.     91

6.2: Woman in the pilot's seat: Carol Danvers (Brie Larson) in control in *Captain Marvel* (2019). Walt Disney Studios.     98

# Chapter 7

7.1: Encountering heterotopic space in *The Thing from Another World* (1951). RKO Pictures.     104

7.2: *District 9* (2009): concentration camp as heterotopia. Sony Pictures.     108

# Chapter 8

8.1: Toho Studio toys with kaiju tropes in *King Kong vs. Godzilla* (1962). Toho Studios.     117

8.2: The monster as metaphor for natural disaster—*Godzilla* (2014). Warner Bros.     123

# Chapter 9

9.1: Elisa (Sally Hawkins) and her aquatic other, the Asset (Doug Jones), recognize their mutual fascination in *The Shape of Water* (2017). Fox Searchlight Pictures.     136

9.2: The Gill Man of *Creature from the Black Lagoon* (1954) makes off with Kay Lawrence (Julia Adams) in "a beautiful moment of love." Universal Pictures.     138

# Chapter 10

10.1: Mixing two-dimensional and three-dimensional textures in Zeman's *Invention for Destruction* (1958). Ceskoslovensky Statni Film. — 143

10.2: Steam technologies proliferate wildly in *Steamboy* (2004). Toho Studios. — 150

10.3: Yu Niang defies giant Western steam-powered machinery (*Tai Chi Zero*, 2012). Huayi Brothers Media. — 159

# Chapter 11

11.1: Tony Stark (Robert Downey Jr.) and the technological sublime: human mastery through techno-scientific creation in *Avengers: Age of Ultron* (2015). Walt Disney Studios. — 171

11.2: Tony Stark (Robert Downey Jr.) constructs his new superhero identity in *Iron Man* (2008). Walt Disney Studios. — 174

# Chapter 12

12.1: In a zoom-out, the world changes around the time traveler Hartdegen in *The Time Machine* (2002). Dreamworks Pictures. — 188

12.2: Hartdegen experiences a new Earth and a new relationship to the world in *The Time Machine* (2002). Dreamworks Pictures. — 190

# Chapter 13

13.1: Tcherny (Andre Benjamin) marvels at how the garden proliferates in outer space (*High Life*, 2018). A24 Films. — 198

13.2: Dr. Dibs (Juliette Binoche) experiments on life in *High Life* (2018). A24 Films. — 203

# Chapter 14

14.1: Preparations for plastic surgery in *Brazil* (1985). Universal Pictures. — 212

14.2: Otto (Emilio Estevez) amid the generic goods at the Pik 'n Pay (*Repo Man*, 1984). Universal Pictures. — 213

14.3: David (Colin Farrell) resists the threat of animal transformation in *The Lobster* (2015). Sony Pictures. — 215

## Chapter 15

15.1: In the digital world of *The Matrix* (1999)—bullet-time effect. Warner Bros.    221

15.2: Low-cost digital spectacle: the Nazi moon base in *Iron Sky* (2014). Walt Disney Studios.    232

## Chapter 16

16.1: The use of cinemicroscopy in *Upstream Color* (2013): the mysterious worm examined. VHX.    244

16.2: Human intimate "entanglement" (Shane Carruth and Amy Seimetz) in *Upstream Color* (2013). VHX.    245

## Chapter 17

17.1: Detective Greer (Bruce Willis) and robotic FBI agents in *Surrogates* (2009). Walt Disney Studios.    255

17.2: The "bright object" of oxygen suddenly spews forth in *Total Recall* (1990). Tri-Star Pictures.    259

## Chapter 18

18.1: The title character (Scarlett Johanson) transforms into a networked being in *Lucy* (2014). Universal Pictures.    267

18.2: At the boundary layer between human and alien in *Arrival* (2016). Paramount Pictures.    274

## Chapter 19

19.1: *Alien Resurrection* (1997) juxtaposes the erotic with maternal tactility, as Ripley (Sigourney Weaver) embraces the android Call (Winona Ryder). 20th Century-Fox.    286

19.2: The android David (Michael Fassbender) adopts the look of T. E. Lawrence (*Prometheus*, 2012). 20th Century-Fox.    288

## Chapter 20

20.1: Cobb (Leonardo Dicaprio) reshapes the city in Limbo (*Inception*, 2010). Warner Bros. 300

20.2: Lucy (Scarlett Johansson) examines the haptic image of her grotesque fingers (*Lucy*, 2014). Universal Pictures. 306

20.3: The disintegration of Justine's subjectivity within the broader ecosphere (*Melancholia*, 2011). Nordisk Film/Magnolia Pictures. 308

# Acknowledgments

Like most books, this one was made possible by a great many friends, colleagues, acquaintances, and supporters of academic enterprise. All of them demonstrated great patience, forbearance, and even humor as we navigated the *Handbook of New Science Fiction Cinemas* through the COVID era. The primary acknowledgment must go to the many colleagues who, when asked to join this project, quickly agreed to contribute their time and talents, probably unaware of how much of both might be involved. While about half of them had worked with me on other efforts, the others took my invitation or importuning on good faith, and I was gratified that they believed I would give their work a worthy stage. I would be most happy to work with them all again. I was also pleased to collaborate with OUP's Norm Hirschy again. After two other projects together, he was a plucky enough editor to consider a third and more complex effort and to shepherd it through a difficult time. Fortunately, he was backed throughout the publication process by several able and detail-minded assistants, Lauralee Yeary and Zara Cannon-Mohammed, while the team at Oxford's Newgen Digital Works affiliate, including project managers Mr. Shunmugapriyan and Ms. Usharani, and copy editor Bob Land ensured a smooth course of production. Together, the Oxford/Newgen group ably kept the project on track and, thanks to their open communications, kept me relatively upbeat throughout the extended review and publication process. In my final years (preretirement) at Georgia Tech when this project was launched, I was fortunate to have the dependable support of Professor Richard Utz, Chair of the School of Literature, Media, and Communication, as well as numerous department members, especially fellow science fiction enthusiasts Lisa Yaszek and Susana Morris. And as ever, I enjoyed the strong support of my wife and fellow scholar, Leigh E. Telotte, with whom I could always discuss whatever problems or satisfactions that came along with the book. Finally, I would like to dedicate this volume to three teachers who helped shape my own critical career and whose inspiration is reflected in this volume: William "Ted" Cotton, Malcolm O. Magaw, and Motley F. Deakin.

# CONTRIBUTORS

**Mark Bould** is Reader in Film and Literature at University of West England, Bristol, and a recipient of both the SFRA Award for Lifetime Contributions to sf Scholarship (2016) and the IAFA Distinguished Scholarship Award (2019). He co-edits the monograph series Studies in Global Science Fiction. His most recent books are *M. John Harrison: Critical Essays* (2019) and *The Anthropocene Unconscious* (2021).

**Levi R. Bryant** is a Professor of Philosophy at Collin College. He is a prominent figure in the speculative realism and object-oriented ontology movements, and the author of a number of seminal texts in these areas, including *Difference and Givenness: Deleuze's Transcendental Empiricism and the Ontology of Immanence* (Northwestern UP, 2008), *The Democracy of Objects* (Open Humanities, 2011), and *Onto-Cartography: An Ontology of Machines and Media* (Edinburgh UP, 2014).

**Gerry Canavan** is an Associate Professor in the English Department at Marquette University, teaching twentieth- and twenty-first-century literature and media. He is the co-editor of *Green Planets: Ecology and Science Fiction* (2014), *The Cambridge Companion to American Science Fiction* (2015), and *The Cambridge History of Science Fiction* (2019), as well as the co-editor of the academic journals *Extrapolation* and *Science Fiction Film and Television*. His first book, *Octavia E. Butler* (2016), was published in the Modern Masters of Science Fiction series by University of Illinois Press.

**Catherine Constable is** a Professor in the Department of Film and Television Studies at the University of Warwick. Her research focuses on film-philosophy, feminist film theory, postmodern theory, and Hollywood cinema. Her publications include the monograph *Adapting Philosophy: Jean Baudrillard and The Matrix Trilogy* and articles such as "Challenging Capitalism: Ethics, Exploitation and the Sublime in *Moon* and *Source Code*" in *Science Fiction Film and Television* and "Surfaces of Science Fiction: Enacting Gender and 'Humanness' in *Ex Machina*," which was short-listed for best article in *Film-Philosophy*. She is currently writing a monograph on female protagonists and the sublime in contemporary science fiction film.

**Matt Denny** is a Teaching Fellow in the Department of Film and Television Studies at the University of Warwick. His research focuses on film theory, postmodern theory, and contemporary Hollywood Cinema, particularly genre cinema. Matt has published work on the aesthetics of action cinema and on intertextuality in representations of the female vampire. His current research explores issues of representation in genre cinema, particularly in relation to images of artificial intelligence in horror and science fiction.

xvi   CONTRIBUTORS

**Gerald Duchovnay**, former department head and Professor of English at Texas A&M–Commerce, is the Founding and General Editor of *Post Script: Essays in Film and the Humanities* (1981–2021). The author of *Humphrey Bogart: A Bio-Bibliography* (1999) and editor of *Film Voices*, a collection of interviews with filmmakers (2004), he has also co-edited (with J. P. Telotte) *Science Fiction Film, Television and Adaptation* (2012) and *Science Fiction Double Feature: The Science Fiction Film as Cult Text* (2015), and has published numerous articles on film, literature, and media.

**Caroline Edwards** is Senior Lecturer in Modern and Contemporary Literature at Birkbeck, University of London where she is Director of the Centre for Contemporary Literature. She is author of *Utopia and the Contemporary British Novel* (2019), editor of *The Cambridge Companion to British Utopian Literature and Culture* (forthcoming), and co-editor of *China Miéville: Critical Essays* (2015) and *Maggie Gee: Critical Essays* (2015). Her research has been featured in broadcasts on BBC One, BBC Radio 4, BBC Radio 3, and the BBC World Service, cited in the *New Statesman* and the *Guardian*, and displayed in a dedicated exhibition at the Museum of London (2017–18).

**Susan A. George**, retired from the University of California, is an independent scholar and librarian. Her work focuses on the construction of gender and the representation of technology in fantastic film and television. Her articles have appeared in such journals as *Science Fiction Film and Television* and *The Journal of Popular Film and Television*, and in various anthologies, including *The Essential Science Fiction Television Reader*. Her work on 1950s science fiction film culminated in her book *Gendering Science Fiction Film: Invaders from the Suburbs*. She is also co-editor of the critical anthology *Supernatural, Humanity, and the Soul: On the Highway to Hell and Back*.

**Joan Gordon** is an editor of *Science Fiction Studies* and a Professor Emerita of Nassau Community College. She has received the SFRA award for lifetime achievement in sf scholarship and writes about feminism and animal studies in sf.

**De Witt Douglas Kilgore** is Associate Professor of English at Indiana University. He is the author of *Astrofuturism: Science, Race and Visions of Utopia in Space* (2003). His recent work includes contributions to *Literary Afrofuturism in the Twentieth Century* and the *Oxford Bibliographies in Cinema and Media Studies*. He is currently working on two book projects: *Galactic Club: Seeking a Postracial Universe in Science/Fiction* and *The Color of the Future: Essays on the Hope of Contemporary Afrofuturism*.

**Thomas Lamarre** teaches Cinema and Media Studies at the University of Chicago. His publications focus on media, thought, and material history, and they include *Uncovering Heian Japan* (2000), on communication networks in medieval Japan; *Shadows on the Screen* (2005), about silent cinema and the global imaginary; *The Anime Machine* (2009), on animation technologies; and *The Anime Ecology* (2018), about animation and infrastructure ecologies.

**Frances McDonald** is Assistant Professor of English at the University of Louisville. Her research and teaching focus on twentieth-century American literature and film, critical theory, and affect studies. She is the author of *Posthumanism: The Modernist Affect of Laughter* (2022), and of work that has appeared in *American Literature, LARB, Post45,* and *The Atlantic*. She is also the co-editor of *Thresholds*, a digital journal for creative/critical scholarship.

**Susana M. Morris** is Associate Professor of Black Media Studies at Georgia Institute of Technology. She writes for the Crunk Feminist Collective blog, is co-editor of *The Crunk Feminist Collection* and *Sycorax's Daughters*, and is the author of *Close Kin and Distant Relatives: The Paradox of Respectability in Black Women's Literature*.

**Angela Ndalianis** is Adjunct Professor in Screen Media and Entertainment at Swinburne University of Technology. Her research focuses on visual effects technologies in the superhero, horror, and science fiction genres. She has published numerous articles and books, including *Neo-Baroque Aesthetics and Contemporary Entertainment* (2004), *Science Fiction Experiences* (2010), *The Horror Sensorium: Media and the Senses* (2012), and the anthology *The Contemporary Comic Book Superhero* (2009).

**Bradley Schauer** is an Associate Professor in the School of Theatre, Film, and Television at the University of Arizona. He is the author of *Escape Velocity: American Science Fiction Film, 1950–1982* (Wesleyan UP, 2017). His work has also appeared in books and journals such as *Film History, The Velvet Light Trap*, and *Cinematography* (Rutgers UP, 2014).

**Lars Schmeink** is an independent scholar of science fiction and has held visiting professorships at the University of Cincinnati and the University of Leeds. In 2010 he inaugurated the Gesellschaft für Fantastikforschung and served as its predicant for ten years. He is the author of *Biopunk Dystopias: Genetic Engineering, Society, and Science Fiction* (2016) and the co-editor of *Cyberpunk and Visual Culture* (2018), *The Routledge Companion to Cyberpunk Culture* (2020), *Fifty Key Figures in Cyberpunk Culture* (2022), and *New Perspectives on Contemporary German Science Fiction* (2022).

**J. P. Telotte** is Professor Emeritus in Georgia Tech's School of Literature, Media, and Communication. Co-editor of the journal *Post Script*, he has published widely on film and television with special emphasis on science fiction and animation. Among his many books are *Animating Space: From Mickey to WALL-E* (2010), *Science Fiction TV* (2014), *Robot Ecology and the Science Fiction Film* (2016), *Animating the Science Fiction Imagination* (2017), and *Movies, Modernism, and the Science Fiction Pulps* (2019).

**Chuck Tryon** is a Professor of English at Fayetteville State University. He is the author of three books, *On-Demand Culture: Digital Delivery and the Future of Movies, Reinventing Cinema: Movies in the Age of Media Convergence*, and *Political TV*, and is a specialist in the impact of digital media on the film industry.

**Sherryl Vint** is a Professor of Media and Cultural Studies at the University of California, Riverside, where she directs the Speculative Fictions and Cultures of Science program. Her books include *Bodies of Tomorrow, Animal Alterity, Science Fiction: A Guide to the Perplexed, Science Fiction: The Essential Knowledge*, and *Biopolitical Futures in*

*Twenty-First-Century Speculative Fiction*. She is an editor of *Science Fiction Studies*, and of the book series *Science and Popular Culture*. She has also edited several books, most recently *After the Human: Theory, Culture and Criticism in the 21st Century*.

**Jeffrey Andrew Weinstock** is Professor of English at Central Michigan University and an associate editor for *The Journal of the Fantastic in the Arts*. He has authored or edited twenty-five books and more than eighty essays and book chapters on such subjects as fantasy, horror, science fiction, and the Gothic. Among his most recent books are *And Now for Something Completely Different: Critical Approaches to Monty Python* (2020, with Kate Egan), *The Monster Theory Reader* (2020), and *Critical Approaches to Welcome to Night Vale: Podcasting between Weather and the Void* (2018). Visit him at JeffreyAndrewWeinstock.com.

# I

# INTRODUCTION

# INTRODUCTION

# CHAPTER 1

## INTRODUCTION: SLANT SCREENS/SLANT SCREENINGS

### J. P. TELOTTE

### I

SCIENCE fiction (sf), and particularly a cinematic sf, has often been described as a kind of universal language. Speaking in icons and a syntax drawn from the realms of science, technology, and reason—a trio of components that Susan Sontag, writing in the late 1960s, placed at the core of the sf film (1966, 216)—it has allowed an element of communication between all industrialized peoples, as it addresses many of the common concerns of those inhabiting modern technological culture. Yet for a lingua franca, sf seems to suffer from two distinct but related problems. First, while much has been written about sf in all of its media modes, we still have some trouble agreeing about precisely what it is; we still lack a universally acceptable definition of the genre, at least beyond Darko Suvin's somewhat timeworn, bounded, yet still useful description of sf as "the literature of cognitive estrangement" (1979, 7). Second, the genre seems to speak in rather diverse ways, as if it were constantly evolving new dialects as it focuses on an ever-widening range of science fictional concerns, not just the readily recognizable subjects like space travel, robotics, and utopian fantasies with which audiences are quite familiar, but also environmental issues, racial tensions, climate concerns, genetic manipulation, gender identity, and so on. This volume seeks to address both problems, not by trying to offer one more definition of the form and then sorting out what is and is not sf, but by considering how these two issues are related and actually part of what makes it such a useful "language" today. That relationship, as the following essays should suggest, shows most prominently in the varying shapes taken by the contemporary sf film in its efforts to speak to and for its global audience.

Like many others who have tackled the difficult task of defining sf, Edward James, one of the genre's primary historians, has tried to account for his subject's multiple dimensions: describing its primary story types, cataloguing its central memes, and exploring its recurring themes. But even in undertaking both a broad and systematic approach to the nature of the genre, its key icons, and how it has used the regime of technoscience as a base for addressing various cultural concerns, he also admits that today, even as sf continues to surge in popularity and presence across a wide range of media, the genre still seems to be very slippery, frequently shifting in its thrusts and imagery, and proving increasingly difficult to define. Addressing this situation, Rob Latham has more recently suggested that, since so much technocultural production seems to share in this idiom, perhaps the real problem of definition today "is to determine *what does not count* as science fiction" (2014, 5). And as Vivian Sobchack has noted, this same sort of slipperiness persists across sf's diverse media appearances—and is particularly obvious in sf cinema where, she claims, there is no "consistent cluster of meanings" that attach to its primary icons (2004, 5). Rather, the very "plasticity of objects and settings" that marks this cinema seems to stand in blatant contrast to "the essentially static worlds of genres such as the western and gangster film," thereby presenting commentators with an inevitably "more ambiguous territory" than we encounter in other genres (Sobchack 2004, 10). It seems that sf is invariably bound up with a variety of other discursive histories, generic formations and reformations, emergences, and (virtual) disappearances that beg to be accounted for.

In response to this situation, James has pragmatically suggested that a possible starting point for discussing the genre—in either its literary or cinematic forms—is acknowledging that "sf is what is marketed as sf: that is a beginning, but no more" (1994, 3). While a simple suggestion, it is not just a case of surrendering to the difficulties of the task or deferring to another frequent and similarly pragmatic approach: the "I know it when I see it" notion. Rather, he is recognizing that sf, especially in the course of its contemporary flourishing, has become something that is largely defined by its place in culture *and* by the ongoing changes that mark—and create a market for—those cultural concerns. Readers and authors, game designers and game players, filmmakers and filmgoers all have simply found sf to be a highly suitable language for expressing their assorted and frequently shifting apprehensions about the world (apprehensions typically linked to the latest, and similarly shifting, scientific and technological developments), and for exploring both its present and future possibilities. Those two issues, of sf's cultural embeddedness and the ongoing changes it reflects, might serve as useful starting points for this handbook—or an admittedly rough roadmap to what remains a surprisingly hard-to-map genre.

While James's comments (like Suvin's definition) are mainly directed at sf literature and its readers, they also echo much of the contemporary conversation about the nature of all popular genres—in literature, television, and especially film—while also reminding us of how those genres have exploited the subject matter of sf. Most obviously, they recall Andrew Tudor's influential, early 1970s discussion of film genres and how they effectively serve their audiences. Drawing largely on the example of the Western, then one of the most popular film forms, Tudor began his assessment of popular genres by describing the obvious, if often side-stepped "chicken-versus-egg" problem that faces every effort at such definition. As he explained, to define a form, one must account for its many characteristics, but to establish what those characteristics are, one must draw on an established canon or group of works. Of course, such a grouping implies that one has already decided which texts fit within the genre under consideration—a decision only made possible by comparing them to a previously determined set of genre characteristics. Given this circular situation, he argued, again quite pragmatically, that undertaking any genre definition ultimately depends on some sort of presumed cultural consensus, a general perception of "what we collectively believe it to be" (Tudor 2003, 7) *at a particular time*, suggesting an inherently fluid situation, while also pointing the way for a variety of more contemporary notions about both genre and ultimately a form like sf.

We can see one of those subsequent developments of genre thinking in Steven Neale's effort to explore more precisely that notion of a "collective" and time-bound appreciation or understanding of genre (see figure 1.1). Neale argued that we might think of every genre as a kind of "inter-textual relay," a point where popular formulas, typically marked by some elements of flexibility in their subjects and actions, intersect with shifting cultural concerns, or what he terms the "trends and values" (2000, 3) of an era. While Neale, like Tudor, acknowledged the cultural positioning of all film genres, and especially their responsiveness to a particular (time-bound) audience's needs and desires—which he grouped under the broad heading of "public expectations"—he also acknowledged that those "expectations" operate in a constant negotiation ("trends")

FIGURE 1.1: The time-bound nature of genres: television as sf icon in 1936's *Things to Come*. United Artists.

with the industrial conditions of film production, especially in the context of the genre-rich and genre-obsessed Hollywood movies (Neale 2000, 4). The result, he suggested, is an ongoing, always evolving pattern of relationships—those "relays," as he terms them—that allows for this inevitably shifting/shifty responsiveness, and that might readily, if somewhat vaguely in the form he offered, open up to more systematic study and understanding. However, such negotiation is itself fraught with ambiguities, for as philosopher Levi Bryant (one of the contributors to this volume) observes, "As is the case with all negotiations, the final outcome or product" practically never coincides with "a pre-existent and well-defined plan" (2014, 19).

While mainly focused on the discourse surrounding television genres, Jason Mittel has sought to nail down some of these ambiguities by focusing more precisely on the actual genre experience. Like Neale, he too assumes that a pattern of "intertextual relations" can be observed between various genre texts and "the cultural operation of genres" (Mittel 2004, 8), but he emphasizes that to better understand those "relations," we need to ground our study in the pervasive "realized practices" (4) of genre consumption—in his test case, those of everyday television viewing. By considering such practices and situating them in the cultural moment, he suggests, we can begin to arrive at a more elaborate accounting: for how "genres shift and evolve" within a cultural context, how they are "actually experienced" (5)—whether on big screens or small, in group experiences or individually, through network-scheduled viewings or self-determined binge streamings—and how those experiences inflect our understanding and *use* of a genre, as filmmakers, as filmgoers, and even as critics. Simply put, Mittel's project aims to explain how a genre like sf, especially in the increasingly varied forms it seems to take as it crosses various sorts of media screens, manages to respond to the concerns and pressures we "actually" experience and with which we continually wrestle.

Similarly suggesting that we might approach genre texts as things that today are variously "experienced"—in film, literature, television, and gaming as well—Mark Bould and Sherryl Vint have effectively linked much contemporary thinking about genre to the specificities of sf, while also striking an interesting note of caution. They too argue that popular genres, and especially sf thanks to its often-hybridized nature, should be seen not as distinct, bounded, and constant formulas, but rather as "fluid and tenuous constructions," produced by all those who are involved in the creation, distribution, and consumption of those genre texts (Bould and Vint 2009, 48)—that is, by everyone involved in the genre "experience." And they suggest that those "constructions" are invariably being shaped and reshaped as our cultural needs and desires dictate, and as those needs and desires undergo their own inevitable mutations over time. On the one hand, Bould and Vint are simply describing what has frequently been noted about sf, that is, the genre's increasing flexibility or adaptability—which, I would argue, is one of the reasons for its great popularity. On the other, they remind us that, much in the manner that Tudor outlined, any sortie at defining a genre like sf will prove to be something of a constant, contingent, and in some ways even frustrating process (recall the chicken and the egg). Acknowledging that fact, while strategically pushing their logic to an extreme but telling conclusion, Bould and Vint note that one might even argue that actually "there is no such *thing* as science fiction" (2009, 50).

Of course, Bould and Vint are not so much trying to send up the genre, to dismiss it in a puff of smoke, as they are trying to link our thinking about both genres and the rather fluid example that is sf. They are simply underscoring the *thing*-oriented nature of much of our genre theorizing, perhaps even of our genre *consumption*, especially when we think about the genre text, as much cultural study has tended to do, as a kind of distraction or palliative—a thing for what ails us at a particular cultural moment. And this thing-y approach has, historically, very often been the case with our thinking about sf. After all, it has long been treated as a distinct form that simply lacks a distinct—or at least satisfactory—name. In the cinema, for example, sf was for many years conflated with horror, a type that was more firmly established in the public consciousness and the semantic elements of which, particularly such icons as the mad scientist and man-made monster, quite readily took on both horrific and science fictional valences. Thus, even as late as the 1951 alien invasion film *The Thing from Another World* (Nyby), studio advertising largely papered over the movie's sf elements, instead posing ominously leading questions, such as "Is It Human or Inhuman? Natural or Supernatural?"—with that issue of the "supernatural" seemingly shutting off the film's sf credentials. And that advertising often printed "*The Thing*" in large red letters seemingly dripping blood—a sure sign of the horrific—while either obscuring or rendering in microprint its additional identifier "from another world" (see Telotte 2019). We might also recall that one of the first rigorous historical treatments of sf cinema went veiled under the title *An Illustrated History of the Horror Film* (Clarens 1967), before being republished, thirty years later, after the genre had attained a new level of popularity (and its author was deceased), as *An Illustrated History of Horror and Science-Fiction Films*.

As for naming, the genre has often been burdened with what Steve Neale nicely describes as "inexact nomenclature" (2000, 121). "Scientifiction," for example, was the popular title championed by sf pioneer Hugo Gernsback, and it would only slowly give way to the simpler "science fiction" over several decades, while leaving an awkward legacy for the cinema thanks to the companion designation "scientifilm," a term that was frequently used in the sf pulp magazines and that lingered well into the 1940s. Yet many critics and commentators even in the 1950s would continue to refer to examples of the genre by such varying terms as "scientific romance," "scientific adventure," "invention story," "pseudo-scientific story," and so on (James 1994, 9–10). Thus, it is hardly surprising that in the film industry, well into the 1950s, the term "science fiction" was just one identifier, used almost interchangeably with various other designations, such as "thriller," "fantasy," "exploitation film," "horror," and "trick picture."[1] More recently, as Lisa Kernan has documented, George Lucas, a filmmaker invariably linked to sf, pointedly "resisted the generic appellation 'science fiction'" in advertising *Star Wars* when it first appeared in 1977, preferring to use the vaguely allusive term "space fantasy" (Kernan 2004, 178). All of these naming efforts were just various ways of marking off that thing-y nature of the form: using (or sidestepping) a particular name to establish boundaries, indicate content to consumers (as well as exhibitors), and declare independence from—or claim intriguing relationships to—other generic kin. However, the variety of those terms further reminds us just how slippery the genre has long seemed to be—as is also evident in the advent of more recent terms and with more recent concerns in mind, titles often applied to literary sf such as "slipstream" or the more broadly allusive and widely embraced "speculative fiction."

Actually more challenge than pronouncement, then, Bould and Vint's question about the thing-status of sf only serves to underscore the genre's elusive nature, and it is a challenge that many others, particularly genre theorist John Rieder, have sought to take up. Even while recognizing the limited and largely rhetorical thrust of their observation, as well as the way that various commentators continue to struggle with defining both the larger notion of genre and the specific instance of something like sf, Rieder has insisted that "constructing genre definitions" remains "a scholarly necessity" (2017, 13), while even suggesting that it is something of a critic/historian's "duty." So in properly dutiful fashion and while ranging widely across literary, televisual, and cinematic versions of sf, Rieder has accepted the challenge "to say what sf is" (15), and in fact does so by integrating many of the rather foundational notions about genres that we outlined earlier, including trying to account for audience expectations and critical practice, along with the inevitable historical changes that have occurred both within and outside of the form.

Thus Rieder argues that sf, like other popular genres (and as distinct from what he terms the larger "classical-academic" forms such as tragedy, comedy, satire, and epic), is "the product of the interaction among different communities of practice," each drawing on its own contemporary "categories" of understanding (2017, 30), and each reflecting different, yet no less significant "motives" for producing and consuming the genre. The inevitable result of this interaction, he suggests, is that sf, like a number of other popular genres, tends to "overlap, change over time, and demand improvisation" (34), especially as it has come to play an increasingly prominent role in what he terms "the mass cultural genre system," that is, the "large-scale commercial production and distribution"—and, I would especially add, consumption—of formulaic fictional narratives (1). An additional result is that a commonly embraced understanding of sf, such as Suvin's simple yet serviceable notion that we think of the genre as a form identified by a pattern of "cognitive estrangement," tends to lose some of its currency, or at least what we might term its precision, in the face of those ongoing, even inter-generic "demands" and "improvisations." We have to wonder whether the methods of "estrangement"—and even how we measure that effect—represent constants, or are they, too, in what often seems an increasingly *strange world*—subject to change?

With such questions in mind, Rieder, much like Tudor, Neale, and Mittel, thus emphasizes that we should incorporate into our sf thinking an awareness of what I keep referring to as the genre's "slipperiness." The result is an inevitable hedging in offering any sort of firm definition or taxonomy simply because of the way such a genre works. In fact, if we follow out his line of reasoning, we might find that a certain fluidity could even be crucial to *how* sf works, as it forces us to *slip* out of our comfort zone to imagine various potentials for difference or change that it offers. In any case, that slippery approach becomes not just, in a return to Edward James's suggestion, a *useful* starting place, but actually a *necessary* one for better thinking about the "historical and mutable" (Rieder 2017, 16) character of sf—a genre that has not only become one of the most popular in the contemporary cinema, but that is also obviously characterized by its Protean possibilities or, put more simply, by its consistent flexibility.

The point of this admittedly selective review of commentary about both genre and sf is to establish the context for what follows here—this volume's own multiple accounts, which, in various ways, try to address what Rieder loosely characterizes as sf's "improvisation." The purpose of the subsequent essays is to follow the lead of these and a number of other critical approaches to genre thinking and especially to contemporary sf cinema, by embracing the "multiple" and "mutable" characteristics that they commonly observe. And I would offer that this volume does so, much as Rieder suggests, rather "dutifully," although not just to clear a path for another—easily frustrated—effort at further refining the notion of genre or to arrive, by volume's end, at a more widely (if only temporarily) satisfying definition of sf. Rather, it takes that highly fluid character *as its problematic*, and as the necessary starting point for exploring sf in the cinema. For sf today is manifestly a multi-form, and therefore also an elusive one that invites our thinking about how open it is to improvisation—or slipperiness. The aim here is to accept that implicit invitation and undertake some of that thinking, capitalizing on this sense of elusiveness to better understand how sf has developed its contemporary character—or, as the body of this collection more properly suggests, *characters*—while starting a mapping process of what seem to be at least *some* of the *multiple* trajectories it is following. If sf is indeed subject to ongoing construction and reconstruction by its many "communities of practice," as all our contributors assume, then these various essays need to follow the lead of those communities, tracking the different paths down which sf—not a "thing," not a bounded formula, but rather a quite shifty "practice"—rather obviously leads. For in tracing those multiple paths and illustrating how a variety of contemporary films enact what Rieder broadly terms "the relation of formal strategies to ideologies" (2017, 170), all linked by a common fascination with science, technology, and the workings of reason, we might better grasp the Protean nature not of a monolithic sf, but, as this volume's title offers, the amazing variety of "New Science Fiction Cinemas."

# II

As I hope the preceding comments have adequately suggested, while sf is arguably the most popular screen genre today, it also has a decidedly multiple character, in fact, far more so than we might observe even during its first great explosion of popularity in the 1950s. Then, hundreds of sf films and numerous television series debuted, although their variety scantly suggests the assortment that characterizes the contemporary genre. In that earlier period alien visitors, adventures in space exploration, and home-grown monsters of various stripes largely dominated the form, and they were most often framed within a context of calamitous events and invasions prompted by and reflecting Cold War traumas. That mixture prompted one of the most famous accounts of the form, Sontag's description of cinematic sf as the "imagination of disaster" (1966, 212), sauced with an "aesthetics of destruction" (216). Of course, contemporary sf looks quite a bit different, and it has the ability to treat these same subjects differently—and certainly more convincingly, thanks to our investment in new technologies for producing more effective instances of what Michele Pierson evocatively describes as "the aesthetic experience of wonder" (2002, 168). It is on the nature and extent of that difference that this section focuses.

Beyond their ability to visualize whatever we might imagine, our sf films have broadened their reach, telling new sorts of stories, some of which might not even have been recognized as sf in an earlier era. For example, we might think of the political/technological power over the human form that is demonstrated in *The Lobster* (Lanthimos 2015), a film cited in several of the essays included here, as culturally nonconforming humans are threatened with the prospect of being turned into animals of their own choosing—such as a lobster; or the paralleling of religious cult activities with the possibility of other-dimensional civilizations glimpsed by a young boy with special powers in *Midnight Special* (Nicholls 2016); or the way in which ancient knowledge and voodoo practice come together and serve as an ultimate counter to both technological and racial repression in *Brown Girl Begins* (Lewis 2017). Such films stretch our conceptions of sf (and even its appropriate subjects), while also providing a popular stage for acting out—and thinking through—a great variety of the important issues that inform what James simply termed the "market" (3): framing contemporary culture's distress over racial disparities, its apprehensions over climate change, the growing desire to give voice to gender, however we choose to define (or redefine) it, acknowledging other spaces than those conventionally sanctioned, along with the other inhabitants of those spaces, searching for the mysterious, even the magical that seems to have become absent in our lives, staking out a space for wondering about other histories and, by implication, other possible futures for human culture.

These and other issues have increasingly found homes in contemporary sf, pointing up just how different it is from much twentieth-century sf, but also just how adaptable a vehicle for our imaginings it can be. Those concerns have gained prominence in the genre not simply by being the subject matter of the moment, as if impelled by the headlines of the day, but rather, as I like to think of it, by angling into the very fabric of earlier sf narratives, refracting and reframing their icons and plots, and, in the process, slanting the genre in the multiple sorts of directions or trajectories that our genre commentators are observing. The result of such cosmic collisions is the development of what I term *slant forms* of sf, shifting a genre that, at least in its early cinematic incarnations, often seemed marked by a relatively limited array of rather conventional icons (spaceships, robots, futuristic cities, scientists—mad and otherwise) and easily charted narrative patterns (the fantastic voyage, the encounter with aliens, the creation of marvelous inventions, etc.), in many new, and for some audiences quite unfamiliar, directions—as it were, knocking the form out of its traditional orbit.

However, I want to use this "slant" conceit not to suggest that there is something not quite right, proper, or really science fictional about our contemporary sf films, but just the opposite: as a way of visualizing how a great many movies have profitably drawn into the sf realm concerns that are often associated with other genres or film types, and to appear, at least to some people, as if they were even vectoring away from the world of traditional sf. Moreover, it provides a terminological way of giving them their due as sf, of not relegating them to the realms of the "subgenre"—a sometimes slighting designation that tends to be all too casually employed. The resulting new sf cinema, if unsettling to some genre traditionalists, is able to so slant, shift direction, or accommodate new material and new thrusts precisely because sf as a genre has itself proven so adaptable; its language of science, technology, and reason so fundamental and appropriate to the modern moment; its general tradition of cognitive estrangement still resonant for a situation in which the changing human world, and across its broad landscape, already—or always—looks and feels quite strange.

The fortunate if also challenging result of such ongoing "improvisation" is that the contemporary cinema seems to confront us not with one dominant flavor of sf but many—or to recall our earlier linguistic conceit, multiple dialects (see figure 1.2). Certainly, this cinema does not even admit of the notion that, as James puts it, there is any sort of "Platonic idea of science fiction" (1994, 1) at work, something to which we might consistently refer or use as a satisfying measure of its "real" generic character and thus to easily mark off any generic changes of the moment. Rather, as this volume's title

**FIGURE 1.2:** SF as improvisational form: sf merges with film noir and a punk aesthetic in *Blade Runner* (1982). Warner Bros.

suggests, the genre recognizes a variety of sf cinemas that all still speak much of that same science fictional "language" to which we are accustomed, but inflect it differently, telling diverse sorts of stories, and challenging audiences to see and interpret them in a variety of new yet also clearly compelling, and we might even say necessary ways.

That challenge is central to this volume's purpose, as the essays collected here aim to describe and follow a number of those new directions or "slants" in order to help map the broad territory of contemporary cinematic sf (particularly because of the inevitably global context of and global audience for these cinemas), and thus to serve as a tentative guide of sorts. While many of these slant forms have found similar prominence in sf literature, this handbook focuses specifically on their cinematic embodiments, in part because they have attracted such attention, with sf film, in its many variations, becoming the most popular—and, as the Marvel Cinematic Universe has repeatedly demonstrated, certainly the most profitable—of contemporary film genres; and also in part because of their common reliance on a powerful regime of digital special effects that is central to both their appeal and promise. These effects not only empower their diversity of visualizations, but also provide us with a point of commonality. Through that regime of effects, they effectively accomplish the common and perhaps the most fundamental goal (and attraction) of all sf: visualizing what might be, visualizing practically anything we might imagine, and in that process illustrating the real possibility for *something else*—whether it is other worlds, other forms of life, or other ways of conducting our lives. In fact, those technologically produced and enhanced utopic, dystopic, or simply heterotopic visions offered by our films also argue for an underlying kinship amid the seemingly multiple trajectories of contemporary sf, suggesting that we need to see them not as representing emerging separate genres, about to, like wayward siblings, go their own ways, but simply as what I have referred to as different slants either emanating from or relatively clustered around a similar, constantly improvised technoscientific core.

Because these different forms of sf film draw on different interests and speak in somewhat different ways, this volume addresses them in several fashions, which I want to describe briefly in this section. Following this introduction is a portfolio of essays exploring a significant, although hardly exhaustive selection of these different forms. Each essay examines a particular film type in terms of its history, current status, and salient texts, while also offering a detailed investigation of one or more of those film texts with an eye to demonstrating how that type of sf most effectively works—and how it might be best understood. Each of these pieces is written by a leading authority, including several recipients of the Science Fiction Research Association's prized Lifetime Contributions award for their impact on the study of sf. These authors bring perspectives drawn from a variety of disciplines (film, literature, philosophy, digital studies), while also providing a properly global perspective, since they represent a number of different nationalities. Following the catalogue of these various slant screens, a second portfolio of articles, drawn from a similarly prominent group of scholars, provides a larger and pointedly theoretically focused vantage. The essays in this section consider some of the dominant critical/theoretical slants that have proven especially fruitful for thinking about contemporary sf. But they also represent the vantages that are enabling these cutting-edge readings, rethinkings, and reformulations of sf that the various film types represent. Each of these pieces thus outlines a particular theory or perspective as it has been broadly used or interpreted (in film as well as in other forms), discusses its specific applicability and especially its *value* for thinking through the "new sf cinemas," and also illustrates that use through the discussion of a contemporary film or group of films. The volume further provides a selective filmography, allowing readers to chart the thinking of our contributors by following up on some of the salient examples of these new sf cinemas. Our collective hope is that the resulting combination will make this collection multiply useful, as it simultaneously and from various vantages addresses a number of overlapping issues—including those of genre formation, the role of cultural studies, film and literary theory, and most obviously, the evolving canon of that ever slippery sf cinema.

# III

Given all that this introduction has said so far about the shifting nature and multiple types that characterize our new sf cinemas, we do run the risk here of suggesting, almost negatively, that the genre might be seen as a fractured, rather than what might be described as a variously vectored form. In fact, looking to the increased multiplication of virtual frames or windows in daily life, Anne Friedberg predicted this sort of outcome in our cultural lives, suggesting that we would increasingly come to "see the world in spatially and fractured frames" (2006, 243). But while Friedberg was simply trying to describe the way in which the contemporary world has become suffused with screens of various sorts, screens that can indeed easily distract and splinter our attention, she also saw an affirmative dimension in this development. In the case of sf, it is one lodged in how different audiences—and filmmakers—have been able to appropriate and elaborate on elements of the larger and always evolving sf supertext to craft stories that have gone untold or to address concerns that have been neglected by older forms of the genre, not to mention the larger culture.

In looking back at other eras we can notice a rather common pattern at work in what Rieder terms the "mass cultural genre system" (2017, 1), one wherein genres come and genres go, as one film form loses its resonance or simply proves unavailing for speaking about our shifting cultural concerns—allowing for another, more evocative type to come to the fore (or be pushed to the fore by various industrial/economic factors) and help us do the necessary working through of cultural issues. But unlike many other genres, sf has demonstrated a remarkable staying power, proving itself flexible enough to develop new voices in which it can speak for and to a great variety of those issues. To determine which of those voices might be especially valuable to users of this handbook, I have surveyed most of the primary journals in sf and cinematic sf studies, including *Science Fiction Studies, Extrapolation, Journal of the Fantastic in the Arts, Science Fiction*

*Film and Television*, and the *Journal of Popular Film and Television*, as well as various press's publication lists. From them I isolated several slants that are consistently drawing attention and that critics, fans, and filmmakers have often identified with titles or at least general descriptors, labels that, in keeping with the adaptable nature of the form, often seem applicable to somewhat different groupings. Thus, this volume's filmography pointedly acknowledges the potential overlap in some of these categories.

Readily demonstrating some of this adaptability or fluidity, the essays included here on Afrofuturist and femspec sf—that is, on works that re-view our reality from Afrocentric and new feminist vantages—show how issues of race and feminist speculation, long repressed or subordinated in mainstream sf cinema, have been able to find an especially evocative space in the genre, while also displacing what has often been seen as a dominant white male perspective driving the form. In fact, both entries in these areas argue that their respective Black and feminist vantages represent necessary changes in trajectory for sf cinema—or as a recent book title quite accurately states the case of femspec sf, "The Future Is Female" (Yaszek, 2018). In similar fashion, the articles on biopunk and ethnogothic film underscore how contemporary punk, Black, and gothic subcultures have been able to contribute strategies for exploring the ways in which different forms of otherness—whether biologically fashioned or ethnically sourced—might become crucial pathways for interrogating the dominant technoscientific world and its inhibiting power structures.

Other useful "fractures" can be found in the articles describing cli-fi, heterotopic, and steampunk film forms. On a very basic level these sf types are closely allied, as they typically address the shape—or potential shape—that our world and its history might take, often serving as protests against or warnings about those shapes. Thus cli-fi films (admittedly, a problematic if popularly used title), with their climatological focus, often play out apocalyptic environmental scenarios, as in the case of the much-lauded Korean work *Snowpiercer* (Bong 2013), with its allegorical description of a constant winter, precipitated by ill-conceived and failed efforts at reversing global warming. Similarly capitalizing on the new digital technologies that have become almost a commonplace marker of the sf cinema, heterotopic films typically visualize other possible spaces and beings, such as the alien world of the Navi found in James Cameron's *Avatar* (2009) or the repurposed Earth and its imprisoned-population-as-energy-source depicted in the *Matrix* film series (Wachowski and Wachowski 1999; 2003; 2021). Steampunk cinema, often described as *retrofuturistic* thanks to its remarkable ability to draw past and future together in a revealing mix, similarly takes on an alternative world character, as it examines a human realm shaped not by electric or atomic power but by steam and/or the industrial processes associated with its use. It thereby illustrates a different path human culture might have followed by taking us on—as contributor Thomas Lamarre's essay in this volume most evocatively puts it, "the adventure of energy"—while also suggesting the presence of alternate paths that humanity might still choose. Whether in its animated forms, as in the case of a Japanese anime feature like *Steamboy* (Otomo 2004), or live action, as in the postapocalyptic narrative *Mortal Engines* (Rivers 2018), steampunk cinema, along with cli-fi and heterotopic film, imagines other possible worlds and other possible directions for our lives, and in the process, does some of the most fundamental work of sf by challenging a set and bounded sense of world and self.

Some of the other sf types given specific treatment here—particularly kaiju, magical realism, and superhero films—are forms that do more than extrapolate the future, the past, or other realms. Rather, they give explicit, if slightly different forms to the common science fictional question of "What if?" The kaiju or giant monster type, for example, asks how humanity might respond if suddenly faced with great creatures, perhaps awakened from their sleep in the earth, as in the case of *Godzilla/Gojira* (Honda 1954) and contemporary spin-offs such as *Pacific Rim* (del Toro 2013), or suddenly arriving from afar, like the titular *The Giant Claw* (Sears 1957) and, apparently, the monster of *Cloverfield* (Reeves 2008). The appearance of these massive creatures completely upsets—rather in the fashion of the COVID-19 pandemic—the very foundations of modern life. Magical realism, especially in the pointedly sf variety demonstrated by Guillermo del Toro's *The Shape of Water* (2017), typically discovers an element of the magical or mystical in the familiar world, something that science has simply overlooked or dismissed as impossible. It then uses that discovery to, in the best sf tradition, "estrange" our view of the everyday, while also suggesting how such an expanded vision could enrich the human experience, redeeming our reality by allowing us to recoup a sense of the *super*natural. A similar sense of the magical is shared by the typical superhero film—and film franchise, as we see in the case of the various Marvel Universe and DC Universe movies—many of them among the most popular in film history. Offering a variation on one of sf's familiar story types, the alternate universe, these films posit a world (or universe) wherein special characters—hulks, wonder women, bat men, super men, guardians of the galaxy all—exist alongside normal human beings, even share some of the same faults and feelings of those normal beings, while lending their "talents" (as heroes tend to do) to that world. In the process, these superheroes, rather like the remarkable figures of kaiju and magical realism cinemas, effectively expand our sense not only of the "known" world, but also of the self, thereby evoking a new sort of science fictional "sense of wonder"—or as Angela Ndalianis's contribution to this collection offers, a new "seven beauties of superhero-science fiction."

Let me also underscore another sort of affirmation that we might find embedded in Friedberg's analysis of our contemporary cultural windows, one that rests in her emphasis on the act of seeing, particularly the ability to see *through* those multiple new windows or screens, to view all that they might open onto or reveal. This emphasis gives reason to the book's other portfolio of articles, each of which details a particular critical/theoretical window that has contributed to our understanding of these new sf cinemas and arguably even helped generate them. To clarify, let me recall in this context noted sf writer and critic Samuel R. Delany's leading remark in his oft-cited essay "About 5,750 Words," that "what makes a given story SF *is* its speculative content" (2009, 1)—a seemingly self-evident point, but one through which he sought to define the genre, at least in its literary form, through its particular linkage of form and content, that is, by the way sf's different-seeming worlds are tied to the seeming difference in how sf works. While Delany's argument that sf form and content are largely indivisible may be suspect, his effort to paint sf as, necessarily, a way of seeing differently remains useful. This simple notion is important for what it says about how we approach sf, and how we employ the various windows, frames, or theories through which we see and read all sorts of texts. It reminds us that when approaching sf, we should always, as Delany nicely puts it, consider "precisely what sort of word-beast sits before us" (2009, 15), lest we forget that "word" and "beast" are conjoined, that how we approach and articulate something becomes part of that "something."

Theory, broadly considered, is indeed one of those "word-beasts," and itself an inherently speculative thing. In some forms it does tend to assume its own answers, leading us along a predetermined path. In others it can prompt naïve perspectives, as in the case of one commentator who, working from a rigidly Marxist perspective, argued that no genre film—and she especially points to the sf film—could ever be "good," because such types can only support "the status quo" (Wright 2012, 60). But in its most useful forms theory allows us to speculate meaningfully—about our texts, our worlds, and ourselves, and thus to be, as Delany might suggest, science fictional. Theory therefore enjoys a kind of symbiotic relationship with speculative fiction, with sf, or at least the sort of fundamental closeness that Rieder might have been sensing when he described what he saw as "a steadily narrowing division separating SF subcultures from literary and academic institutional communities and practices" (2017, 114). As it opens new windows onto the world of sf, contemporary theory thus both contributes to the sf vision and is, in fact, enriched by it, as we might see evidenced by Donna Haraway's leading commentary in her "Cyborg Manifesto," wherein she argues that "the boundary between science fiction and social reality is an optical illusion" (1991, 149). Following her lead, we might consider how Haraway's work echoes in such contemporary films as *The Machine* (James 2013), *Ex Machina* (Garland 2014), and *Alita: Battle Angel* (Rodriguez 2019), all of which seem as if they were trying to illustrate, embody, or advance her notion of a "cyberfeminism," even as her writing repeatedly draws on cinematic inspirations, as in the case of the cyborg film *Blade Runner* (Scott 1982) to which she has often turned to illustrate her theories.[2]

Examining some of those windows, as well as their impact on sf cinema, thus becomes the fitting task of the various essays that make up the "New Slants" section of this volume. Among the articles we might especially note are those that, like Haraway's work, provide different angles on the changing nature of the human, such as the pieces on bioethics, feminist materialism, queer theory, and posthumanism. Bioethics, as theory, considers how life itself has come to be seen as a material resource, one that industry has often tried to mine for profit, although at the core of that perspective lies a challenge for the proper uses of that resource—uses that align with humanistic ideals rather than trading in a base economic valuation. It is a challenge at the core of a film like Claire Denis' *High Life* (2019), with its narrative about convicts, viewed by society as expendable human capital, being forced into space experiments to help preserve a dying Earth. A similar ethical thrust underpins recent work in feminist materialism, which fashions speculative practices that call attention to material conditions—economic, psychological, social—involved in the cultural production of gender, and that ultimately challenge those conditions. Such conditions are especially well illustrated in the experimental sf film *Upstream Color* (Carruth 2013), which describes how imbedded parasites allow for human control and produce a kind of false subjectivity, as they colonize human memories and ultimately even identities. Focused on the broader issue of all gender as culturally constructed, queer (and to some extent postgender) theory often sees gender itself as a kind of science fictional subject, a speculative project or cultural performance that questions the traditional role and even the need for gender designation in contemporary society. Ripe for such analysis is a film such as the postapocalyptic *Mad Max: Fury Road* (Miller 2015), with its lead character, Imperator Furiosa (Charlize Theron), pointedly marked by conflicting gender imagery, while other characters, such as Rictus Erectus (Nathan Jones), provide almost comiclike gender performances that invite their dismissal or defeat, while also prompting audiences to reexamine their own sense of what constitutes the "normal." It might be said, admittedly with a note of irony, that an even larger human perspective is at work in posthuman studies, a field that considers the dissolution of the rational human agent—along with a primary narrative focus on the human—as a center of value. It is a pattern that we can see being literally played out in the slow transformation of the character Wikus (Sharlto Copley) of the South African film *District 9* (Blomkamp 2009) into an alien "prawn," even as the alien figure Christopher (Jason Cope) gradually comes to the forefront of the narrative and engages audience sympathies with his promise of returning to "save" Wikus, and by implication perhaps humanity itself from its many failings.

Another grouping of critical categories, which includes the Anthropocene, speculative realism (or object-oriented ontology), and utopianism, provides us with perspectives on the environment broadly considered—both natural and constructed—and its increasingly central place within our speculative narratives. The Anthropocene perspective perceives the current geological period as uniquely marked by human alterations in or impacts on the planet's natural systems, a situation destructively detailed in Dean Devlin's *Geostorm* (2017), which describes how a faulty global climate-control system eventually produces a worldwide storm of apocalyptic consequence. The Anthropocene essay in this volume describes how such narratives effectively investigate this sort of detrimental human impact on the land, seas, and atmosphere to both project and possibly ward off its consequences. The vantage offered by the new field of speculative realism suggests that all things, including the human, might most effectively be approached in similar fashion—as like objects, equally open to speculation. Applied to film, this notion affords a new way of examining both character and content, a leveling out that demands a resetting or new determination of values, focused on what contributor Levi Bryant characterizes as "affordances and constraints," as we examine our sf speculations about other times, other social orders, and other possible worlds. A similar fascination with "other" possibilities has always been at the core of utopian theory, although current efforts in this vein have evolved new approaches to cultural texts that move away from examining the simple depiction of utopian worlds—such as the typical city-of-tomorrow, rendered iconically in films like *Metropolis* (Lang 1927) and *Things to Come* (Menzies 1936)—to emphasize instead the very breadth of that utopian impulse.

Thus the chapter on utopian thinking explores how such a vantage can uncover new formal, political, and generic implications in such seemingly different contemporary sf texts as *Cloud Atlas* (Wachowskis/Tykwer 2012), *Lucy* (Besson 2014), and *Arrival* (Villeneuve 2016).

A final set of essays focuses on what we might term "special filters" through which audiences see and appreciate contemporary sf cinemas (see figure 1.3). Because the digital permeates all aspects of sf cinema today, from subjects and production to distribution and exhibition, it also opens new windows for "digital thinking" through that cinema, and especially for examining how that cinema generates an aesthetics of wonder through what Garrett Stewart has described as its "pixel reflex" (2020, 2). Reading sf cinema from the vantage of the digital, whether in the context of its production of new/old worlds such as *Jurassic World* (Trevorrow 2015) or its pointed investigation of actual digital environments, as in the various *Matrix* films, allows us to rethink new structures of community, new models of self, and new ways of being in a world that is itself increasingly defined by digital creation and interaction. A similarly self-aware approach, cult criticism approaches the sf film from the perspective of a community of viewers who share a kind of irrational love or appreciation for particular qualities or aspects of film, elements that often run counter to conventional cinematic expectations and evaluations—including our sense of what makes for a "good" film. As is evidenced by the cults surrounding such mainstream texts as *Donnie Darko* (Kelly 2001) and *Firefly/Serenity* (Whedon 2002-03, 2005), as well as those appreciative of cultural outliers like *Plan 9 from Outer Space* (Wood 1959), *Repo Man* (Cox 1984), and *Iron Sky* (Vuorensola 2012), sf provides an especially fertile ground for such studies, allowing for a productive

**FIGURE 1.3:** Digital effects visualize a world that might be: the world-train of *Snowpiercer* (2013) careens through a climate catastrophe. CJ Entertainment.

investigation of one of the enduring characteristics—and appeals—of the genre: its depiction of otherness, including what might be characterized as an industrial otherness, as these films are often made and find their audiences outside the conventional channels of the established film industry.

While thoroughly embedded in the cultural, all of these critical slants, just like the many slant-genre forms we have identified, challenge many of our conventional notions about sf, as they place the genre within a variety of shaping cultural and industrial processes (or beyond them), and as they demonstrate how those processes are effectively adapting/adopting the estranging voice of sf. The result, almost paradoxically, is that through their very variety and with their seemingly different identities, these slants also provide something of a challenge—to both readers and moviegoers. While focusing throughout this volume on the different science fictions that are emerging and on the different perspectives that currently inform both our critical thinking and our sf, the essays gathered here also point to notions that lie beyond simple national or even immediate cultural concerns. Particularly, they might help us to recognize something larger at work in the new sf cinemas and the thinking that attends them—a strong sense of the various boundaries that always seem to trouble our conceptions. For they insistently question how we conventionally think about the human, about the human space or place, and also about the narrative "place" of genre and its own often binding, limiting conventions. This is an important, even necessary recognition that, I would hope, imparts a kind of unity to the various articulations of difference and alterity that form this book's general structure and give it purpose.

## Notes

1. This collection of various namings was drawn from reviews of a single sf film, *Dr. Cyclops* (Schoedsack 1940), cited in the "Critics' Quotes" column of the movie trade paper *Motion Picture Daily* (1940). Similarly shifting designations can be found throughout the trade papers even into the mid-1950s, by which time sf is often thought to have cemented a more consistent sense of identity with the moviegoing public.
2. For a detailed discussion of this interweaving of theory and cinematic practice, especially in the context of cyberfeminism, see Short 2005.

## Works Cited

Bould, Mark, and Sherryl Vint. "There Is No Such Thing as Science Fiction." In *Reading Science Fiction*, edited by James Gunn, Marleen Barr, and Matthew Candelaria, 43–51. New York: Palgrave Macmillan, 2009.

Bryant, Levi R. *Onto-Cartography: An Ontology of Machines and Media*. Edinburgh: Edinburgh UP, 2014.

"Critics' Quotes." *Motion Picture Daily*, April 22, 1940, 8.

Delany, Samuel R. "About 5,750 Words." In *The Jewel-Hinged Jaw: Notes on the Language of Science Fiction*, 1–16. Rev. ed. Middletown, CT: Wesleyan UP, 2009.

Friedberg, Anne. *The Virtual Window: From Alberti to Microsoft*. Cambridge, MA: MIT P, 2006.

Haraway, Donna. "A Cyborg Manifesto: Science, Technology, and Socialist-Feminism in the Late Twentieth Century." In *Simians, Cyborgs and Women: The Reinvention of Nature*, 149–81. New York: Routledge, 1991.

James, Edward. *Science Fiction in the Twentieth Century*. Oxford: Oxford UP, 1994.

Kernan, Lisa. *Coming Attractions: Reading American Movie Trailers*. Austin: U of Texas P, 2004.

Latham, Rob. "Introduction." In *The Oxford Handbook of Science Fiction*, edited by Rob Latham, 1–19. Oxford: Oxford UP , 2014.

Mittel, Jason. *Genre and Television: From Cop Shows to Cartoons in American Culture*. London: Routledge, 2004.

Neale, Steve. *Genre and Hollywood*. London: Routledge, 2000.

Pierson, Michele. *Special Effects: Still in Search of Wonder*. New York: Columbia UP , 2002.

Rieder, John. *Science Fiction and the Mass Cultural Genre System*. Middletown, CT: Wesleyan UP, 2017.

Short, Sue. *Cyborg Cinema and Contemporary Subjectivity*. London: Palgrave Macmillan, 2005.

Sobchack, Vivian. "Images of Wonder: The Look of Science Fiction." In *Liquid Metal: The Science Fiction Film Reader*, edited by Sean Redmond, 4–10. London: Wallflower Press, 2004.

Sontag, Susan. "The Imagination of Disaster." In *Against Interpretation*, 212–28. New York: Dell, 1966.

Stewart, Garrett. *Cinemachines: An Essay on Media and Method*. Chicago: U of Chicago P, 2020.

Suvin, Darko. *Metamorphoses of Science Fiction: On the Poetics and History of a Literary Genre*. New Haven, CT: Yale UP, 1979.

Telotte, J. P. "What Is This *Thing*? Framing and Unframing the Science Fiction Film." *Film History* 31, no. 1 (2019): 56–80.

Tudor, Andrew. "Genre." In *Film Genre Reader III*, edited by Barry Keith Grant, 3–11. Austin: U of Texas P, 2003.

Wright, Judith Hess. "Genre Films and the Status Quo." In *Film Genre Reader IV*, edited by Barry Keith Grant, 60–68. Austin: University of Texas Press, 2012.

Yaszek, Lisa, ed. *The Future Is Female! 25 Classic Science Fiction Stories by Woman, from Pulp Pioneers to Ursula K. Le Guin*. New York: Library of America, 2018.

# 11

## THE SLANT SCREENS OF NEW SCIENCE FICTION CINEMA

# II

## THE SLANT SCREENS OF NEW SCIENCE FICTION CINEMA

# CHAPTER 2

................................................................................

# AFROFUTURIST CINEMA

................................................................................

## DE WITT DOUGLAS KILGORE

THE challenge of cinematic Afrofuturism is to create an Afrocentric iconography, a future in which the people and cultures of the African Diaspora direct human prosperity. This visionary cinema typically exchanges mimetic renderings of Black trauma for pictures of homecoming and liberation. Afrofuturist filmmakers use the resources of science fiction (sf) to tell stories that range from a contemporary realism slightly altered by a single novum, to near futures in which dystopian conditions spark a revolution, to futuristic heterotopias divorced from a valorizing whiteness. Whatever their narrative strategy, these filmmakers are allied in placing Black people at the center of visionary fantasies that remake how we see things to come.

Afrofuturist films imagine a yet to be in which Black communities flourish and afford a space for new values. They do so by engaging an oppressive history with a powerful array of creative resources to visualize the descendants of the African Diaspora as unencumbered human beings. Afrofuturist cinema thus moves away from the color-blind (whitewashed) futures offered by films like Stanley Kubrick's *2001: A Space Odyssey* (1968) or Stephen Spielberg's *Minority Report* (2002), and instead asks us to see a post-Black condition that might emerge in futures where the social construct of race has shifted from current usage. However, the awkward neologism "post-Black," often used to evoke such futures, should not distract us from recognizing how important it is to Afrofuturism's effect that we recognize its characters and worlds as "Black." Supporting that thrust, a distinguishing feature of these Black worlds is their reference to folklores and theologies outside of a Christian Europe.

Sf has often been theorized in a way that erases such mythic or religious knowledge from future practice—an exclusion that traditionally distinguishes the genre from fantasy or horror. The effect is to disentangle the rational-not-yet from the irrational days of yore, the future from the past. In practice, this habit often translates into future stories that are sanitized extrapolations of a deracinated twentieth-century America, as in the case of *Minority Report*, a film that benefited from the advice of futurologists primarily concerned with forecasting recent trends in urban design and advertising. This technocratic focus avoids wrestling with the persistent myths and presences that haunt human knowledge and that inform how we organize our sciences, technologies, and social relations. By including such knowledge, Afrofuturism narrows, as it also sometimes ignores, the customary gap between sf and fantasy.

Thus, the visual vocabulary of Afrofuturist cinema forms a kind of science fantasy that imports African legends and divine presences into current and future technoscapes, allowing circuit boards, computers, extraterrestrials, and space travel to coexist with summonings, witchcraft, and Panther gods. This combination opens space for a rethinking of sf's metaphysical as well as its cultural coordinates, thereby helping break the genre out of what Helen Young calls a "habit of whiteness" that has been a traditional feature of anglophone fantasy (2016, 10–11). In film this shift means moving Black and brown bodies; their experiences, concerns, and interiority; their humanity out of what Isiah Lavender III calls the "blackground" and centering them on the screen (2011, 19–20).

While much Afrofuturism has been focused on art in literature and music, this focus reflects the very different economics of the media involved. It is simply more expensive to make even a modestly budgeted motion picture than to produce a novel or compose an album. However, as an art form and a public medium, film can reach audiences unavailable to the other arts. While fiction is vital for working through the details of character and philosophy, and music essential for creating soundscapes that evoke important emotional architectures, film adds the power of composed images and evocative sound. Film art builds, photographs, and dramatizes what has been imagined on the page. This complement is especially important when film seeks to capture the fantastic or invented worlds of sf, since it makes our dreams part of a productive reality. The momentary heterotopias of film production—whether on location or in studio—thereby create diegetic worlds that reshape our perception of the real and the possible. How, then, do we define Afrofuturism as a distinct mode or genre of cinematic production? Are there particular patterns, characters, or visual signifiers that make a motion picture more or less Afrofuturist? For the purposes of this survey, I want to suggest three general types of Afrofuturist film that we might label: thin, strong, and revolutionary.

*Thin* Afrofuturism is characterized by films that focus on an African American or Black character in a narrative that makes no specific reference to racial or cultural difference, but that also does not isolate him or her from Black or biracial families and friends. These films, however, are usually not in direct conversation with Afrofuturist themes or tradition. They follow customary assumptions about the kinds of pictures that entertain the broadest possible (assumed to be predominantly white) audiences. Films that typify this category are *A Wrinkle in Time* (DuVerney 2018), *After Earth* (Shyamalan 2013), and *I, Robot* (Proyas 2004).

A movie centered on an African or African American who is strongly identified with and emmeshed in some aspect of African Diasporic culture may be considered an example of *strong* Afrofuturism. These films speak more directly to the Black experience (history) and illuminate futures occupied by thriving, politically ascendant Black communities. While technology may be an important part of these narratives, a neat separation between science and magic is rarely assumed. This conjunction allows for the weaving of non-Western mythologies, religions, and supernatural presences into human life. Reckoning with past oppression (slavery)—and drawing a redemptive message from the experience—is important, even necessary to present action and a sense of destiny. Some films that exemplify this mode are *Afronauts* (Bodomo 2014), *Black Panther* (Coogler 2018), *Brown Girl Begins* (Lewis 2017), *Destination: Planet Negro* (Willmott 2013), *Hello, Rain* (Obasi 2018), *The Last Angel of History* (Akomfrah 1996), and *Janelle Monaé: Dirty Computer* (Donono and Lightning 2018).

*Revolutionary* Afrofuturism emphasizes revolt against racist oppression. Filmmakers working in this mode accept the Frederick Douglass adage that "power concedes nothing without a demand" (1999, 412). In some of these films, revolution is strongly inflected by the need to challenge the patriarchal nature of racial control. Such films are the strongest statements of an Afrofuturism that seeks change through militant political activity. Examples of films suggesting this direction are *Born in Flames* (Borden 1983), *Welcome to the Terrordome* (Onwurah 1995), *Robots of Brixton* (Tavares 2011), *Space Is the Place* (Coney 1974), and *The Brother from Another Planet* (Sayles 1983).

When we define a film as Afrofuturist, we also need to consider whether the term covers any futuristic or fantastic feature that has a Black person as the central character or only those films written, directed, or produced by a Black filmmaker. Are such films as *The World, the Flesh and the Devil* (MacDougall 1959, starring Harry Belafonte), *Demolition Man* (Brambilla 1993, featuring Wesley Snipes), *Supernova* (Hill1 2000, with Angela Bassett), *The Adventures of Pluto Nash* (Underwood 2002, starring Eddie Murphy), *The Matrix* (Wachowskis 1999, featuring Lawrence Fishburne), and *Men in Black: International* (Gray 2019, with Tessa Thompson) automatically considered Afrofuturist because African American actors play leading roles in them? This question might be sharpened by briefly considering the career of one of the top Black actors, Will Smith.

In the mid-1990s Smith emerged as the first Black film star for whom sf became an important genre. This development was significant not only for marking the commercial viability of a Black actor in the motion picture industry, but also for demonstrating sf's marketability. However, slapping an Afrofuturist label on Smith vehicles such as *Men in Black* (Sonnenfeld 1997), *I, Robot* (Proyas 2004), *I Am Legend* (Lawrence 2007), and *Hancock* (Berg 2008) seems too casual. These projects leveraged his particular urbane Blackness to disturb the habitual whiteness of their sources. However, their links to Black experience and futurity are limited to and by Smith's persona. Afrofuturism, as defined by Dery (1994), Nelson (2002), Womack (2013), and Kilgore (2014), is often filtered through an individual's experience, but almost always as a participant in a vibrant Black communitas. The adventures of Smith's mostly deracinated heroes elides Afrofuturism's social character. Thus, Smith and similar Hollywood stars (such as Denzel Washington and Halle Berry) often represent a Blackness that has been integrated into futurities in which whiteness is still the norm. In short, their roles could be undertaken by white actors with little to no revision. The Black star's job is simply, as Adilifu Nama offers, to make the future look cool (39). If this is Afrofuturism, it is of a kind that is only skin deep.

In this same vein, should we label an sf movie produced by a Black film company Afrofuturist? The Will Smith–produced adventure *After Earth* (2013), directed by M. Night Shyamalan, offers one answer to this question. The film is set in an apocalyptic future when humanity has had to leave its native planet and colonize another world. It is based on a story by Smith and was produced by his company, Overbrook Entertainment. It meets the sociopolitical goal of having people of color (Black and Asian) in important roles both in front of and behind the camera. *After Earth* establishes a future in which Black people not only survive but thrive in roles critical to the destiny of humanity. The centrality of Smith's character and that of his on-screen family—marked by their obvious wealth and unquestioned social prominence—disrupt the visual conceit that the human future is hegemonically white (in décor or personnel).

However, *After Earth* is a military sf/survival/coming-of-age/father-son story that avoids deep engagement with the sociocultural lives of its characters. It is enough that they are courageous against an alien threat and that the father/son bond is healed. The social effect of the tale and its resolution are very similar to that established by Robert A. Heinlein's *Starship Troopers* (1959), the novel at the foundation of much modern military sf. The film shares that novel's political strategy of making the military a colorblind heterotopia that defends the human species against alien predators (see figure 2.1). However, while Heinlein elides his lead character's Filipino heritage for much of the book, *After Earth* makes the race of its heroes plainly visible. The result is, admittedly, a kind of progress.

FIGURE 2.1: "Thin" Afrofuturism offers a Black hero (Will Smith) in a conventional space adventure: *After Earth* (2013). Sony Pictures.

Ava DuVerney's *A Wrinkle in Time* is another example of what might be called thin Afrofuturism. The motion picture seems to appear on Afrofuturist lists for three reasons. First, DuVerney is the first Black woman to direct a multimillion-dollar mainstream motion picture. She came to the project from an award-winning career in independent cinema with films that directly address the Black experience. Her task was to bring that background to adapting a classic of children's literature, Madeleine L'Engle's *A Wrinkle in Time* ([1962] 1978). A second reason is that Black and brown actors make up the majority of the film's cast. This gesture updates the exclusively white worlds (real and fantastic) of the source novel. The original white family is recast as interracial with a Black mother and a white father. Thus, Meg Murry (Storm Reid), the protagonist, is biracial, and her mother, Mrs. Murry (Gugu Mbatha-Raw), is Black. A third reason is that the otherworldly Mrs. Which and Mrs. Who are played by Oprah Winfrey and Mindy Kaling. As Black and brown women who are culture industry stars, they lend a kind of box office credit to their parts as celestial beings who represent the forces of cosmic good. Recasting these iconic children's literature characters as Black and brown women is a deliberate attempt to allow non-white viewers to do more than "racebend" themselves into popular narratives (Warner 2015).

However, the problem with defining this film as fully Afrofuturist is a function of both the source and its adaptation. L'Engle was concerned with the relationship between science and faith, the nature of good and evil within a Christian eschatology. While her staging of these concerns in a story invoking time and space travel offers a vision of the future, it is not a foretaste of a multiracial human destiny, much less one that gives direct agency to Black people as a cultural group. DuVernay's adaptation complicates the story's social vision with the casting noted earlier, but only through the visual signs supplied by a wider variety of physiognomy and skin tones. Because the screenplay hews fairly faithfully to the L'Engle novel, it does not allow for the alternate sociocultural content of a strong Afrofuturism. We see, instead, a deracinated middle-class mise en scène as the playground on which human destiny unfolds. As a result, the film can be read as strongly Afrofuturist only by extrapolating outward from its explicit interest in Meg's individual salvation. The film aims neither to offer itself as a metaphor of the Black experience nor to tease out attractive images of that experience as the foundation of things to come. Instead it places Black and brown women into a social world that implies a deracinated cosmos. If this science fantasy implies a *particular* future vision, it is one in which space travel may be undertaken in a Manichean universe divided between a light and a dark that do not map onto racial signifiers. The result is a liberal, pluralist vision, but not one that is Afrofuturist in either a cultural nationalist sense or in the way of *Black Panther*'s United Nations–style cosmopolitanism.

The films that I would define as strongly Afrofuturist come from both independent and mainstream cinema. They range from short films, available either at film festivals or online, to features produced by independent companies for art houses (such that remain), to Disney-produced blockbusters. Naturally, these films do not all take the same approach, nor do they represent a coherent school with a manifesto that orients their aesthetics in a single direction. While a film like *Space is the Place* can be read as foundational, a more recent work such as *Dirty Computer* is in more direct conversation with an Afrofuturist cinematic tradition. And while films like *Destination: Planet Negro* and *Black Panther* are recent enough for their filmmakers to be aware of precedent, they are more directly in tune with literary, historical, and comic book sources than with an sf tradition that includes Black artists as paradigmatic influences. Even *Dirty Computer*, which is strongly linked to a Black musical tradition that includes Afrofuturist pioneers such as Sun Ra and George Clinton, derives its futurist iconography from the utopian/dystopian models established by Fritz Lang's *Metropolis* (1927) and George Lucas's *THX 1138* (1971). This link is not a fault but an inevitable feature of a cinematic Afrofuturism that speaks from within a hybridized Western culture that Black artists have played a critical role in creating. Thus, the performing arts of the African Diaspora are strongly influenced by European-derived practice as they recover and enact their African roots.

In an effort to define Afrofuturism as a particular mode of Black artistic production, Mark Dery cited John Sayles's *Brother from Another Planet* (1984) and Lizzie Borden's *Born in Flames* (1983) as films that offer "glimpses" of Afrofuturism (1994, 182). These films are, in fact, examples of a revolutionary Afrofuturism, as productions that directly engage alternatives to liberal ways of seeing sociopolitical change. Dery, however, overlooks John Coney's *Space Is the Place*, an earlier effort inspired by the figure and performances of Sun Ra, the adventurously philosophical jazz composer. I would argue that *Space Is the Place* has a solid claim as a foundational expression of Afrofuturist cinema. While it shares with Dery's examples the distinction of being written and directed by a white production staff, it is profoundly shaped by its star performer's style and sensibility.

*Space Is the Place* is a generic hybrid: documentary, music video, and narrative film. It puts into play the past and the present in terms of a speculative future. It is "cinematic" in the sense that it uses its generic furniture to tackle several big issues: racism in America, the "unreality" of Black people in a world system predicated on white supremacy, sociopolitical tensions within the Black community, the possibility of a second chance in "outer space," the opening of that chance through new science and technology, the following of a philosophical (even religious) alternative to mainstream preoccupations, and the potential for living beyond capitalism. It supports these issues by creating an expansive mise en scène that frames its documentary-style treatment of 1979 Oakland, California, with extraterrestrial and allegorical settings. The film also prompts us to consider the place and role of Black women in Sun Ra's messianic yet patriarchal vision. As a representative of Sun Ra's philosophical radicalism, the film establishes an Afrofuturism that is not only a rejection of whiteness, but also of the institutional inequalities that place some Black men in positions of attractive, exploitative authority.

*Space Is the Place* gives us two heroic icons in a dichotomous struggle between racial good and evil. Sun Ra represents the good—he is introduced as an elaborately costumed celestial visitor, the Savior/messiah who will take the Black community to a world free of white people. But he is also the rebellious musician whose creativity and inventiveness can literally blow away the tawdry spectacle of Black exploitation. His power is made evident by the practical effects of a nightclub scene. "Sonny" (Sun Ra's character) is opposed by the Overseer (Ray Johnson), a Superfly-style pimp figure who operates within the Black community while exploiting it. They duel for the community's soul by way of a chess game, in a gesture that seems deliberately evocative of the match between Death and the Knight in Ingmar Bergman's *The Seventh Seal* (1957). Afrofuturism, as modeled here, is a mode firmly grounded in the African American urban past (a 1940s jazz sequence), the present (the situation in Oakland during the 1970s), and a fantastic extraterrestrial alternative. The chess sequence elevates the Black community's fate from the quotidian to the mythic: a struggle between good and evil, life and death, slavery and freedom that projects human destiny into outer space.

Sun Ra's rebellion against the white-dominated world (the now) is encapsulated in his flat declaration, "I hate your reality." This sentiment makes him, arguably, the most radical Afrofuturist in any form. It is important to note that the film's revolutionary Afrofuturism advocates the overthrow of white authority and its Black disguise in the white-suited Overseer by escaping from it. Thus, it has some resonance with the Garveyite back-to-Africa movement of Sun Ra's youth. In his life and art, the musician made little concession to the ways of the American mainstream. His disinclination to accommodate his vision to what he saw as a corrupt history is the catalyst for his thought and his art, so his Astro-Black myth suggests a new way of seeing the future.

Through its form and content *Space is the Place* establishes cinematic Afrofuturism as a revolutionary vision, appealingly grounded in its refusal of any stricture levied by hegemonic whiteness or a grounded Afro-American social realism. However, that very quality makes it a hard act to follow, partly because Blacks are immersed in, and thus vulnerable to a Western tradition wherein African American art makes sense. An Afrofuturism that does not take account of this system, that excises the history we know for a potpourri of popular Egyptology mixed with alien abduction conspiracy and a generous dollop of mystical reading, is hard to digest. So, what have later Afrofuturist filmmakers salvaged from the ruins of Sun Ra's eccentric, mid-twentieth-century vision? How has Afrofuturism built on Sun Ra's initiative while correcting some of his faults, and how has it offset what Dery identifies as white "ownership" of the future (1994, 180)?

In the decades after *Space Is the Place*, Afrofuturist cinema has followed two productive streams: first, short and feature films by independent American and international producers, destined for festivals, art houses, and streaming services; and second, mainstream features produced by the Hollywood system. There is more work in the first category than the second. Important independent productions include the aforementioned *The Brother from Another Planet*, *Born in Flames*, and *Space Is the Place*, while more recent titles include *Destination: Planet Negro*, *Dirty Computer*, *Afronauts*, *Hello, Rain*, and *Brown Girl Begins*. As I have here defined an Afrofuturist cinema, the second category holds just two films, *A Wrinkle in Time* and *Black Panther*. Each of these categories contains examples of *thin* and *strong* Afrofuturism. Revolutionary Afrofuturism is, so far, the exclusive province of independent cinema.

Against the background of these thinly Afrofuturist films, we might better view some in the strong camp. One such effort is Kevin Willmott's *Destination: Planet Negro*, which draws inspiration from the low-budget sf film serials, like those from Republic Pictures in the 1930s and 1940s. The film's Afrofuturism is premised on a historical vision that links the current reformed racial landscape to a more oppressive past, as it offers a satire of twenty-first-century racial politics. Its political vision is buttressed by its portrayal of a Black community whose imbrication in the ordinary political, religious, educational, military, and scientific regimes of the American nation makes unity against racist oppression difficult. Willmott's satire resolves these debates in favor of the powers and perspectives offered by liberal, science-based education. His characters time travel from a black-and-white early twentieth century to a full-color twenty-first in which "we are beautiful." The present that "we" inhabit is revealed as the Afrofuture that scientific explorers of the past sought when they first escaped planet Earth. *Destination: Planet Negro* reinforces the idea that cinematic Afrofuturism is a fundamentally optimistic, speculative modality. While a Black future might not be utopia, neither is it a dystopia in which whiteness has uncontested dominion. Willmott's film gives us a comic Afrofuturism that cuts against the pessimism inherent in Sun Ra's mid-twentieth century dreams of escape from planet Earth as a viable solution, as the actions and knowledge of Black people make an Afrofuture manifest here on earth.

Following *Destination: Planet Negro*'s lead, four strong Afrofuturist films were released in 2018: *Brown Girl Begins*; *Hello, Rain*; *Janelle Monaé: Dirty Computer*; and *Black Panther*. This body of work indicates that we may be approaching a tipping point in terms of the kinds of cinematic storytelling that can attract consistent investment and distribution. They demonstrate, in any case, that Black filmmakers have achieved a visibility reminiscent of blaxploitation in the 1970s and the 'hood films of the 1990s. The economic, even critical success of films from these moments made the film industry's portrayal of race less monolithic—thereby giving crucial support to the careers of some Black filmmakers—but did little to change its white habits. However, *Black Panther*'s economic, critical, and social successes have sparked hope that the film industry is entering a new age of "diverse" production. Thus, an Afrofuturist film came to represent an industry seeking solutions to the racial inequalities spotlighted by the #OscarsSoWhite movement. Unfortunately, it is impossible to predict whether these films herald the habitual production of films in which Black people are the agents of things to come.

*Brown Girl Begins* boasts an impeccable pedigree from contemporary sf. It is based on Caribbean-Canadian author Nalo Hopkinson's award-winning 1998 novel, *Brown Girl in the Ring*. Modestly budgeted, this Canadian film is set in a dystopian future when poor, mostly Black people have been exiled from Toronto. This future Canada's struggles with class, race, and gender have metastasized. However, while Black women are at the margins of power, their concerns and abilities are privileged. The film's initial focus is on its young protagonist's desire to escape her grandmother's plan that she take her mother's place as the vessel of a supernatural power, even though such a personal surrender could help save her people. A standard treatment of the hero's journey would follow her from rural poverty and superstition to big-city Toronto where she would fulfill her dreams and, perhaps, find the knowledge needed to enlighten her community. Thus, the poor Black immigrant would find salvation in the predominantly white domed city, but in deference to its source—and testimony to the director's sensibility—*Brown Girl Begins* avoids this mythic convention. Instead, the film's hero overcomes her fear, reconciles with her grandmother, and becomes a priestess who hosts her family's power. By reconciling with her ancestral/cultural background and making peace with its spiritual powers, she opens the way to a better future for her community, one that will be different from that defended by the towers of the domed city. Like Julie Dash's *Daughters of the Dust* (1991), *Brown Girl Begins* gives us a world in which the lives and affinities of Black women make the future possible.

*Janelle Monaé: Dirty Computer* is a similarly strong Afrofuturist production. This "emotion picture," as it is termed, is committed to a vision that breaks with conventions about how race, queer sexuality, and gender roles may be presented in an Afrofuturist text. *Dirty Computer* is like Sun Ra's *Space is the Place* in that its effect is catalyzed by viewers recognizing its central character as a well-known musician. And like the older film, it may simply be read as a vehicle for its star's celebrity. However, while *Space* is awkwardly shaped around concert footage, *Dirty Computer* uses its fictional sections as the frame for music videos shot in support of an album. The videos are presented as the dreams or memories of Monaé's on-screen avatar, Jane 57821, who must endure reeducation in a dystopian reality. While more than a celebrity vehicle, the film does exploit Monaé's standing as a multimedia star to authorize a liberation from traditional racial/sexual identities.

Casting a Black woman as the intellectual and emotive center of an sf narrative shifts how we perceive its depicted social relations. Monaé's initial reputation as a musician/performer was through the concept albums that constitute her *Metropolis Suite*. In that music and the supporting tours, she presented herself as an android, Cindi Mayweather, rebelling against the limits and prejudices of the society that made her. Until recently, cinematic cyborgs have almost always been cast as white. This was the case even if their situation can be read as metaphorically Black, as has been argued for the replicants of *Blade Runner* (Scott 1982) (Bukatman [1997] 2008, 76). Other examples are the "robot Maria" (Bridgette Helm) in *Metropolis* and "Andrew" (Robin Williams) of *Bicentennial Man* (Columbus 1999). The former is a tempestuous threat to an explicitly racialized social order controlled by men; the latter's great goal is to become "human" as a white upper-class male. Their difference from white humanity is subcutaneous, their danger only revealed by the "shocking" exposure of tubes and wires. Their status as replacements for labor categorized by ethnicity, race, class, or gender is a metaphorical contrivance that either ducks the issue or a cognitive estrangement that produces sympathetic engagement. In either case, to be "human" as a Black woman is not the default condition of these sf works. Even though such films are almost always constructed to affirm some normative definition of human being, Blackness is customarily excluded as part of the design, as we also see in *A.I.: Artificial Intelligence* (Spielberg 2001). The implication is that by metaphorizing race through humanoid aliens or manufactured beings, sf transcends our customary racial stratification. Thus, as Scholes and Rabkin have argued, the genre renders "race comparatively unimportant" (1977, 189). Unfortunately, this postracial liberalization may leave whiteness in the position of an unmarked universal.

Monáe's incarnation of the android aspires to the human condition but not to a state of heteronormative whiteness. Her innovation is to remake herself, a Black woman (while her sexuality undergoes redefinition, her race and gender do not), as an android in a way that foregrounds racial difference. Her imitation of life is the traditional impression drawn from the skin and one that can be read at a glance—open access panel not required. Thus, her android's goal is not to become unremarkable, but to lead the way to a self and a communitas free from heteronormative/monoracial convention. Through her Mayweather persona, we see an experiment in shifting the sf "icon of the robot" (Wolfe 1979, 151–83) away from simply transcending or representing Black abjection. It may now refer to both a particular Black woman and the communities to which she belongs.

Monáe's android performance was the first stage of her transformation into an Afrofuturist figure who can represent (a subaltern) humanity. In *Dirty Computer* she is an ordinary woman, as the Mayweather character has done its work. As a famous actor Monáe now carries with her the parts that she has played, allowing audiences to see each new iteration as a variation that layers and amplifies her central themes: difference, joy, community, love. Her performance of a Black woman revolting against a dystopian near future signals a general embodiment that is quite different from the universality accorded to speculations foregrounding either white survival or destruction. Jane, along with her dream selves, is the Black icon who is the best hope for a human future. Since she is also part of a polyamorous ménage à trois, when Jane escapes from brainwashing with her male and female lovers, it is also clear that she represents a reproductive future in which Black people both survive and change. While Monáe's Afrofuturism shares a great deal with that of Sun Ra's *Space Is the Place*, her work thus represents a break from that movie's patriarchal vision. The crucial difference between the two is bound up with her generation's public embrace of LBGTQ+ identities as both valuable and ordinary—her future authorized by the portrait of a young artist who is a queer Black woman.

Of the four *strong* Afrofuturist films released in 2018, Marvel's *Black Panther* is the most visible and successful. It became the highest-grossing North American film in 2018, the ninth-highest-grossing film of all time, and the highest-grossing motion picture by a Black director. The film's success has intervened in old debates about the position of Blacks in a white-dominated film industry and helped bring the hitherto countercultural term Afrofuturism into popular usage. But much of that impact is specifically due to a cinematic design that renders the fictional African nation of Wakanda as a technoscientific utopia: a fictive-historical alternative to the racist colonialism that informs the global inequalities of our world. By placing an attractively pictured pan-African elite in its foreground, the film takes the spectacle of cinematic sf away from unconscious investment in futurisms founded in the history of white colonizers, while offering a strikingly novel iteration of a conventional adventure film.

The Marvel Cinematic Universe, a media franchise focusing on superhero adventures, gave *Black Panther* a generic, economic, and distributive platform unavailable to prior Afrofuturist films. While this cinematic sandbox has presented itself as socially diverse, its films have largely followed Hollywood convention as vehicles for heroic white masculinity. As the only film of twenty-two (and counting) that prioritizes an Afrocentric narrative, *Black Panther* is unique, although not isolated within its generic container. The events in this film are shown to have social and political impact in the larger franchise "universe," as the ideals held by its characters and the actions they take are understood to have consequences for individuals, races, and nations beyond this narrative. As a result, *Black Panther*'s Afrofuturism is positioned as a difference that can change the social investments of the universe produced by Marvel Studios. The film's social and political goals reside in dramatic images—of people, their machines, buildings, and the landscapes they occupy—that seek to change how we see the world.

Martin Green argues that the adventure narrative, as a literary genre, has conventionally served as the imaginative script for a world system controlled by a white Western male elite (16). However, *Black Panther* revises that practice by moving exotic adventure's supporting players into roles of heroic privilege, thus reforming the genre's habit of whiteness. In its place, the film's adventurous mise en scène correlates to a world system that will be changed by the existence of a super-scientific African civilization. That novum heralds a clear political hope: that humanity can turn away from the failures and corruptions of white Western hegemony and toward the wiser, more humane course found in an Afrofuturist vision.

Admittedly, *Black Panther* is fairly conservative in that it immediately draws our gaze to an attractive male who seeks revenge for himself, as well as salvation for his people. The protagonist T'Challa (Chadwick Boseman) is a superhero, scientist, military tactician, and spiritual leader, and an extremely wealthy bachelor. Historically, racist beliefs would have made assigning any of these qualities to a Black character almost impossible. However, T'Challa is African (Wakandan) and not African American. As such, the character is an outsider to the history of slavery, segregation, and civil disenfranchisement that informs African American identity. That remove gives *Black Panther*'s African American writers room to complicate the film's political message by doubling its title character. Screenwriters Ryan Coogler and Joe Robert Cole created two Black Panthers: T'Challa, the African king, and Erik "Killmonger" Stevens/ N'Jadaka (Michael B. Jordan), his American cousin. The narrative's politics is staged primarily as a duality between these two Panthers who both, in their different ways, seek revenge and salvation. Both sit on the Wakandan throne and hope to use its vast technoscientific might to transform the world we know, so our sympathies are initially divided between the two. Whom do we follow—the royalist African Panther from Wakanda or the revolutionary American Panther from Oakland, California? The film thus offers two different images of political hope.

This reading of the film's two leads, one a superhero, the other a supervillain, helps illustrate the difference Afrofuturism makes in the construction of the adventure narrative. It allows the filmmakers to intervene directly into the history of race and racism in America, as well as the way we now live. They make a virtue of the historical fact that the comic book character's creation, by Jewish American comics creators Stan Lee and Jack Kirby in 1966, is entangled with the emergence of the Black Panther Party (BPP) that same year. The Coogler and Cole screenplay ties the Wakandan Royal family to that history by making one of its heirs a member of that revolutionary movement in Oakland. While the choice to make the villainous Killmonger a child of that political movement is not unproblematic, it should not divert attention from the film's sympathetic introduction to the conditions that inspired the BPP's call for militant action in defense of the Black community. This situation allows the film to stage a political conversation/debate that takes place within the Black community. We are not asked to choose between a white futurity and a Black not yet, but instead, are given two sympathetically rendered kinds of Black political speculation about our problems and how to solve them, with the film making clear that both human salvation and the fate of a free world hinge on how that debate is resolved. It is a remarkable feature of our moment, since this kind of political focus remains a culturally radical idea, yet it is expressed in the immensely popular form of sf.

By examining the settings that frame its characters we can also better appreciate how *Black Panther*'s strong Afrofuture is realized. One of the film's most striking sequences is the moment when three of its protagonists, T'Challa, Nakia (Lupita Nyong'o), and Okoye (Danai Gurira), fly across Wakanda's frontier and over its border. The scene opens with their ship cruising over the African heartland, revealing a vast rolling landscape enlivened by streaming herds of wild game and energetic natives on horseback, keeping pace. This aerial view recalls the century-old convention of wildlife or exploration photography sponsored by institutions like the American Museum of Natural History or television's *Mutual of Omaha's Wild Kingdom* (1963–1988, 2002–2018). It is an image that celebrates the power to survey non-Western landscapes, to cast them as a lush, primal past, flying before the advance of civilization, ripe for exploitation. But that image is recontextualized when the Wakandan jet speeds into an ultramodern Africanesque cityscape. The light changes, the music swells, and Okoye smiles proudly while saying, "Sister Nakia, my prince, we are home." As they (and we) survey the beautiful, bountiful land and then, after breaking through the verdant illusion, Wakanda's futuristic capital, T'Challa responds, "This never gets old." We have discovered the real Wakanda, with the wild kingdom revealed as simply a disguise, a stereotype the world is all too willing to accept. It hides Wakanda's true nature from a system that is, by contrast, a nightmare. This sequence is one of the film's iconoclastic moments. As the standard image of Africa as an exotic safari land is broken, Wakanda's golden city, Birnin Zana, takes its place (see figure 2.2). The audience is encouraged to love this new vision of Africa as much as its people do.

Following this bravura introduction, a street-level tour of the Wakandan capital underscores the social order that T'Challa and Nakia defend. Coogler's camera escorts us through a vibrant street scene that demonstrates the wealth and diversity of the Wakandan way. Costume designer Ruth E. Carter's designs are elegantly contemporary, while referencing a variety of African traditions (Baer 2019). High-tech communication and transport technologies facilitate the more tactile face-to-face of marketplace encounters. The scene is an attractive prospect: the kind of busy, peaceful commerce common on the utopian side of cinematic sf. This scene is an invitation to both the film's Afrotopian diegetic and its gloss on pan-African politics. The density and dynamism of

FIGURE 2.2: "Strong" Afrofuturism and a new image of Africa: entering Wakanda in *Black Panther* (2018). Walt Disney Studios.

its lighting, scene, and costume design lend the place a momentary reality—the kind of realism that might persuade us to seek its imitation in our own world.

In an interview, Hannah Beachler, *Black Panther*'s production designer, recounts the work that went into crafting this aura of authenticity for the film's Afrofuturist Wakanda (Flatow 2018). Her research produced a five-hundred-page "bible" that gathered impressions from several African nations, especially of their landscapes and traditions. Her design program included cultural and historical rationales for the existence of particular buildings, such as the Hall of Records, and landscape features, like Warrior Falls (Edwards 2018, 56, 65). While these places are prominent in the film's mise en scène, they are neither named nor explained; they are just there, as backgrounds and settings, as frames of reference for character and action. They inspired the filmmakers, but also entertain the curious eye with what seems a thickly detailed reality. Beachler's postfilm descriptions of the nature and function of particular buildings, interior designs, and landscape features satisfy that curiosity, providing us with the rationale for an Afrofuturist design philosophy. Moreover, it is an artistic gesture that insists on displacing the entrenched images of wild Africa with an ultramodern, Afrofuturist spectacle. In this way the film seeks to replace the colonial-era fantasy of Africa as "dark," a place that can only represent the past instead of the future or—following Chinua Achebe's critique of Joseph Conrad's *Heart of Darkness*—a projection of the darkness abiding in the white mind (Achebe 1988, 247–48). It is this potential that T'Challa wishes to preserve and that Nakia hopes to share.

The political import of *Black Panther*'s Afrofuturist iconography is highlighted by T'Challa's later speech at the United Nations. The scene broadcasts an ascendant Africa. The king stands before the Assembly in modern garb, flanked by attractive women whose form-fitting attire is equally contemporary. In this scene costume designer Carter chose not to dress the characters in the traditional clothing we see elsewhere in the film. She hints at a non-European civilization by placing a subtly patterned scarf atop T'Challa's suit and augmenting the accompanying women's (two of whom are important players in the adventure) sheath dresses with equally subtle jewelry. This strategy marks Wakanda as a forward-looking nation that lives within the global circuit of fashion, not apart from it. Carter's stylistic choices help place Wakanda in a space similar to that occupied by Japan, a modern global power whose habits are in conversation with but not directed by the West. The image, therefore, underpins the proposal that T'Challa makes in his speech: that Wakanda has things to give to the world that could change it for the better.

The African monarch's speech also lays out a vision of world peace sponsored by the Wakandan people. The incredulity with which his statement is greeted by a white European diplomat is a sign of contemporary geopolitical reality. Independent, postcolonial Africa remains subject to low expectations under the current world system. What we have been trained to expect from its nations is postcolonial bad news. While their peoples may have rich histories and vibrant cultures, they are always cast as "developing," never arrived. This is the stereotype that Wakanda has used as protective cover, hiding its wealth from an avaricious, racist hegemony. Since Wakanda has used the West's own prejudices against it, it is no wonder that the white diplomat cannot see what is in front of him: the illusion is too powerful. But breaking that illusion, removing that disguise, is the goal of an Afrofuturist vision.

It is also a goal that drives a number of *Black Panther*'s other shifts in our customary cinematic vocabulary. Its delineation of African women, the non-European elements of its soundtrack, its vivid rendering of a non-Christian religion and mythology are all part of an exercise in world-building that lends stability to the film's pan-Africanist Afrofuturism. While there may be faults in the film's political vision or technical execution, its robust realization of Afrofuturism has shown mass appeal and acceptance. This success means that, as a cinematic mode, Afrofuturism has moved from the margins of artistic expression to the center of popular entertainment, where it might be replicated and monetized in ways that we can both foresee (in merchandising and sequels) and yet cannot imagine. What is clear is that, whether revolutionary, strong, or thin, the new Afrofuturist cinema embodies a politico-aesthetic that allows for a vibrant mode of cultural (re)invention. It produces an iconography that breaks with clichéd white salvationist future visions. As other contemporary filmmakers, like Kibwe Tavares (*Robots of Brixton*, 2011) and Sama'an Ashrawi (*Afrofuturism*, 2017), exploit online distribution channels with their own efforts in this vein, additionally diverse narratives should appear in which the African Diaspora argues for a flourishing human future.

## WORKS CITED

Achebe, Chinua. "An Image of Africa: Racism in Conrad's *Heart of Darkness*" (1975). In Joseph Conrad, *Heart of Darkness*, 251–61. New York: Norton, 1988.

Baer, April. "Designer Ruth E. Carter on Creating Oscar-Winning Looks for *Black Panther*." OPB. September 8, 2019. https://www.opb.org/radio/article/black-panther-costume-design-ruth-carter-interview/.

"*Black Panther* (film)." *Wikipedia*. Accessed August 1, 2020. https://en.wikipedia.org/wiki/Black_Panther_(film).

Bukatman, Scott. *Blade Runner*. London: British Film Institute, [1997] 2008.

Dery, Mark. "Black to the Future: Interviews with Samuel R. Delany, Greg Tate, and Tricia Rose." In *Flame Wars: The Discourse of Cyberculture*, edited by Mark Dery, 179–222. Durham: Duke UP, 1994.

Douglass, Frederick. *Frederick Douglass: Selected Speeches and Writings*. Ed. Philip S. Foner. Chicago: Lawrence Hill Books, 1999.

Edwards, Graham. "Black Panther." *Cinefex* 158 (April 2018): 54–82.

Flatow, Nicole. "The Social Responsibility of Wakanda's Golden City." *CityLab*. November 5, 2018. www.citylab.com/life/2018/11/black-panther-wakanda-golden-city-hannah-beachler-interview/574420/.

Heinlein, Robert A. *Starship Troopers*. New York: Ace Book, [1959] 1987.

Kilgore, De Witt Douglas. "Afrofuturism." In *The Oxford Handbook of Science Fiction*, edited by Rob Latham, 561–72. New York: Oxford UP, 2014.

———. "Welcome to Wakanda: Reforming African Adventure the Marvel Way." *Paradoxa* 25 (2013): 228–54.

Lavender, Isiah, III. *Race in American Science Fiction*. Bloomington: Indiana UP, 2011.

L'Engle, Madeleine. *A Wrinkle in Time*. New York: Farrar, Strauss and Giroux, [1962] 1978.

Le Guin, Ursula K. *The Language of the Night: Essays on Fantasy and Science Fiction*. New York: Berkley, 1982.

Nama, Adilifu. *Black Space: Imagining Race in Science Fiction Film*. Austin: U of Texas P, 2008.

Nelson, Alondra. "Introduction: Future Texts." *Social Text* 71 (Summer 2002): 1–15.

Scholes, Robert, and Eric S. Rabkin, *Science Fiction: History-Science-Vision*. New York: Oxford UP, 1977.

Warner, Kristin J. "ABC's *Scandal* and Black Women's Fandom." In *Cupcakes, Pinterest, and Ladyporn: Feminized Popular Culture in the Early Twenty-First Century*, edited by Elana Levine, 32–50. Champaign: U of Illinois P, 2015.

Wolfe, Gary K. *The Known and the Unknown: The Iconography of Science Fiction*. Kent, OH: Kent State UP, 1979.

Womack, Ytasha L. *Afrofuturism: the World of Black Sci-fi and Fantasy Culture*. Chicago: Lawrence Hill Books, 2013.

Young, Helen. *Race and Popular Fantasy Literature: Habits of Whiteness*. New York: Routledge, 2016.

# CHAPTER 3

## BIOPUNK FILM

### LARS SCHMEINK

RIDLEY Scott's *Blade Runner* (1982) is widely considered to be one of the most essential cyberpunk films; its depiction of a bleak, dystopian world of the twenty-first century has become iconic to science fiction (sf) cinema. Interestingly, though, its central novum is not the virtuality of computer worlds or the mechanical augmentation of human bodies that we find in many other cyberpunk films. Instead, the film discusses the creation of replicants, a new slave race of genetically engineered posthumans with superior abilities. In this focus, *Blade Runner* can be considered an early example of what Brian McHale refers to as the "'bio-punk' sub variety of cyberpunk SF" (1992, 257). Biopunk, like cyberpunk, negotiates how technologies in a late-capitalist globalized economy can push life toward becoming a commodity. Both are integral to trans- and posthumanist[1] discourses—concerned with technoscientific developments and their influence on life—with biopunk preferring "bio-engineering techniques—cloning, genetic engineering" (McHale 1992, 255) over cybernetic or biomechanical technologies as the driving force. Biopunk is, so to speak, the younger sibling of cyberpunk—whose cyberspace motif and promise of bodily transcendence took the limelight during the 1980s and 1990s—that has now largely displaced its predecessor and become a cultural formation in its own right.[2]

## BIOLOGY IN SF FILM

Dealing mainly with genetic engineering as a posthuman technology, but also other biotechnologies such as tissue culture and cloning, biopunk cinema has its origin in a broader range of sf biological themes, among them "evolution, genetics, and comparative or exobiology" (Parker 1977, 11). As Helen Parker argues, genetic sf typically "develops two major premises": life might be haphazardly altered through the "genetic accident" or it might be purposely changed by "genetic alteration" (1977, 35). And even though early sf authors and filmmakers could not fully understand the science behind genetics, examples such as H. G. Wells's *The Island of Doctor Moreau* (1896) anticipated the science, depicting the potential and consequences of the alteration of life. In the film adaptation *Island of Lost Souls* (Kenton 1932), human-animal hybrids are not created through genetic splicing but by surgical intervention, with the film's main premise showcasing how a posthumanity is denied subject status by Western bio-medical science and treated merely as an object of study. In such early sf films science largely functioned as a backdrop or gimmick. But as audiences became more knowledgeable and sophisticated, so did the cinematic depictions of biological science. In his *Biology Run Amok!*, Mark C. Glassy cites several Universal horror movies, from *Son of Frankenstein* (Lee 1939) to *House of Dracula* (Kenton 1945), that use images of blood cells to provide supposed explanations of how their various monsters are the product of mutations in their biology (2018, 17–19).[3] A similar pattern appears in many creature features of the 1950s, wherein biological alterations resulting from nuclear radiation or chemical spills produce the monstrous ants of *Them!* (Douglas 1954) or the giant lizardlike creature *Gojira/Godzilla* (Honda 1954). And in the early 1970s films such as *The Andromeda Strain* (Wise 1971) and *The Omega Man* (Sagal 1971) explored the outbreak of deadly viruses and their effects on society, although mainly as a resource to generate horror effects.

Beginning in the 1970s, successes in genetic engineering (such as the gene-splicing and recombinant DNA experiments at Stanford University) began to garner more attention in the press. But not until the breakthrough successes in genetic engineering and cloning (such as the Vacanti mouse and Dolly the sheep), and especially the sequencing of the human genome, did the marginal depictions of biology in sf shift into the cultural formation that is biopunk. As I have argued elsewhere, "With the turn of the twenty-first century, the genetic has become not just a theme in sf, but rather a cultural formation that transcends the borders of the literary genre and establishes itself in mainstream culture" (Schmeink 2016, 9). Biopunk film, then, is not so much a standalone cinematic genre, as it is a specific medial expression of the larger cultural formation that finds expressions for discourses of the genetic in different cultural productions.[4]

In the mid-1990s, genetics and other biotechnologies started to emerge in mainstream sf film with more regularity. Fittingly, a new adaptation of *The Island of Dr. Moreau* (Frankenheimer 1996) shifted the creation of human-animal hybrids from surgery to genetic splicing, while both *Jurassic Park* (Spielberg 1993) and *Species* (Donaldson 1995) explored how new forms of life could be grown from DNA samples, gathered from either ancient dinosaurs or invasive aliens, respectively, with similarly unsettling results for the currently dominant human species. While these films cross over into the action-adventure form, their use of genetics as a central motivator for the action maintains their sf thrust, while suggesting a shift in which the public views some forms of scientific progress as potentially threatening.

Dealing with how biotechnological innovation threatens or promises (depending on one's viewpoint) to change life itself, biopunk is a slant of sf cinema that has become more and more central in the twenty-first century and that has easily moved beyond the action subgenre. In fact, biopunk, understood as a larger cultural formation, is allowing filmmakers to explore posthumanist discourses in a variety of cinematic genre forms, negotiating controversial issues such as genetic or racial purity, transhumanist ascendency, self-improvement and life extension through medical interventions, the social construction of identity, biological or genetic determinism, and many others. In the following sections, I discuss examples drawn from a range of generic expressions to showcase how biopunk's adaptability has allowed it to intervene significantly in posthumanist discourse. The films have been chosen from a wide and exemplary variety of narrative types which the cultural formation of biopunk uses: *Gattaca* (Niccol 1997) plays to the tradition of noir crime, *Code 46* (Winterbottom 2004) is a melodramatic love story, *Antiviral* (Cronenberg 2012) is an avant-garde film with body horror elements, and *The Girl with All the Gifts* (McCarthy 2016) evokes the mass-market category of zombie films. While these films could all be read as coming from different film categories, deploying a wide and disparate range of generic conventions, framing them within a biopunk slant of sf cinema allows us to examine their similarities and intimate connection to key issues of genetic posthumanism (i.e., genetic determinism, biopolitical control over bodies, the commodification of life, and a posthuman becoming "after the human"), seeing them as part of a cultural formation that is becoming ubiquitous in contemporary society.

## GENETIC DETERMINISM IN *GATTACA*

In *The DNA Mystique*, Dorothy Nelkin and Susan Lindee argue that DNA has become a "cultural icon" with a strong "plasticity and openness to interpretation" (2004, xii), which allows us to articulate almost any purpose through the cultural interpretation of DNA. Its most essential meaning, of course, is as a carrier of "information that helps to form living cells and tissues" (Nelkin and Lindee 2004, 2), providing a data set that determines how life itself is expressed and embodied. But seeing humans as mere expressions of genetic code severely limits how we view the human experience. Such a vantage becomes a form of genetic determinism that "reduces the self to a molecular entity, equating human beings, in all their social, historical, and moral complexity, with their genes" (Nelkin and Lindee 2004, 2).

The biopunk film *Gattaca*, for example, seizes upon the idea of genetic determinism and its implementation of a "new eugenics" (Kirby 2000, 197) of gene editing, not only to eliminate unwanted or defective genes but also, as Lee Silver (1997) argues, to become "*GenRich*" (4) and "enhance a variety of physical and mental attributes" (6) beyond what "*Naturals*" (4) are born with. The film thus depicts a highly stratified society in which the GenRich or "valids" are genetically edited before birth and ubiquitous genetic screening leads to discrimination against those born unenhanced, the "in-valids." In this context, all social life hinges upon the genetically determined potential of each individual—schools, jobs, housing, personal relationships. In-valids become the serving underclass of a society that is focused on the genetically enhanced potential of the valids. The film is focused on the in-valid Vincent (Ethan Hawke), who does not accept his genetic limitations and seeks to gain acceptance in the astronaut program by taking on the identity of Jerome Eugene Morrow (Jude Law), a professional swimmer who has severed his spine in an accident and now hopes to sell his DNA. Vincent needs Jerome in order to pass the near-constant DNA testing that bars him from participation in society, as "validity" is granted upon one's genes and the probabilities they determine. For Vincent, a blood test reveals "neurological condition, 60% probability, ... attention deficit disorder, 98% probability, heart disorder, 99% probability, early fatal potential, life expectancy: 30.2 years." These probabilities not only determine Vincent's self-image but all social interactions (see figure 3.1). Though the film points out that discrimination based on genetic screening is illegal, the practice has become common and is widely used.

But as David Kirby and Laura Gaither (2005) point out, it is not only the in-valids who see themselves as determined and limited by their genome, as "*Gattaca* portrays a world in which genomic alterations take the form of unyielding expectations. In fact, *Gattaca*'s society has so much confidence in the predictive power of genomics that their culture revolves around these expectations. In *Gattaca*'s world, no one is free to pursue a life of his or her own choosing" (269), although, as Vincent explains, when "a member of the elite falls on hard times, their genetic identity becomes a valued commodity for the unscrupulous. One man's loss is another man's gain." Valids feel a similar pressure to succeed and fulfill societal expectations, as Jerome reveals, when he shows Vincent a silver medal from a sports competition: "Jerome Morrow was never meant to be one step down on the podium. With all I had going for me, I was still second-best." Genetic engineering, it seems, does not guarantee success, and the pressure exerted by society is beyond any person to live up to.

**FIGURE 3.1:** Trapped in the patterns of DNA, Vincent (Ethan Hawke) of *Gattaca* (1997). Sony Pictures.

Other scenes support this reading. For example, Irene (Uma Thurman) is similarly designed to excel, but her heart condition excludes her from becoming an astronaut. Even though she was designed for accomplishment, she does not measure up to expectations: "An acceptable likelihood of heart failure. That is what the manual says." And similar genetically determined expectations drive the competition between Vincent and his brother Anton (Loren Dean), who loses a game of distance swimming to the invalid: "This is how I did it, Anton. I never saved anything for the swim back." Vincent's individual will to succeed, the film suggests, makes him exceed his own limitations. The fact that the unenhanced Vincent can pass and outperform any expectations, while enhanced characters fall short of expected standards, illustrates the film's rejection of genetic determinism, although the critique is troubled insofar as the film advertises the success of individual resistance to systemic changes. While genetic engineering in this biopunk future has changed what constitutes the human in terms of biology, it has not fundamentally changed discrimination and evaluation on the basis of perceived difference or challenged the concepts of a systemic social stratification. Instead, while *Gattaca* comments on ethical issues involved in genetic determinism, genetic engineering, and eugenics, the film also upholds humanist ideals about the individual's ability to push beyond their limitations when faced with technology threatening the boundary of life itself.

## GENETIC CONTROL IN *CODE 46*

*Code 46*, a film by British director Michael Winterbottom, takes the idea of genetic determinism a step further, interconnecting biopolitical control with the geopolitics of border control. Biopunk here showcases the Anthropocene via the impact of genetic technologies on a globalized world. The film depicts a future in climate crisis, with humanity either living in the corporately controlled environments of megacities or the lawless slums cropping up in the deserts between them. The story follows William (Tim Robbins), an investigator for supranational corporate entity The Sphinx on his trip to Shanghai, where he is supposed to find the person responsible for forging illegal "papeles"—a form of insurance coverage for travel to other cities. The Sphinx is a corporate-government hybrid responsible for the mobility of its citizens/customers based on genetic testing. Depending on genetic (in)compatabilities, The Sphinx grants or rejects the right to travel or permanent residence. William discovers that Maria (Samantha Morton) is smuggling papeles for people stuck in a rigid bureaucratic system, falls in love with her rebellious spirit, and spends a night of passionate intimacy. Later, he learns that Maria has been sent to a clinic, a newly formed pregnancy terminated, and her memory wiped, because of what is known as a Code 46 violation. Maria and William are genetically closely related without their knowledge, as a result of in vitro fertilization and cloning in the world of *Code 46*. Clinging to his affair, William elopes with Maria, prompting another violation of genetic laws and their arrest. While William's memory is now wiped and he returns to his family, Maria is banished to live out the rest of her life in the desert.

Through William and Maria's relationship, *Code 46* doubly comments on genetics becoming a factor in global control. The film appropriates the Oedipus myth in biopunk terms to reveal the breaking down of the biological barriers in familial relations and identity, especially through the practices of in vitro fertilization and cloning. William, as we see, has no direct link to his parents—he was born in vitro and does not know his genetic origins. When he has Maria's DNA tested, he learns that she has the DNA of his mother: "A 100% match. Genetically identical." When asked how that is possible, the geneticist explains, "Anything's possible in vitro. Your mother was one of a batch of twenty-four cloned fetuses. This hair is from your mother or one of her genetic twins." The film thus makes clear that biotechnologies such as cloning and genetic engineering blur, if not eliminate, previously drawn boundaries in our biological and social interactions. From the perspective of current social standards, as Brian Goss argues, "putting a 21st-century imprimatur on the ancient myth," William is "troubled upon learning that he committed incest" (2007, 69). Biopunk here uses the incest taboo to reveal how our conception of "family" might be determined by cultural frameworks, not through genetic proximity. The lived experience of raising a child, the shift in power distribution (mother to son), and the marked difference in age determine what we consider a parental relation, not the shared genome.

In the film, posthuman technologies in reproductive medicine, cloning, in vitro fertilization, and genetic engineering place everyone at risk for unwittingly breaking this taboo, thus making it necessary to enforce biopolitically any Code 46 violations and genetically screen potential partners. Thus, the biological control over partnership and reproduction shifts to the technocratic corporate-government entity that enforces the violation: The Sphinx. However, The Sphinx not only reads the genetic information for biopolitical control; it also "writes" upon the body and its genes any policy deemed necessary. That "writing" is accomplished through viruses that are genetically programmed to change human behavior and social interaction, such as providing William with empathy for his investigative work and making him better able to manipulate other people to get the information he needs.

Further emphasizing the work of biopower here, a virus is also used to prohibit any further violations of Code 46. To achieve that end, Maria is injected with a prevention virus that forces a physical reaction to any intimacy with William; as we learn, "It's involuntary. It's like adrenaline. When you're . . . scared of something, your body gets ready to run away." In fact, the virus is so potent that Maria's body violently rejects contact with William, so while she wants to be with him, she must be physically forced to have "consensual" sex. Even after that act, the virus controls Maria, compelling her to file a report about her own "crime"; as Goss observes, "Maria's subsequent, postcoital ambivalence is underscored in her (stilted and zombified) call to the authorities to report the Code 46 violation" (2007, 69). The genetically created virus forces The Sphinx's will on Maria and enacts genetic control over her body, thus powerfully showing that meaningful biological interactions between humans, whether birth or sex, are now subject to what

Brian Baker calls "anatomo-politics," or the governing of "the human body reproduced by technical means" (2015, 124).

Moreover, the film not only comments on biopolitical control but also links the genome to issues of governance in times of globalization. By issuing papeles as travel documents that indicate a status of legal sojourn in a specific region, as well as insurance against risk, The Sphinx becomes a sovereign entity that can limit human subjectivity, "not only in terms of physical mobility, but also in terms of the constitution of an informatic body (DNA as code) which may be read as a deterministic understanding of future risk" (Baker 2015, 116). The film makes clear that who gets to live inside the borders of the cities and who gets expelled are choices completely dependent on the genetic screenings that The Sphinx enforces, just as are movements between the regions. But beyond the individual story of Maria and William, we are unable to verify the justification for this genetically enforced biopower. Instead, we are left with the promotional slogan "The Sphinx knows best" as a cypher for possible genetic incompatibilities or defects that can restrict people's mobility (both physically and socially). And the film seems to stress the system's correct demarcation of people as genetically flawed or at risk. For example, one of Maria's customers is Damien (David Fahm), a researcher who wants to travel to Delhi but is denied permission by The Sphinx due to a possible health risk. While he leaves with a forged papele, we later learn that he died from "Ward's disease," to which most people in Delhi are immune. Repeatedly, the film points out that the technocratic system of The Sphinx is in place for good reason, even when it argues that people should be exiled from life in the cities. And while the system replaces current forms of racism or sexism—that is, more traditional biases for evaluating people based on physical markers such as skin color, gender, age, and so on—it merely transforms inequality into subordination based on genetic code, with a corporate entity acting as the sovereign of a social system based on genetic determinism.

## COMMODIFICATION OF LIFE IN *ANTIVIRAL*

Interestingly, in *Code 46*, even though The Sphinx is a private corporate entity, the notion of capitalist exploitation is rather opaque, which is unusual for biopunk films. In most cases, when genes are seen as biological containers of data, biopunk narratives seem to suggest that genes can and will be exploited, patented, marketed—that the building blocks of life will be subsumed by capitalism. As Steven Shaviro explains, such appropriation means that "Everything without exception is subordinated to an economic logic, an economic rationality. Everything must be measured, and made commensurable, through the mediation of some sort of 'universal equivalent': money or information" (2015, 29). Any and all objects, processes, and goals are thus typically placed within this framework of economics—and with DNA as a form of information, life itself thus becomes a commodity. Thierry Bardini refers to this development as the "genetic phase" (2011, 127) of global capitalism. He claims that the abstract category of code, whether biological or informational, is a twenty-first-century substitute for money: "Hereditary information, as the invention of genetic capital, now grounds absolutely the general equivalence in the realm of the living, making us living money in essence" (11). Seen in this context, the code of the human genome easily becomes a kind of currency that might be traded among capitalist shareholders.

A biopunk film that fully embraces this commodification of genetic material is Brandon Cronenberg's *Antiviral*, which explores our contemporary celebrity culture and extrapolates it to reveal a society fixated on consuming celebrity down to its DNA. In *Antiviral*, fans inject themselves with viruses and diseases that had afflicted celebrities, they eat flesh grown from celebrity tissue cultures, or engage in any number of services surrounding a posthuman desire for becoming-celebrity. In the film Syd March (Caleb Landry Jones) is a sales representative of the Lucas Clinic, which deals in high-end viral, bacterial, and fungal infections from a list of exclusive celebrity properties. To make extra money, Syd engages in illegal copying of diseases and selling them on the black market. When superstar Hannah Geist (Sarah Gadon) contracts a rare and ultimately deadly virus, Syd steals the virus by injecting it into his own body, thus unwittingly becoming the center of a conspiracy-fueled corporate war involving celebrity clinics, underground entrepreneurs, and grief-stricken fans who create a run on Hannah Geist products.

The "bioeconomy" of *Antiviral* is highly stratified in terms of its product, with "muscle cells taken from celebrities, cultivated in cell gardens and turned into a foodstuff, and sold as 'astral bodies' in butcher shops and restaurants" (McQueen 2016, 222) at the lower end of the spectrum. Here, biopunk draws close to body horror, as the film explores the impact of a capitalist mode of consumption on human and animal life, in the process pushing the envelope on the acceptability of a commodification of life. In fact, the film deliberately purges animals from its diegetic world altogether and instead highlights celebrity meat consumption (in butcher shops, restaurants, etc.). In doing so, the narrative seems to be "transgressing the human-animal divide and expressing a normalization of the levelling of certain humans with certain animals selected for consumption" (Üçoluk Krane 2020, 8). In contrast to animal consumption—and highlighting the techno-utopian haunting that occurs here—the celebrity is not slaughtered but rendered as cells grown separately from the original body in a form of simulacral replication: a genetic duplication of an original without variation that never produces the original celebrity body.

Folded into this vision is the notion that tissues artificially grown from celebrities' bodies are not considered to have the same life value as when they constitute an autonomous human being—a devaluation that is also true for animals' lives. But Syd expresses an unease with this argument that is grounded in human exceptionalism: the question of whether any consumption of human tissue constitutes a form of cannibalism. The butcher responds that "these are just muscle cells. It all depends on whether the human being is found in its materials. . . . We'll see what happens when we go from growing celebrity cell steaks to growing complete celebrity bodies." Cells independent from recognizable embodiment are not seen to constitute life, but under the logic of capitalist subsumption merely express a certain informational value—as quantified in measures such as nutrition, price, and weight. Bio-artists Oron Catts and Ionat Zurr offer a contrasting view, arguing that the "biomass of disassociated living cells and tissues" used in laboratories for experimentation or discarded by medical procedures needs to be understood as "the Extended Body," an "amalgamation of the human extended phenotype and tissue life—a unified body for disembodied living fragments" (2006). The astral bodies depicted in *Antiviral* are just such an "extended body"—as the film uses biopunk's blurring of biological boundaries to renegotiate what we perceive as part of ourselves, what constitutes life. By normalizing human meat consumption as part of genetic capitalism, the film pushes our conceptions of how the genome can become commodified, prompting a resistance to these posthuman technologies and a critical rethinking of our own subjectivity.

In contrast to the low-end astral bodies, *Antiviral* opens with a sales pitch for "boutique biocommodities, the apex of the bioeconomy" (McQueen 2016, 228), available at the Lucas Clinic, where Syd harnesses the desire to consume celebrity in more sophisticated ways for affluent clients. Short media clips of Hannah Geist and other celebrities advertise the genetic capitalist products as a kind of posthuman life, created from viral contagion and host cells, but also a commodification of life without its historic trace of destruction. As Mark Hansen puts it, "The destructive impact of viruses is effectively suspended: far from destroying the bodies of their hosts, viruses form new 'bodies'" (2001)—a logic that echoes in genetic capitalism where the host body can produce ever more "new bodies" to be marketed and a new hybrid subjectivity; or as Syd explains, "a biological communion between you and Aria [another celebrity]. From her body to your body, from her cells to your cells."

The reduction of the human host becomes even more prominent at the end of the film, when Syd introduces the "Afterlife Capsule," a pressure chamber that keeps the cells of Hannah's body alive as a system "perpetuated, even expanded beyond what existed during her lifetime." Hannah's cells have been cultivated into a biotechnological system of feeding tubes and harvesting racks that allow for experimentation with new viral infections exclusively available for the customers. Hannah thus becomes an Extended Body, more "disembodied living fragments" (Catts and Zurr 2006) than embodied human subject, always becoming-with new viral bodies. As Syd explains to interested customers, "From the perspective of the virus, the human being is irrelevant. What matters is the system that allows it to function." The Afterlife Capsule, then, represents the future of human life through the lens of a genetic capitalism that sees human DNA as just another commodity.

# Posthuman Alternatives in *The Girl with All the Gifts*

Biopunk negotiates the boundaries of what constitutes life and how we evaluate it, from individual bodies as presented in the body horror of *Antiviral* to the larger scale of the Anthropocene, where "the human" as a whole is in jeopardy through our own interventions in the genetic makeup of life. One such radical future or new world has been introduced in recent zombie narratives, which are commonly viewed "as apocalyptic because we believe that we are watching either the slow breakdown or the catastrophic destruction of human society" (Christie 2011, 61–62). But that view is anthropocentric and one could just as easily see zombies taking over the world as the coming of a new, different form of world. From this vantage, zombie fictions become posthuman alternatives, perfect screens against which we might define our human subjectivity. In the biopunk context, zombies represent the *not*-human, a form of posthuman life, part of a "grey ecology," as Jeffrey Jerome Cohen argues, a "thriving of life in other forms" (2017, 382). From this nonanthropocentric view, life itself is both bigger and smaller than the human scale imagines, for a grey ecology "propels us beyond our finitude, opens us to alien scales of both being (the micro and the macro) and time (the effervescent, barely glimpsed; the geologic, as life proceeds at a billion-year pace)" (383). And zombie films, such as Colm McCarthy's *The Girl with all the Gifts*, introduce us to such an alien scale of life in a world no longer made for the human.

In the film, a variation of the ant-attacking fungus *Ophiocordyceps unilateralis* has broken the species barrier and infected humans. As with these famous "zombie-ants," the fungus wraps around the brain of the human host and then hijacks its behavior in order to spread the infection. The result is that humans are turned into "hungries," wanting to spread the fungal infection via fluid transfer by biting. At this stage, hungries are simply zombies by another name, undead in their loss of human agency, at the same time a vegetative fungal host and an urge-driven hunting animal. As in most zombie fiction, the film focuses on a group of survivors that tries to maintain barriers against the infected, while researchers seek a possible cure to return humanity to a "normal" state of affairs.

However, the film blurs the line between the "normal" human and infected non-human by problematizing the category of hungry: it introduces a group of children who are kept in an underground army facility and form the basis of the cure research. The children are prisoners, tested and probed, taught in a classroom, watched for behavioral clues, and even sliced open to reveal their inner workings. They are referred to both as "freaking abominations" and as "test subjects," and while the film is careful to build sympathy for them—showing them as clever, articulate, and even loveable—they are later revealed to be something other than human. When exposed to a human scent, the kids lose control of their cognitive ability and change physically into a more vivid form of hungry. Dr. Caldwell (Glenn Close), the lead scientist and main representative of the old-world order, believes the children represent another stage of the fungal infection, so she coldly studies and dissects them in her battle to restore a stable, human-centered world. She believes the fungus uses the kids as hosts to mimic human behavior, that these are not really children but rather a form of the fungus "present[ing] as children" and driven by the fungal urge to reproduce.

Melanie (Sennia Nanua), the titular character and one of the children, is not aware of the fungus and loses control to her hungry urges (see figure 3.2). Over the course of the film, she learns to control these urges by keeping fed and by building a tolerance for

FIGURE 3.2: Children as "freakish abominations" in the postapocalyptic *The Girl with All the Gifts* (2016). Warner Bros.

human smells. She realizes that she was infected in her mother's womb and has thus developed symbiotically with the fungus. The film emphasizes this nonhuman character through the trope of a monstrous birth, coldly described by Dr. Caldwell: "The mothers were ... empty. Cored. All their organs devoured. ... The embryos they were carrying took the infection as well, through the placenta. They ate their way out."

Interestingly, the scene where Melanie is told about her origin is intercut with scenes of her teacher and mother-surrogate Helen Justineau (Gemma Arterton) checking for threats and discovering a hungry, whose actions suggest that Caldwell's analysis might be faulty and based on her anthropocentric approach. The hungry does not act according to Caldwell's fungus-in-the-driver-seat-theory, which predicts it should simply be inactive and vegetative until prompted by sensory input. Instead, it repeatedly slams into a locker it is chained to, wanting to break free. While this behavior, like Melanie's existence ("unlike anything that has ever existed before"), is unprecedented, it does not prompt Caldwell to see Melanie for what she is—essentially a new species. Caldwell's anthropocentric ideal of an uncontaminated existence is not based in scientific reality, since "*we have never been humans proper* by fact of our constitution by microbial and amoebic intelligences that are not *our own*" (Wallin 2016, 64, italics in original). But seeing Melanie as a nonhuman other serves an ideological purpose, keeping intact the concept of human exceptionalism and the hierarchies that go with it.

In the film's final conflict, old and new world orders clash after Melanie discovers more children like herself. Caldwell tries to persuade Melanie to sacrifice herself, to let herself be dissected for a cure, thereby saving the remaining humans. While Caldwell argues for compassion, that the survivors—among them Ms. Justineau—deserve a long and safe life, she ignores the fact that Melanie might deserve the same thing. In the end, Melanie realizes that self-sacrifice would not just be her death, but the death of all the children symbiotically living with the fungus. Caldwell's hierarchy of what constitutes valuable life would remain intact, never allowing for co-habitation or mutual understanding. In this instance the film uses biopunk's themes of genetic mutation and the breaking down of categorical distinctions of life to reveal that Caldwell's thinking is rooted in concepts of human exceptionalism and essentiality, while ignoring the grey ecology that surrounds her. When Melanie recognizes her posthuman subjectivity and the toxic hierarchies at work, she stakes out a new position, arguing, "Why should we have to die for you to live?" As Victoria Carrington has pointed out, the symbiotic children might be thought of as posthuman, a break with the old concepts that forces us to realize, "This is a world coming to an end. There will be no recovery. Melanie and the other zombie children, evolved to thrive in the changed conditions, will build a new order" (2016, 32).

It is thus fitting that Melanie, to usher in the new, literally just needs to strike a match and light up the fungal pods containing spores all around her. When they open in the blaze, they infect every human remaining in the world, leaving the symbiotic children to inherit it. When confronted with Melanie's decision, Sergeant Parks (Paddy Considine), the film's stand-in for the old world's military authority, recognizes that "it's over, it's all over." But what he doesn't understand is that the world is only ending from his anthropocentric worldview. From a biopunk perspective, measured in the time scale of a grey ecology, Melanie rightly assumes the opposite: "It's not over. It's just not yours anymore."

# CONCLUSION

Biopunk film is ideally suited to address any and all of the above-mentioned scenarios that have been ushered in by the new world of biotechnology. It throws into stark relief the consequences of orienting our lives around the concept of genetic determinism and what this perspective entails for our understanding of subjectivity and identity. It addresses the inequalities inherent in genetically enforced biopolitical control, and it maps them onto the geopolitical challenges that we similarly face. We might not be able to solve these issues separately, since genetic engineering is poised to change our social hierarchies because it also easily impacts food supplies, species diversity, and climate change. However, a biopunk cinema can teach us about the fallout from allowing a broadly construed genetic capitalism to reign unchecked, making the genome part and parcel of our bioeconomy and dealing in patented life forms. Moreover, the concept of biopunk lets filmmakers reveal a world in which a posthuman becoming is already changing the fundamental structures of life, wherein the human is no longer an exceptional or essential category. In all, biopunk films seek to negotiate a fundamental change in our existence and play an essential part in the discourses of what it means to be human in the genomic age.

## NOTES

1. There are many definitions of either term, but most scholars agree that both deal with concepts that move "beyond the human." The central distinction between them is that

transhumanism argues that technology will eventually allow humans to evolve beyond their current limitations—an argument based on humanistic ideals of exceptionalism and individualism. In contrast, a critical posthumanism sees the human as always already being hybrid and relational, thus defining subjectivity in communication and companionship with animals, machines, world, and all life. For more on these distinctions, see Schmeink (2016, 29–46), Braidotti (2013), and Herbrechter (2013).

2. Instead of conventional generic forms, both cyberpunk and biopunk have been described as "cultural formations," meaning that their motifs, themes, and structures have become dispersed through culture in general, found in practices as well as a variety of products. For more details see Foster (2005, xiv) and Schmeink (2016, 21).

3. David A. Kirby in "The New Eugenics in Cinema" (2000) provides an overview of early biological sf, largely focused on eugenics as a practice in film.

4. For a longer discussion of the advantages of using "cultural formation" over genre regarding both cyberpunk and biopunk, see McFarlane, Murphy, and Schmeink (2020).

## WORKS CITED

Baker, Brian. "'Here on the Outside': Mobility and Bio-politics in Michael Winterbottom's *Code 46*." *Science Fiction Studies* 42, no. 1 (2015): 115–31.

Bardini, Thierry. *Junkware*. Minneapolis: U of Minnesota P, 2011.

Braidotti, Rosi. *The Posthuman*. London: Polity, 2013.

Carrington, Victoria. "The 'Next People': And the Zombies Shall Inherit the Earth." In *Generation Z: Zombies, Popular Culture and Educating Youth*, edited by Carrington et al., 21–36. New York: Springer, 2016.

Catts, Oron, and Ionat Zurr. "Towards a New Class of Being—the Extended Body." *Artnodes: Intersections between Arts, Sciences and Technologies* 6 (2006). https://raco.cat/index.php/Artnodes/issue/view/4308.

Christie, Deborah. "And the Dead Shall Walk." In *Better Off Dead: The Evolution of the Zombie as Post-Human*, edited by Deborah Christie and Sarah Juliet Lauro, 21–36. New York: Fordham UP, 2011.

Cohen, Jeffrey Jerome. "Grey: A Zombie Ecology." In *Zombie Theory: A Reader*, edited by Sarah Juliet Lauro, 381–94. Minneapolis: U of Minnesota P, 2017.

Foster, Thomas. *The Souls of Cyberfolk: Posthumanism as Vernacular Theory*. Minneapolis: U of Minneapolis P, 2005.

Glassy, Mark C. *Biology Run Amok! The Life Science Lessons of Science Fiction Cinema*. Jefferson, NC: McFarland, 2018.

Goss, Brian Michael. "Taking Cover from Progress: Michael Winterbottom's Code 46." *Journal of Communication Inquiry* 31, no. 1 (2007): 62–78.

Hansen, Mark. "Internal Resonance, or Three Steps Towards a Non-Viral Becoming." *Culture Machine* 3 (2001). http://svr91.edns1.com/~culturem/index.php/cm/article/view/429/446.

Herbrechter, Stefan. *Posthumanism: A Critical Analysis*. London: Bloomsbury, 2013.

Kirby, David A. "The New Eugenics in Cinema: Genetic Determinism and Gene Therapy in *Gattaca*." *Science Fiction Studies* 27, no. 2 (2000): 193–215.

Kirby, David A., and Laura A. Gaither. "Genetic Coming of Age: Genomics, Enhancement, and Identity in Film." *New Literary History* 36, no. 2 (2005): 263–82.

McFarlane, Ann, Graham J. Murphy, and Lars Schmeink. "Cyberpunk as Cultural Formation." In *The Routledge Companion to Cyberpunk Culture*, edited by Anna McFarlane, Graham J. Murphy, and Lars Schmeink, 1–3. New York: Routledge, 2020.

McHale, Brian. *Constructing Postmodernism*. New York: Routledge, 1992.

McQueen, Sean. *Deleuze and Baudrillard: From Cyberpunk to Biopunk*. Edinburgh: Edinburgh UP, 2016.

Nelkin, Dorothy, and M. Susan Lindee. *The DNA Mystique: The Gene as a Cultural Icon*. Ann Arbor: U of Michigan P, 2004.

Parker, Helen. *Biological Themes in Modern Science Fiction*. Ann Arbor: UMI Research, 1977.

Schmeink, Lars. *Biopunk Dystopias: Genetic Engineering, Society, and Science Fiction*. Liverpool: Liverpool UP, 2016.

Shaviro, Steven. *No Speed Limit: Three Essays on Accelerationism*. Minneapolis: U of Minnesota P, 2015.

Silver, Lee M. *Remaking Eden: How Genetic Engineering and Cloning Will Transform the American Family*. New York: Avon, 1997.

Üçoluk Krane, Ece. "Eat Me(at): Phantom Animality and Spectral Posthumanism in *Antiviral*." *Continuum* 34, no. 1 (2020): 1–15. DOI: 10.1080/10304312.2019.1675589.

Wallin, Jason J. "Into the Black: Zombie Pedagogy, Education and Youth at the End of the Anthropocene." In *Generation Z: Zombies, Popular Culture and Educating Youth*, edited by Victoria Carrington et al., 55–70. New York: Springer: 2016.

# CHAPTER 4

·············································································

# CLI-FI CINEMA

·············································································

## MARK BOULD

THE term "climate fiction," or "cli-fi," was coined in 2007 by Dan Bloom to describe "works of art and storytelling that deal with climate change and global warming concerns" (Thorpe n.d.). Much of the ensuing commentary on this narrative type has displayed a curious determination to establish cli-fi as a distinct form, rather than a variety of science fiction (sf). This seems to stem from either an unfamiliarity with or a calculated misrepresentation of sf, thus perpetuating inaccurate stereotypes of the genre. For example, Bloom (2013) distinguishes this "new kid on the literary block" from the rest of sf by claiming that "rather than look outward at the stars and the cosmos, cli-fi looks inward, at our warming planet, this third rock from the sun, a planet in trouble." This rhetorical move substitutes a part of sf—an image, say, of space operas such as the *Star Wars* and *Star Trek* series—for the whole of the genre, as if the equally long traditions of utopian, dystopian, satirical, apocalyptic, near-future, and contemporarily set sf addressing current scientific concerns and social trends simply do not exist.[1] In fact, Bloom describes Hamish MacDonald's cli-fi novel *Finitude* (2010) in terms indistinguishable from sf, as he notes that "the setting is a city much like London in some un-named country much like Britain in the near future, where all hell breaks loose and a group of people search for a safe haven, against all odds" (Thorpe n.d.). Somehow this description is meant to show that cli-fi provides "a platform for writers and film people to explore the future, not in a sci-fi but in a cli-fi way" (Thorpe n.d.). But only by suppressing vast swathes of the genre can Bloom boost cli-fi as something quite different from sf.

Similar blind spots, confusions, and misconceptions recur in much popular coverage of literary climate fiction. For example, in a *Guardian* article subheaded "Not Sci-Fi but Cli-Fi," Sarah Holding (2015) declares her love for sf, which she then demonstrates by conflating it with magical realism and exemplifying it with children's fantasy novels such as E. Nesbit's *Five Children and It* (1902), C. S. Lewis's *The Lion, the Witch and the Wardrobe* (1950), and Philippa Pearce's *Tom's Midnight Garden* (1958). In order "to participate imaginatively in rewriting our future," Holding claims it was necessary to free herself from sf's constraints and embrace the fact she was actually writing cli-fi, the distinctiveness of which apparently hinges on its ability to tackle "a wide range of compelling issues: shortages of food, fresh water, oxygen or dry land, giving us vivid depictions of the human consequences of climate change" (2015). Only then, apparently, did she think she could write something that was distinguishable from much actually existing sf.[2]

Two other *Guardian* articles have offered similar faulty efforts at genre distinction. In one, John Abraham (2017) notes that cli-fi "often present[s] real science in a credible way," arguing that "good" cli-fi "has to have some real science in it" and "get the science right," while also being "fun to read." Indeed, what makes this new development "so important," he claims, is that "we can unintentionally learn real science" from it (Abraham 2017). He seems unaware that he is echoing key formulations and (improbable) claims made by Hugo Gernsback as he sought to establish sf in American pulp magazines in the 1920s. In the second piece, Rodge Glass (2013) states, "Unlike sci-fi, cli-fi writing comes primarily from a place of warning rather than discovery," a claim promptly undone by the image of sf he conjures being entirely composed of such premonitory examples as *District 9* (Blomkamp 2009) and George Orwell's *Nineteen Eighty-four* (1949): "There are no spaceships hovering in the sky; no clocks striking 13. On the contrary, many of the horrors described seem oddly familiar" (Glass 2013). More significant, though, is his acknowledgment of the cultural politics of cli-fi, as he ponders whether it is "a term designed for squeamish writers and critics who dislike the box labelled 'science fiction.'" Such misgivings certainly resonate with the actions of publishers and marketers who regularly and unsubtly discriminate in this way so as to maintain engrained and presumably profitable distinctions between "literary" and "genre" fiction.[3]

This tendency is also evident in some book-length criticism. Despite engaging with sf in some detail, Adam Trexler's (2015) otherwise invaluable mapping of Anglophone climate change novels is shaped by an implicit hierarchy that privileges the literary novel. While Amitav Ghosh acknowledges sf's capacity for depicting planetary events, he then sets aside such "generic outhouses" (2016, 24) to focus on what he calls the "serious literary novel," even though he considers it inherently incapable of representing climate change. In contrast, Andrew Milner and J. R. Burgmann (2020) draw no distinctions in terms of supposed literary quality and happily subsume cli-fi into the broader field of sf. They argue that it should be considered an sf subgenre "rather than a distinct and separate genre" because "its texts and practitioners—writers, readers, publishers, film directors, fans, etc.—relate primarily to the SF selective tradition" and "articulate a structure of feeling that accords centrality to science and technology, in this case, normally climate science" (2020, 25, 26). Despite this assumption, Milner and Burgmann share Trexler's and Ghosh's curious literal-mindedness about what constitutes fiction about climate change, and leave little room for the oblique, the symbolic, the partial, or the estranged.[4]

Work on the cinema of climate change has not been quite so constrained, perhaps because very few films might be considered as overtly cli-fi. For example, E. Ann Kaplan's *Climate Trauma* (2016) focuses on dystopian and apocalyptic films with literary pedigrees and arthouse credentials, such as *Children of Men* (Cuarón 2006), *Blindness* (Meirelles 2008), *The Road* (Hillcoat 2009), *Take Shelter* (Nichols 2011), and middlebrow documentaries like *Manufactured Landscapes* (Baichwal 2008) and *Into Eternity* (Madsen 2010), alongside the pulpier mainstream *The Happening* (Shyamalan 2008) and *The Book of Eli* (Hughes brothers 2010). However, not one of these examples says "anthropogenic climate change" aloud, and none of them easily fits into the climate fiction models noted by Trexler, Ghosh, or Milner and Burgmann. The same is true of the films analyzed in Jennifer Fay's *Inhospitable World* (2018), which range from Buster Keaton's *Steamboat Bill, Jr.* (Reisner 1928) to Zhangke Jia's *Still Life* (2006).

## THE DAY AFTER TOMORROW AND BEYOND

Probably the most famous and financially successful example of overt cli-fi sf is *The Day After Tomorrow* (Emmerich 2004). Inspired by Art Bell and Whitley Strieber's *The Coming Global Superstorm* (1999), it depicts the global-warming-induced partial collapse of the Larsen B ice shelf, which reduces ocean salinity and thus disrupts the North Atlantic Ocean Circulation. There are unprecedented snowfalls in New Delhi, giant hailstones pummel Tokyo, and multiple simultaneous tornadoes level Los Angeles. Three arctic superstorms ravage the northern hemisphere. The sea level in Nova Scotia rises twenty-five feet in seconds, and air drawn from the upper troposphere causes temperatures to plummet, covering Europe in fifteen feet of snow. As the eye of each storm passes, local air temperature drops 10 degrees per second, all the way down to −150° Fahrenheit. In the midst of these calamitous events, paleoclimatologist Jack Hall (Dennis Quaid), who has warned about climate change for years, undertakes a cross-country trek to rescue his son Sam (Jake Gyllenhaal), who is sheltering in the New York Public Library (figure 4.1). Widely derided over its questionable science,[5] the film has also been criticized for its "reliance on dramatic exaggerations, instant consequences, and dazzling special effects; its absolute apocalyptic premise and . . . its emphasis on

**FIGURE 4.1:** A frozen New York harbor in *The Day After Tomorrow* (2004). 20th Century-Fox.

the individualistic heroic actions of the male protagonist" (Willoquet-Maricondi 2010, 49). While valid, such criticisms are actually neutral observations about the dominant narrative and affective strategies of Hollywood blockbusters. Seen in that context, it is unsurprising that, after all the death and destruction of a world-changing catastrophe, the father brings his son safely home to mother, the apparently broken family is healed, the boy gets the girl, and heteronormative society is saved. While the crisis opens up intriguing cracks through which a potentially different future is glimpsed—Americans crossing the Rio Grande into Mexico are labeled illegal immigrants, and the United States swiftly cancels all Latin American debt in exchange for Mexico opening its borders to American refugees—the film's ending shuts down these possibilities, as the United States survives, displaced southward but otherwise apparently unchanged.

Such political, aesthetic, and imaginative shortcomings are not just a matter of unthinkingly reiterating ideological and affective norms. They also derive from a dominant narrative form—itself a product, articulation, and mediation of ideological and commercial imperatives—in which an equilibrium is disrupted until a new equilibrium is achieved. This pattern is established early in the film. Asked if the weather anomalies will continue to get worse, Hall replies that there will be a major climate shift; and the film ends with disruptions calmed as the high albedo of the ice-and-snow-covered northern hemisphere reflects more solar radiation back into space, resulting in a new, if differently stable climate (and family unit). Of course, in the face of real climate chaos (see Lynas 2008; Wallace-Wells 2019), this conventional narrative structure can be dangerously misleading.[6]

However, by prosecuting Roland Emmerich's film in such terms, critics overlook the important role it played in increasing popular awareness of climate change (see Leiserowitz 2004; Hart and Leiserowitz 2009). And criticism that chooses to privilege a niche ecocinema "more likely to reflect an independent and experimental approach to production, play at film festivals, art houses, and on public television" (Willoquet-Maricondi 2010, 48) disdains the mass audience in several ways. It assumes that such viewers will encounter only this particular representation of climate change, or that they are incapable of understanding it in relation to other representations, including nonfiction accounts. It assumes that *The Day after Tomorrow* has a singular meaning straightforwardly accepted by the entire audience (except, of course, those critics who see right through it). And by failing to embrace climate change movies across a range of modes and formats, it assumes that the climate education of a middle-class, middlebrow liberal audience is the key that will unlock the systemic changes necessary to avert further climate destabilization.

While *The Day After Tomorrow* continues to be a problematic touchstone in accounts of cli-fi cinema, anthropogenic climate change has been part of sf story worlds since at least the 1990s in films such as *Highlander II: The Quickening* (Mulcahy 1991) and *Split Second* (Thompson 1992). This date can be pushed back even further if one interprets the dystopian settings of films such as *Soylent Green* (Fleischer 1973) and *Blade Runner* (Scott 1982) from our current state of knowledge. Furthermore, if one dates the Anthropocene to the post–World War II "Great Acceleration," one can legitimately include films featuring the climatological effects of nuclear weapons, from *On the Beach*

(Kramer 1959) and *The Day the Earth Caught Fire* (Guest 1961) to *Mad Max 2* (Miller 1981) and *Hardware* (Stanley 1990). More recently, weather and climate have become a more pronounced element of sf worldbuilding. Manhattan is mostly beneath the sea in *A.I.: Artificial Intelligence* (Spielberg 2001), although the suburbs seem unfazed. Weather is weaponized in *The Matrix* franchise (Wachowskis 1999–) and *Geostorm* (Devlin 2017). Geoengineering goes badly wrong in *Snowpiercer* (Bong 2013), and the Sun runs down in *Sunshine* (Boyle 2007), both producing frozen futures. The Earth is scorched by some unspecified catastrophe in *The Road* (Hillcoat 2009) and by dragons in *Reign of Fire* (Bowman 2002). Thanks to climate change, interdimensional *kaiju* now find the planet habitable and thus worth invading in the *Pacific Rim* films (2013, 2018), while the solution to a global energy crisis unleashes monsters in *The Cloverfield Paradox* (Onah 2018). *Resident Evil: The Final Chapter* (Anderson 2016) reveals that climate change was one of the key reasons the Umbrella Corporation designed and released the T-virus, intending to massively reduce the human population, but instead unleashing a zombie plague. In *The Day the Earth Stood Still* (Derrickson 2008), aliens decide the only way to save the planet from the sixth great extinction is to make humans extinct; the eponymous AI in *Avengers: Age of Ultron* (Whedon 2015) comes to a similar conclusion, as does Thanos in *Avengers: Infinity War* (Joe and Anthony Russo 2018), although his plan is to destroy half of all life in what he considers a dangerously overpopulated universe.

The range of potential examples extends even further if one includes films featuring just one or two of the elements of climate destabilization, such as rising temperatures, desertification, wildfires, droughts, famine, food shortages, rising sea levels, floods and inundations, extreme weather events, depleted water resources, loss of agricultural production (land, crops, and livestock), ocean acidification, reduced air quality, particulate pollution, respiratory ailments, mutated diseases, relocated disease vectors, new viruses, epidemics, pandemics, contagion, resource conflicts, climate conflicts, economic collapse, species extinctions, and so on. And it extends even further still if one moves beyond the most literal-minded definitions of what it means to represent climate change.[7] But besides cataloguing examples, it might be useful to identify three tendencies found among them: strategic realism, techno-utopianism, and apocalyptic environmentalism.[8] They tend to appear together, combined in different measures and with different emphases, and as Randy Schroeder notes, all three "collapse into a more foundational view that mistakenly takes capitalism as the final horizon that circumscribes available terms for thinking the world" and perpetuate "a deeper, more persistent . . . malaise: the celebration of the species and the reification of the perceptible world in which that species must survive" (2017, 356–57).

# STRATEGIC REALISM

Strategic realism focuses on "geopolitical maneuverings" and "potential political and economic tensions that will inevitably arise as countries pursue their individual" interests

(Szeman 2017, 59). In reality—and in fiction—this realpolitik involves such things as "military intervention," "economic agreements between states," "the creation of new trade and security arrangements," and, at the same time, efforts to disentangle and isolate the nation-state from any international interdependence that might render it vulnerable (59). The nation-state, rather than being globalized into insignificance, is instead revealed to be "the central actor in the drama of the looming disaster . . . engag[ing] in often brutal geopolitical calculations in order to secure the stability of national economies and communities" (59). Strategic realism thus sees the climate crisis not as an existential threat to humans and other species, but as a problem requiring action, so that when the deck chairs on the sinking ship are rearranged, the particular nation-state manages to "preserve or enhance" its geopolitical status in a "political future that will look more or less like the present" (60).

Reading *The Day After Tomorrow* against the grain reveals elements of this attitude in the United States' southward relocation (and its complete abandonment of the entire American population farther north than the fortieth parallel, along with Canada). To anyone familiar with the US history of settler-colonialism (see Dunbar-Ortiz 2014; Wolfe 2006), there is something chilling in the way the president in this film refers to Americans being "guests" of Mexico and then, moments later, talks about bringing the Manhattan survivors "home." Indeed, borders and walls are often key images in cli-fi. For example, in *Pacific Rim* (del Toro 2013), world leaders (mistakenly) defund the Jaeger program: rather than fighting invading *kaiju*—the beneficiaries and avatars of climate change—with human-piloted giant robots, they construct monumental coastal defenses to keep them from rampaging ashore. Similarly, all zombie films, regardless of whether they include climatological changes, have some kinship to cli-fi movies because any depiction of a massive, mobile, and "unwanted" population will likely be seen, on some level, as about climate refugees, of which there may be as many as 250 to 500 million by 2050. Among others, *Doomsday* (Marshall 2008), *World War Z* (Forster 2013), and the conclusion of *I Am Legend* (Lawrence 2007), emphasize massive militarized walls that often echo historical and contemporary colonial structures, along with other, lesser barriers, which recall the fortifications of both settler homes and suburbs against the movement of displaced and excluded others.

*Monsters* (Edwards 2010), recalling both *It Happened One Night* (Capra 1934) and *Apocalypse Now* (Coppola 1979), combines a gap-year sensibility with a touristic gaze and desktop special effects. American photojournalist Andrew Kaulder (Scoot McNairy) is tasked, unwillingly, with bringing his boss's daughter, Sam Wyndem (Whitney Able), safely home from a perilous warzone. Six years earlier, a NASA probe, carrying specimens of life from elsewhere in the solar system, crashed in the US/Mexico borderlands. To the south of a hastily constructed but genuinely massive defensive border wall, a quarantine zone extends two hundred kilometers into northern Mexico. Regularly bombed by US and Mexican air forces, it is the breeding ground of amphibious aliens. Their bioluminescent eggs grow, like fungi, on tree trunks, which eventually topple into the river to be swept out to sea, where the eggs hatch. The adults—which resemble giant cephalopods, but with some tentacles rigid like spider legs so that they can walk—return inland to spawn. While not incurious creatures, they are indifferent to humans.

The landscape through which Kaulder and Sam flee north combines beautifully lush nature with a postcatastrophic sense of a world without humans (versions of this emptied-of-us Earth appear in *The Wild Blue Yonder* [Herzog 2005], *WALL-E* [Stanton 2008], *After Earth* [Shyamalan 2013], and *Oblivion* [Kosinski 2013]). Derelict buildings, torn apart by monsters or ordnance, rise intermittently among the trees, like traces of a dead civilization, and echo the Mayan temple that Sam discovers in the jungle near the Texas border. Elsewhere, cars and boats are lodged in treetops, as if deposited there by a tidal surge or some other sudden inundation. Such imagery further identifies the aliens as climate change metaphors, while the inferable backstory suggests agreements between the neighboring nations about establishing a quarantine zone, coordinating military action, and designating anyone found there as *homo sacer*, that is, people who can be killed with impunity. In realpolitik terms, it is unsurprising that this kill zone is located outside of the United States and that, along with the wall, it signals a simultaneous effort to disentangle and distance the United States from Mexico. These geopolitical calculations, however, are not wholly successful. When Kaulder and Sam cross into Texas, they find the nearest border town flattened, as if by a hurricane, but hiding out in a deserted gas station they witness a pair of aliens roaming free, demonstrating that a border wall is inadequate protection from the creatures, as well as the climate catastrophe they represent.

# TECHNO-UTOPIANISM

Techno-utopian discourse imagines a potentially—but not too—different future produced by scientific and technological innovations. It is shared by those who wish to maintain the status quo—wringing their hands as they kick the crisis into the future where some vague notion of progress supposedly ensures that the problem will be easier to solve—and by those actively working to create the means of gradual, reformist transformation. Lacking what Fredric Jameson calls "a conception of systematic otherness," both routes lead at best to the "bad utopia" of the more-or-less unchanged same (2014, 36). Thus, *Downsizing* (Payne 2017) offers as a solution to the ecologically unsustainable American lifestyle not a radical reorganization of economic and social life but miniaturizing people. Reduced to a mere five inches in height, Americans can continue to enjoy consumer luxury while using just a fraction of the resources. As Imre Szeman notes, techno-utopian discourse "imagines scientific innovations that are in perfect synchrony with the operation of the capitalist economy" (2017, 61), obviating any "need for radical ruptures or alterations in political and social life" (62).

In this sort of account, scientific innovation hates a public policy vacuum, so the "flow of scientific discovery" is seen as resolving "the energy and environmental problems we have produced for ourselves" (Szeman 2017, 62). Technological solutions are always "just around the corner, always just on the verge of arriving," and they never fail to "arrive just in time" (62). *The Martian* (Scott 2015), in which accidentally abandoned astronaut Mark Watney (Matt Damon) must survive 560 days in Mars's uninhabitable biosphere, is a succession of overlapping lethal countdowns, each halted by a just-in-time solution. *Gravity* (Cuarón 2013), in which mission specialist and novice astronaut Ryan Stone (Sandra Bullock), sole survivor of an orbital disaster, must make her way from shuttle to space station to partly disabled reentry vehicle to space station to reentry vehicle to Earth, has a similar structure. While neither film is overtly about climate change, both rehearse techno-utopian logics while emphasizing energy and environment. In contrast, *Interstellar* (Nolan 2014) is set on an ecologically devastated Earth, slowly turning into a global dust bowl as a nitrogen-breathing blight destroys entire species of plant crops and reduces atmospheric oxygen. Billions have died, and humanity is barely more than one failed harvest from extinction. Former astronaut Joseph Cooper (Matthew McConaughey) pilots a desperate mission through a wormhole in search of a habitable new world on which *Homo sapiens* as a species might be preserved, even if the entire terrestrial population must die out. However, the film discloses a five-dimensional posthuman civilization, bending time and space so that Cooper can send a message through time to his genius daughter, Murphy (Mackenzie Foy, Jessica Chastain, Ellen Burstyn at different ages), so that she can complete the equation that will enable the surviving remnants of humanity to control gravity and thus move off-world, ensuring not only their survival but also the evolution of their five-dimensional posthuman descendants.

The success or failure of the techno-utopian film typically depends on the felt adequacy of the solution it proposes to the problem on which it is predicated, as a film like *Elysium* (Blomkamp 2013) demonstrates. It opens with a tracking shot over derelict houses and dry, dusty land, over favela rooftops and on through smoke-filled air to a central business district whose crumbling skyscrapers have been squatted, salvaged, and retrofitted into precarious homes for a predominantly Latinized population of color. This is Los Angeles in the late twenty-first century, and the world is a planet of such slums—diseased, polluted, and overpopulated. Following this opening sequence is a slowly rotating view of the Earth from space. Blue seas limn familiar coastlines; to the north, there is the green of Europe and, to the southwest, a diminished Sahel and depleted Sudanian savanna; the middle of the image, though, is the brown of advancing desertification, stretching from west to east Africa, through the Middle East, Central Asia, the Indian subcontinent, and across to China's Pacific coast. But not everyone is trapped in this warmer world; as we are told, "Earth's wealthiest inhabitants fled the planet to preserve their way of life." The virtual camera pans away from this pale blue (and brown) dot to a massive space habitat, and then inward and down through Elysium's metallic superstructure to glide in over the green and pleasant, vaguely Mediterranean-looking super-suburb of the ultra-rich (figure 4.2). This techno-utopian gated community in the sky is the product of a strategic realism, exercised by the upper class who have extracted themselves from national and terrestrial entanglements and created a nation-state with which to dominate geopolitics from on high. Desperately sick people from Earth intermittently attempt to gain access to Elysium's high-tech medical facilities—capable not

FIGURE 4.2: A "strategic realism": the controlled space habitat of *Elysium* (2013). Sony Pictures.

only of curing all diseases, but also of reversing aging processes and regenerating limbs and organs—but their "undocumented shuttles" are shot down. The extrapolation of contemporary US political discourse around socialized medicine and immigration, especially from Latin America, could not be less subtle.

Rebels fit Max (Matt Damon), a factory worker lethally dosed with radiation in an industrial accident, with a powered exoskeleton so that he can storm Elysium and upload into its central computer a program extending citizenship—and thus healthcare—to all humans. In a stirring finale, Elysian medical shuttles are dispatched to relieve the suffering of the planetary sick, but not even the spectacle of sacrificial white masculinity can conceal the absurdity of it all. Any system that might be used for awarding citizenship can just as easily be employed to revoke it. Providing medical care for billions of people would require not just more shuttles and med-bays than Elysium could possibly possess, but the complete restructuring of the economy. And focusing on medical treatment leaves intact the ecological, economic, social, and political causes of disease, illness, and injury. Such a techno-utopian Band-Aid leaves the forces driving anthropogenic climate change untouched, and even the film's director Blomkamp has recognized that, despite the potential of this imagined future, "the story [was] not the right story" (Ryan 2015).

## APOCALYPTIC ENVIRONMENTALISM

The discourse of apocalyptic environmentalism is most common among environmentalists and on the political left. It recognizes the inertia of existing political systems faced by an unfolding catastrophe and the inability of the market to address crises other than as (crisis-deepening) opportunities to enhance profitability, usually through accumulation by dispossession (see Harvey 2004). Such narratives confound every attempt to maintain the status quo, or something close to it; at the same time, nation-state and supranational action (and inaction) in defense of the status quo squash any possibility of making the fundamental political, social, and economic transformations needed. Embracing the fact of anthropogenic climate upheaval shades over into embracing the disaster itself. Finally, the conundrum about it being easier to imagine the end of the world than to imagine the end of capitalism is resolved as they become, for humans at least, the same thing. Bypassing the proletariat, capital becomes its own gravedigger.

*Snowpiercer* makes this point overtly. Revolution breaks out on the train bearing the last surviving humans on an endless journey around the frozen world, but the finale opts instead for a catastrophic derailment. The closing shots are deeply ambivalent. Only a seventeen-year-old Asian girl, Yona (Ko Asung), and a five-year-old African American boy, Tim (Marcanthonee Reis), emerge from the wreckage. They are dressed in furs, but have no supplies or apparent means of continued survival; meanwhile, in the distance, there is a polar bear, that icon of global warming concerns that has somehow survived, implying that despite the purported lifelessness of the planetary wasteland, there is an ecosystem that may or may not also sustain human life. But that is all it offers. In similar fashion, the revolutionary conclusion of *Mad Max: Fury Road* (Miller 2015) sees a tyrant toppled and his stockpiled water released freely to the benefit of all. However, this potent image of a new world dawning struggles to overcome the preceding two hours of gorgeous, excessive depictions of empty desert.

*Homo sapiens* (Geyrhalter 2016) is a feature-length, experimental, slow-cinema documentary. With an average shot length of over 25.25 seconds and a stationary camera, it comprises nothing but long and medium-long shots of derelict and deserted spaces, of man-made structures fallen into ruin and overrun by nature, of abandoned objects hidden away in a world from which we have departed. Roads and streets, rail tracks and streetcar tracks, corridors and rooms are shot so as to emphasize lines of perspective converging on a vanishing point at the center of the screen, often within the rectangle of a window or doorway, implying receding depths unplumbed by the camera eye. Long lenses ensure consistent depth of field, so that those lines into infinity—or sometimes the way light falls across the image—provide the only visual clues as to what within the shot might be significant. But they are red herrings, as nothing in the frame is more significant than anything else. The only movements are chance ones within the static frame: wind blowing against a blind or lifting some garbage into the air, a frog hopping unexpectedly across a mall floor, pigeons and gulls fluttering within buildings, and flies. The soundtrack is empty apart from ambient noise: water dripping, wind gusting and flurrying, birds calling and singing, insects buzzing.

There are no humans in the film, just traces that we were once here, in these mostly rural, suburban, and exurban spaces. Pylons and telegraph poles stretch into the distance. There are boats run aground on dried-out riverbeds, discarded crates of gas masks, abandoned military vehicles and vessels, pillboxes and bunkers. There are panoptical spaces and nonplaces, such as prisons, airports, and shopping malls. There are bowling alleys, offices, grocery stores, a crematorium, a church, a nightclub in a former church, a swimming pool, a seaside roller coaster, theaters, the interior of a power plant. Also appearing—if we recognize them—are the Buzludzha Monument, Fukushima, the Hashima Island offshore mining facility, a salt lake in Argentina. Potted plants persist inside buildings. Grass, shrubs, flowers, and trees grow through man-made surfaces. Nature is often green, vibrant, doing fine without us, but the film also includes pallid dead trees, shrouded in salt.

*Homo sapiens* shows a world without humans, both derelict and beautiful, and as indifferent as *Monsters'* monsters to absent humans and our relics. As it made its way around the international film festival circuit in 2016 and then onto DVD, reviewers repeatedly described it as science-fictional. They said it depicted the world as if aliens (or the Rapture) had abducted us all, or a plague had wiped us out (but left no corpses). The director, Nikolaus Geyrhalter, allows that audiences might interpret it in this way, but denies that he intended it as a postapocalyptic vision. And mostly he is right to do so. All of the footage is real—actually existing locations that already look this way, right now in the middle of climate catastrophe. This is not a vision of the future, but of the recent and still unfurling past. Indeed, the sole—but utterly transformative—sf intervention into Geyrhalter's record of the nonhuman, of late-capitalist construction collapsing, is to be found in the soundtrack. It consists of archival and specially recorded ambient sound added in postproduction because his efforts to record synchronized a-human sound while filming failed: it repeatedly proved impossible to keep out all human-produced noise, demonstrating how ubiquitous and persistent the eponymous species' pollution has become.

# Not Knowingly Cli-Fi

We might think of all film footage ever shot on Earth as an archive of the visually indiscernible concentration of atmospheric $CO_2$ and other greenhouse gases. Invisible, intangible, it can be recorded but not easily represented. While anthropogenic climate destabilization is happening everywhere all the time and is continuously accelerating, most narrative forms favor the discrete event, a punctual moment with a beginning and an end. Such a rupture in the status quo typically includes among its consequences either a fresh status quo or a return to the status quo ante, thus ensuring the continuity of the subject and of a position from which the world can be narrated. Therefore, while it is relatively easy to represent a changed climate as the background to a narrative about something else entirely—see, for example, *A.I.: Artificial Intelligence*—it is much more difficult to represent climate change itself as a narrative's principal concern. There seem to be three basic options: global catastrophe thrillers along the lines of *The Day After Tomorrow*; films in which the climate-changed setting is so inescapably woven into the story that it can never be mistaken for mere backdrop, such as *Nuoc 2030* (Minh 2014), which slowly discloses how Vietnamese farmers now live in the coastal waters above their former fields; or films that depict localized events but contextualize them as part of larger climate upheavals, such as *Aniara* (Kagerman and Lilja 2018), in which a colony vessel fleeing climate change suffers its own concatenation of closed-system breakdowns and catastrophes; or *The Age of Stupid* (Armstrong 2009), in which an archivist (Pete Postlethwaite) of surviving human knowledge, all alone in a repository in the ice-free Arctic of 2055, watches what are effectively six embedded documentary shorts about in-progress climate change from back "when we could have saved ourselves."

An alternative that might avoid the reductive simplifications required by a block-buster budget *and* the awkwardness of needing to enframe a chamber piece is to condense climate change into a ruptural event that can function metaphorically.[9] For example, in *Arrival* (Villeneuve 2016), a dozen alien spaceships suddenly appear in the skies, each half a kilometer high and hovering ten meters or so above the surface. Featureless, they look like elongated eggs or pebbles, although later they are shown to have a concave side and to also function on a horizontal axis, making them more like mushroom caps. Their advent changes everything. Each craft contains a pair of seven-limbed "heptapods," their appearance blurred by their swirling atmosphere. They resemble banyan trees, standing up on thick roots, but also squid. Their mottled flesh ripples and swells. They have no recognizable faces. Their seven limbs are flexible, like cephalopod tentacles, and their movements aquatic, but at other times their limbs seem jointed, like those of a crustacean. At least one of their tentacles ends in a seven-digited "hand" that opens out to look like the underside of a starfish. From these "palms" they spray jets of an inklike substance to write in the air and on the transparent barrier wall that separates their "aquarium" from part of the meeting chamber designed for humans. Their orthography—each symbol or logogram is a complex circle with a shifting structure, some sections dense with fractal detail—is temporally nonlinear, as if a human were to write a sentence using both hands and working simultaneously from both of its ends. Their written language is semasiographic rather than glossographic—that is, the logograms express meaning directly, rather than approximating the sound of their spoken language, which itself evokes whale song and didgeridoos, the chittering of a Predator, the knocking root stumps of a BBC Triffid. These familiar aliens—as their nonlinear orthography implies—also perceive time differently, experiencing it simultaneously (like *Interstellar*'s five-dimensional posthumans). At the same time, though, their spoken language (presumably) enables them also to experience time sequentially, like humans.

The heptapods' aquatic and tidal zone associations—which include the darkly clouded skies into which their ships seem to dissolve on departure—imply a world submerged. At the same time, their weird physical design and alien mode of apprehension locate them firmly within the literary tradition derived most centrally from H. P. Lovecraft's cosmic horror, revived and revised around the turn of the millennium by such New Weird writers as China Miéville and Jeff VanderMeer. The tentacular and the mycological are central to their respective imaginaries, and both open up the

Weird's usefulness for thinking about anthropogenic climate change.[10] In fact, Donna J. Haraway (2016)—while deliberately misspelling Lovecraft's "Cthulhu," so as to disavow his racism, classism, and misogyny while hopefully deriving a more useful term from the etymological roots of his most famous cosmic entity—argues for renaming and reconceptualizing the "Anthropocene" as the "Chthulucene." Seen in this context, the heptapods invite a climate change interpretation.

The twelve heptapod vessels hover above China, Greenland, Japan, Pakistan, Sierra Leone, Sudan, the United Kingdom, the United States (over Montana), Venezuela, the Indian Ocean west of Australia, and two Russian locations. This eschewal of cinematic aliens' typical fascination with iconic architectural signifiers seems to suggest an indifference to human social arrangements, but the disregard of human geopolitics initially implied by these locations is more apparent than real. Their seemingly random distribution—although in relation to both the entire globe and landmasses, far from random—exposes the limits of the film's geopolitical imaginary. Construing the world in terms compatible with the most hawkish US-centrism, the film repeatedly depicts China, Russia, Pakistan, Sierra Leone, Sudan, and Venezuela as hot-headed and inherently violent, and the United States as fundamentally committed to peace, cooperation, rationality, and reasonableness, only reluctantly turning to military options. The US soldiers who commit the only act of violence against the aliens in the film—detonating a bomb smuggled aboard the Montana spaceship, killing a heptapod—were acting on their own. They do not represent the nation-state or its military, and there is no suggestion that the chain of command bears any responsibility for its subordinates' actions. Before questions of culpability can even be raised, the Chinese General Shang (Tzi Ma) orders the spaceship in Chinese airspace to leave or face military action, and the film treats this as the primary cause of the culminating crisis, especially as others follow China's lead.

It is therefore unsurprising that whenever the film recognizes that these aliens echo terrestrial colonial history, the American characters do not refer to their own settler-colonialism but cite the British in India, the Germans in Ruanda (now Rwanda), and the genocide of Australian aborigines. The sole trace of America's own colonial past is a young girl glimpsed playing in a cowboy costume. While colonial structures extending into the present and future might seem a long way from climate change concerns, it is worth recalling that one inception date for the Anthropocene currently under stratigraphic consideration is the European "discovery" of the Americas. In the 150 years after Columbus landed, the violence of exploitation, extraction, and settlement eradicated approximately 50 million indigenous people. Consequently, the jungle reclaimed agricultural land so fast that its increased uptake of atmospheric $CO_2$ can be observed in early-seventeenth-century ice cores. Furthermore, any consideration of capitalist-colonialism's responsibility for the massive concentration of atmospheric greenhouse gases and other biosphere damage reveals starkly differential culpabilities for climate change—and the extent to which capitalist-colonialism already determines an unequal distribution of climate change consequences. It is the frame within which strategic realism here plays out.

In *Arrival*, the US military recruits linguistics professor Louise Banks (Amy Adams) to interpret the heptapod language and astrophysicist Ian Donnelly (Jeremy Renner) to head the science team. International efforts to communicate with the aliens proceed slowly. With this monumental task stymied by different national agendas and security state paranoia, tensions mount globally. Well before she feels ready, Banks is compelled to ask the heptapods their purpose in coming to the Earth. The answer seems to be "Offer weapon," and although it is just a possible interpretation of "visitors friends heptapods offer give donate award technology apparatus method humanity man woman host person," it nonetheless prompts Shang's ultimatum and the breakdown of communications between national scientific teams and governments. The heptapods present Banks and Donnelly with a massive information download, one-twelfth of all heptapod knowledge, but none of it understandable until integrated with the other eleven-twelfths shared at the other sites. But that download is not really the "technology apparatus method" they have brought; it is the written heptapod language itself. Learning to think in heptapod rewires the human brain so that time is perceived simultaneously rather than sequentially. The flashbacks that have so disturbed Banks—to her daughter's childhood, adolescence, and slow death from a rare disease, to her estrangement from her (unseen) husband—have actually been flash-forwards. She will marry Donnelly, give birth to and raise the palindromically named Hannah (at different ages, Jadyn Malone, Abigail Pniowsky, and Julia Scarlett Dan), and later she will tell Donnelly of her foreknowledge of their daughter's debilitating illness and death, even though she also knows that this revelation will drive him away.

As peril mounts, Banks also flash-forwards eighteen months to a celebration of global unification at which Shang thanks her for phoning him at a vital moment with a message that persuaded him not to attack the aliens, thus saving the world. He then tells her what she would have needed to know to have been able to do that. Back in the present, she phones him, changes his mind, and saves the world. The twelve parts of heptapod knowledge are shared, the aliens depart a united Earth, safe in the knowledge that three thousand years hence, when they desperately need humanity's aid, we will have survived to provide it.

The film's play with complex temporalities is another invitation to climate change interpretation. It transforms the "alien invasion" scenario into something more closely resembling a hyperobject. In Timothy Morton's conceptualization, a hyperobject is so vast as to be "almost impossible to hold in mind" (2013, 58), so "massively distributed in time and space relative to humans" that it functions on "profoundly different" scales and temporalities than those with which we are familiar (1). It showers us with effects and affects but withdraws from comprehension, just as we might see rain falling but cannot perceive the climate. The weather is but "a flimsy, superficial appearance," a "mere local representation" of a phenomenon so extended in space-time that it shatters simple illusions of linear cause-and-effect (104). In *Arrival*, what might seem like a punctual moment (aliens arrive in the skies) and a contained narrative (the story of humanity and the heterosexual couple united) is also situated within two larger histories: the ensuing decade or more of Banks, Donnelly, and Hannah's lives, and an interspecies history that extends for millennia. To even begin to grasp such a hyperobjectal encounter—to see that this is not an alien invasion—Banks must perceive time in such a way that linear causality, the very essence of narrative, gives way to chaos, complexity, and nonlinearity. She must herself become at least partly nonhuman, as she moves beyond the circumscriptions of modernity, of the monadic subject in a reified world, and opens herself to the web of life (see Moore 2015).

*Arrival*'s play with temporality tricks the (first-time) viewer into seeing Banks through rather tired clichés of maternal trauma, and then pulls the rug out. What we have actually been witnessing is her subjection to the early stages of Pretraumatic Stress Disorder—a term E. Ann Kaplan (2016) uses to describe the psychological effect of the perpetual sense of impending dread produced by the slow violence of climate change, even before it happens. At the same time, Banks is faced with an impossible dilemma—whether she should have a child, despite knowing her daughter will suffer and die young—that sooner or later we must all confront, and not just on the scale of the family. The film asks how we might proceed into a future of foreknown catastrophe.

Roy Scranton (2015) contends that anthropogenic climate change narratives present us with the opportunity to learn—not as individuals but as a civilization—how to die. The end of modernity's conceptual and material limits might, he suggests, produce new ways of being in the world, but *Arrival* of course cannot show us that. Conventional narrative closure demands heterosexual union, and the film can make only vague gestures at the notion of global unity (and in the brief glimpse we have of it, nothing much seems changed). But the moment at which Banks makes her first breakthrough in communicating with the heptapods is suggestive. With only the old carbon culture standby—a canary in a cage—to assure her that the air in the alien ship is safe to breathe, Banks sheds her hazmat suit. She risks her life to demonstrate her trust that the aliens will maintain the habitable environment they have created. She strips away protection to show her face, her humanity, herself to the heptapods. This moment recalls Jacques Derrida's (2008) reflections upon finding himself standing naked before his cat—the object of a nonhuman animal's gaze—and his subsequent dissolution of the culturally erected barrier between human animals and nonhuman animals in order not to eradicate but to pluralize difference. But as Banks steps forward, the moment transforms. She raises a hand to touch the transparent wall separating her from the heptapods. It looks like a hailing gesture, but is at least equally an attempt at contact; she reduces the separation between species to the environmental barrier upon which both their lives depend. Behind her, Donnelly discards his hazmat suit. Their relative positions gender-role reverse those of the naked man and woman pictured on the plaques attached to the Pioneer 10 and 11 space probes. But here the woman, not the man, is the active figure (even as their remaining clothing conceals the secondary sexual characteristics the Pioneer nudes link to gender); it is Banks who makes kin in this Cthulhu scene. While this image might still normalize white people, a global minority, as representative humans, even Donnelly realizes that he is not the protagonist.

In these ways at least, *Arrival*, an sf film that is categorically not cli-fi, is also very clearly about climate change. It reminds us that in attempting to find the limits of cli-fi, and to turn it to useful transformative ends, one must do more than merely look at the obvious, self-declared examples. As Fredric Jameson has argued, criticism must rewrite the "text in terms of a particular master code" (1989, 10). And as the anthropogenic crisis continues to develop, our contemporary cinema reminds us that we have a duty to engage with its cultural texts through the fresh lenses it produces—to better understand our culture, our politics, and our world, and to revolutionize them.

## Notes

1. While Bloom concedes cli-fi's indebtedness to sf and admits that it might be considered an sf subgenre, he buries both points amid his declarations about this "new" genre.
2. Intriguingly, her cli-fi *SEAbean* trilogy (2013–14) reads precisely like sf that has failed to escape the gravitational pull of classic children's timeslip fantasies.
3. Ironically, Bloom's neologism is modeled on "sci-fi," a term coined in the 1950s by sf fan Forrest J. Ackerman, who was trying to make the genre sound as trendy and cutting edge as "hi-fi." However, it has long been despised by sf aficionados as a juvenilizing and derogatory term, and is often now used to designate and deride "bad" sf—a category into which many instances of "literary" cli-fi might fit.
4. To which they add overt exclusions of geogenic, xenogenic, and astrogenic climate change fiction such as Maggie Gee's *The Ice People* (1998), N. K. Jemisin's *Broken Earth* trilogy (2015–17), and Liu Cixin's *Remembrance of Earth's Past* trilogy (2006–10). They also, like Trexler (and Peter Christoff), are keen to find the "magic bullet" novel, that is, the cli-fi equivalent of Neville Shute's *On the Beach* (1957), often credited with shifting cultural attitudes about nuclear weapons (see Baker 2012).
5. For example, global warming does threaten the circulation of ocean currents that currently keep the climate of such countries as the United Kingdom relatively temperate, but the cooling consequences are unlikely to be so sudden or so drastic.
6. A better sense of the accelerating upheavals through which we are already living might be obtained by watching the *Sharknado* movies back-to-back but skipping the last ten minutes of the sixth one (see Bould 2021).
7. I have yet to encounter a cli-fi novel that could not just as easily be described as sf. However, while the catastrophic flooding in, for example, *Beasts of the Southern Wild* (Zeitlin 2012), *Lost River* (Gosling 2014), *Crawl* (Aja 2019), and *Parasite* (Bong 2019) clearly addresses how race, class, gender, or predatory finance determine climate change's consequences, even those with fantastic elements are difficult to subsume into sf.
8. I am indebted to Imre Szeman (2017), who uses these terms to describe the three commonly occurring narratives that dominated political and media framings of the peak-oil crisis and speculations about transitioning to a postcarbon economy.
9. Or one might envision a ruptural location, as in the zones of *Stalker* (Tarkovsky 1979), based on Boris and Arkady Strugatsky's *Roadside Picnic* (1972), *Annihilation* (Garland 2018), based on Jeff VanderMeer's 2014 novel of the same name, and *The Color Out of Space* (Stanley 2019), based on H.P. Lovecraft's "The Colour Out of Space" (1927).
10. See, for example, Miéville's "Covehithe" (2011) and "Polynia" (2014), both collected in *Three Moments of an Explosion: Stories* (2015), and VanderMeer's *Southern Reach Trilogy* (2014), beginning with *Annihilation*.

## WORKS CITED

Abraham, John. "CliFi—A New Way to Talk about Climate Change." *The Guardian*, October 18, 2017. https://www.theguardian.com/environment/climate-consensus-97-per-cent/2017/oct/18/clifi-a-new-way-to-talk-about-climate-change.

Baker, Brian. "*On the Beach*: British Nuclear Fiction and the Spaces of Empire's End." In *Future Wars: The Anticipations and the Fears*, edited by David Seed, 144–60. Liverpool: Liverpool UP, 2012.

Bloom, Dan. "How and Why 'Sci-Fi' Gave Birth to 'Cli-Fi.'" *Teleread*, December 13, 2013. https://teleread.com/how-and-why-sci-fi-gave-birth-to-cli-fi/index.html.

Bould, Mark. *The Anthropocene Unconscious: Climate Catastrophe Culture*. London: Verso, 2021.

Christoff, Peter. "The End of the World as We Know It." *The Age*, January 15, 2008. https://www.theage.com.au/national/the-end-of-the-world-as-we-know-it-20080115-ge6luy.html.

Derrida, Jacques. *The Animal That Therefore I Am*. Ed. Marie-Louise Mallet. Trans. David Wills. New York: Fordham UP, 2008.

Dunbar-Ortiz, Roxanne. *An Indigenous Peoples' History of the United States*. Boston: Beacon Press, 2014.

Fay, Jennifer. *Inhospitable World: Cinema in the Time of the Anthropocene*. New York: Oxford UP, 2018.

Gee, Maggie. *The Ice People*. London: Richard Cohen, 1998.

Ghosh, Amitav. *The Great Derangement: Climate Change and the Unthinkable*. Chicago: U of Chicago P, 2016.

Glass, Rodge. "Global Warning: The Rise of 'Cli-Fi.'" *The Guardian*, May 31, 2013. https://www.theguardian.com/books/2013/may/31/global-warning-rise-cli-fi.

Haraway, Donna J. *Staying with the Trouble: Making Kin in the Chthulucene*. Durham, NC: Duke UP, 2016.

Hart, Philip Solomon, and Anthony A. Leiserowtiz, "Finding the Teachable Moment: An Analysis of Information-Seeking Behavior on Global Warming Related Websites during the Release of *The Day After Tomorrow*." *Environmental Communication* 3, no. 3 (2009): 355–66.

Harvey, David. "The 'New' Imperialism: Accumulation by Dispossession." *Socialist Register* 40 (2004): 63–87.

Holding, Sarah. *SeaBEAN*. Cowes: Medina Publishing, 2013.

Holding, Sarah. *SeaWAR*. Cowes: Medina Publishing, 2014.

Holding, Sarah. *SeaRISE*. Cowes: Medina Publishing, 2014.

Holding, Sarah. "What Is Cli-Fi? And Why Do I Write It?" *The Guardian*, February 6, 2015. https://www.theguardian.com/childrens-books-site/2015/feb/06/what-is-cli-fi-sarah-holding.

Jameson, Fredric. *The Political Unconscious: Narrative as a Socially Symbolic Act*. London: Routledge, 1989.

Jameson, Fredric. "The Politics of Utopia." *New Left Review* 25 (2004): 35–54.

Jemisin, N. K. *The Fifth Season*. New York: Orbit, 2015.

Jemisin, N. K. *The Obelisk Gate*. New York: Orbit, 2016.

Jemisin, N. K. *The Stone Sky*. New York: Orbit, 2017.

Kaplan, E. Ann. *Climate Trauma: Foreseeing the Future in Dystopian Film and Fiction*. New Brunswick, NJ: Rutgers UP, 2016.

Leiserowitz, Anthony A. "Before and after *The Day after Tomorrow*: A U.S. Study of Climate Change Risk Perception." *Environment* 46, no. 9 (2004): 22–37.

Lewis, C. S. *The Lion, the Witch and the Wardrobe*. London: Geoffrey Bles, 1950.

Liu, Cixin. 三体. Chongqing: Chongqing Press, 2008. Translated by Ken Liu as *The Three-Body Problem*. New York: Tor, 2014.

Liu, Cixin. 黑暗森林. Chongqing: Chongqing Press, 2008. Translated by Joel Martinson as *The Dark Forest*. New York: Tor, 2015.

Liu, Cixin. 死神永生. Chongqing: Chongqing Press, 2010. Translated by Ken Liu as *Death's End*. New York: Tor, 2016.

Lovecraft, H. P. "The Colour Out of Space." *Amazing Stories* 2, no. 6 (September 1927): 556–567.

Lynas, Mark. *Six Degrees: Our Future on a Hotter Planet*. London: Harper Perennial, 2008.

Miéville, China. "Covehithe." *The Guardian*. 22 April 2011. https://www.theguardian.com/books/2011/apr/22/china-mieville-covehithe-short-story.

Miéville, China. "Polynia." *Tor.com* (1 July 2014).

Miéville, China. *Three Moments of an Explosion: Stories*. London: Macmillan, 2015.

Milner, Andrew, and J. R. Burgmann. *Science Fiction and Climate Change: A Sociological Approach*. Liverpool: Liverpool UP, 2020.

Moore, Jason W. *Capitalism and the Web of Life: Ecology and the Accumulation of Capital*. London: Verso, 2015.

Morton, Timothy, *Hyperobjects: Philosophy and Ecology after the End of the World*. Minneapolis: U of Minnesota P, 2013.

Nesbit, E. *Five Children and It*. London: T. Fisher Unwin, 1902.

Pearce, Philippa. *Tom's Midnight Garden*. Oxford: Oxford University Press, 1958.

Ryan, Mike. "New *Alien* and *Chappie* Director Neill Blomkamp on *Elysium*: 'I F*cked It Up.'" *Uproxx*, February 26, 2015. https://uproxx.com/movies/neill-blomkamp-elysium-alien/.

Schroeder, Randy. "Getting into Accidents: Stoekl, Virilio, Postsustainability." In *Petrocultures: Oil, Politics, Culture*, edited by Sheen Wilson, Adam Carlson, and Imre Szeman, 355–76. Montreal: McGill-Queen's UP, 2017.

Scranton, Roy. *Learning to Die in the Anthropocene: Reflections on the End of a Civilization*. San Francisco: City Lights, 2015.

Shute, Neville. *On the Beach*. London: Heinemann, 1957.

Strugatsky, Boris, and Arkady. Пикник на обочине. *Avrora* issues 7–10, 1972. Translated by Antonina W. Bouis as *Roadside Picnic*. New York: Macmillan, 1977.

Szeman, Imre. "System Failure: Oil, Futurity, and the Anticipation of Disaster." In *Energy Humanities: An Anthology*, edited by Imre Szeman and Dominic Boyer, 55–70. Baltimore: Johns Hopkins UP, 2017.

Thorpe, David. "Interview: Dan Bloom on CliFi and Imagining the Cities of the Future." *Smart Cities Dive*. Accessed September 11, 2022. https://www.smartcitiesdive.com/ex/sustainablecitiescollective/interview-dan-bloom-clifi-and-imagining-cities-future/1037731/.

Trexler, Adam. *Anthropocene Fictions: The Novel in a Time of Climate Change*. Charlottesville: U of Virginia P, 2015.

VanderMeer, Jeff. *Annihilation*. New York: Fourth Estate, 2014.

VanderMeer, Jeff. *Authority*. New York: Fourth Estate, 2014.

VanderMeer, Jeff. *Acceptance*. New York: Fourth Estate, 2014.

Wallace-Wells, David. *The Uninhabitable Earth: A Story of the Future*. London: Penguin, 2019.

Willoquet-Maricondi, Paula. "Shifting Paradigms: From Environmentalist Films to Ecocinema." In *Framing the World: Explorations in Ecocriticism and Film*, edited by Paula Willoquet-Maricondi, 43–61. Charlottesville: University of Virginia Press, 2010.

Wolfe, Patrick. "Settler Colonialism and the Elimination of the Native." *Journal of Genocide Research* 8, no. 4 (2006): 387–409.

# CHAPTER 5

........................................................................................

# ETHNOGOTHIC FILM

........................................................................................

### SUSANA M. MORRIS

JORDAN Peele's 2017 film *Get Out* has most of the hallmarks of contemporary Gothic cinema. City dwellers drive into the country for a weekend with family. Although there is no dilapidated estate, the isolated pastoral setting is more uncanny than idyllic. The presence of creepy servants, kidnapping, hypnotic trances, and body swapping—all set to a brooding score—mark the film's participation in the Gothic tradition, mingled with some science fiction (sf) elements. Eventually, the hero discovers a sinister plot and is lucky to make it out alive. However, the specifics of the film's plot twist distinguish *Get Out* from its mainstream peers: its primary villains are wealthy whites who covet Black bodies for everything from their sexual prowess to their vision. While a racial concern is no stranger to Gothic cinema, the specter of nonwhite contagion, rather than white supremacy, has most often been the cause of terror in such films. However, *Get Out* reworks the relationship of race and the Gothic in sf cinema—and it is not singular in doing so.

Indeed, Peele's film participates in a broader tradition, that of the ethnogothic narrative. Ethnogothic works, whether film, literature, or fine art, reflect a longstanding tradition in Black cultural production that reframes the monstrous other often found at the heart of traditional Gothic works. While Black people have often been associated with abjection, deviation, and the uncanny, creators from across the African diaspora have repurposed the Gothic as a way to engage with existential questions without resorting to tired tropes of menacing Blackness. Artists Stanford Carpenter and John Jennings initially created the term "ethnogothic" to describe the proliferation of ghosts, vampires, hauntings, and the supernatural in Black diasporic literatures, with Jennings noting that "the black body has been an index for other spaces. White folk think of our bodies as heterotopic places and then we internalize that" (Black Speculative 2020). Based in the work of Michel Foucault (1994), the notion of heterotopia describes certain cultural, institutional, and discursive spaces that are somehow "other": disturbing, intense, incompatible, contradictory, or transforming. This sense of otherness has often been depicted as synonymous with Blackness. To that end, Jennings further defines the ethnogothic as "Primarily speculative cultural narratives that actively engage with negatively affective racialized and/or intersectionally situated psychological traumas via the traditions of Gothic tropes and technologies such as: the grotesque other, body horror, haunted spaces, the hungry ghost, the uncanny, the doppelgänger, fictional historical artifacts, and multivalent disruptive tensions between the constructions of memory, history, the present, and the self" (Jennings 2016). Thus, the ethnogothic tradition recasts the specter of Blackness as a tool for illuminating the myriad horrors of white supremacy and anti-Blackness.

This essay describes how, while drawing heavily on a Gothic tradition commonly associated with horror, ethnogothic cinema can also be understood as a particular type of sf cinema that merges Gothic material with a Black cultural context. Ethnogothic works thereby produce iconographies that emphasize a haunting sense of otherness, while disrupting heterotopic depictions of Blackness. These ethnogothic narratives have found a significant presence in contemporary sf cinema, as they effectively explore part of a cinematic tradition that exploits tensions between the so-called natural world and the supernatural world, calling into question Western epistemologies of reality. The infusion of Gothic elements in Black sf cinema helps to illuminate the horrors of science and technology, as seen in the racist pseudoscience used to snatch the bodies of Black victims in Peele's *Get Out*. To that end, ethnogothic cinema often interrogates the contemporary techno-scientific world—as both a repressive power and a power that might be turned upon itself—to sketch the complexly haunted identity of the Black experience, while also challenging a sense of white supremacy. This chapter charts several key elements that recur across ethnogothic films: the uncanny environment, the supernatural, African diasporic traditions such as Vodou or hoodoo, and the rejection of Blackness as abject. These characteristics reflect a distinct tradition that distinguishes ethnogothic cinema from both its Gothic peers and from traditional sf and fantasy traditions.

Ethnogothic films—from *Blacula* to *Get Out*—have existed for decades. Along with ethnogothic literature and art, they constitute a rejection of heterotopic Blackness in favor of more complex renderings of race, particularly in the West. In *African American Gothic*, Maisha Wester suggests that the African American Gothic tradition "reveals the archetypal depictions of racial, sexual, and gendered others as constructions useful in the production of white patriarchal dominance" (2012, 2). When Black cultural producers engage in the Gothic tradition, they do so not just to appropriate a dominant discourse, but to illuminate and denaturalize that dominant discourse; as Wester asserts, "African American texts go beyond merely inverting the color scheme of the gothic trope—blackened evil that torments whiteness—to destabilizing the entire notion of categories and boundaries" (2). Likewise, Ytasha Womack notes how the "presence of ghost stories and hauntings in black literature and art" is "a way of dealing with cultural trauma" (2013, 103). More specifically, she suggests that Black sf cultural production, such as ethnogothic cinema, rejects the visual gatekeeping of Western cinema and resists anti-Black themes (132–33). Thus, ethnogothic cinema rarely includes hapless virgins running through haunted houses, chased by dark, monstrous villains, as one might expect to find in conventional Gothic cinema. Instead, it troubles dominant tropes of the uncanny and the abject so that those deemed monstrous must deal with the specter of themselves and those who dehumanize them.

## GOTHIC ANTECEDENTS

The Gothic has been a staple in Western popular culture since the publication of Horace Walpole's *The Castle of Otrantro* in 1764. Since the advent of film, the Gothic has been reworked in myriad ways, ranging from *Nosferatu* (Murnau 1922) to *Rebecca* (Hitchcock 1940) to *Rosemary's Baby* (Polanski 1968), to more contemporary depictions such as *The Others* (Amenabar 2001) and *The Orphanage* (Bayona 2007). Gothic cinema has proven extremely malleable, easily translating elements from architecture, literature, and music into its narratives. As Nick Groom notes,

> Black-and-white *chiaroscuro* was ideally suited to Gothic style from the outset—indeed films are watched in the dark—but it was also the demand for new cinematic experiences such as camera trickery, special effects, fantastical scenery, and monstrous figures that merged in a new visual style. In the cinema, the Gothic was deliberately updated and came to represent a distinct aesthetic. And just as the Gothic Revival reflected imperial attitudes and anxieties in Victorian Britain, Gothic cinema was likewise shaped by social and political forces. Great War trauma in Weimar Germany, mid-century isolationism in America, and British attempts at cultural renewal after the Second World War: all motivated the medium. (2012, 123–24)

These works are often preoccupied with a monstrous other that represents social fears and anxieties, particularly about taboo subjects such as sexuality, queerness, race mixing, and the like. Jack Halberstam suggests that "in the Gothic, crime is embodied within a specifically deviant form—the monster—that announces itself (de-monstrates) as the place of corruption" (1995, 2). Take, for instance, Bram Stoker's 1897 novel *Dracula*. Dracula embodied fears of the specter of Eastern European influence on its Western neighbors and was especially cautionary about deviant sexuality that could destroy chaste Western womanhood. The film adaptations of *Dracula* toyed with these same themes, updating the monstrous fears to match the anxieties of the time, prompting Stephen T. Asma's comment that "the monster is more than an odious creature of the imagination; it is a kind of *cultural category*, employed in domains as diverse as religion, biology, literature, and politics" (2009, 13, italics in original).

Reflecting this cultural fixation, a significant portion of Gothic film is preoccupied with anxieties about sexuality, particularly queerness, and race, especially Blackness. As Halberstam suggests, "If race in the nineteenth-century Gothic was one of many clashing surfaces of monstrosity, in the context of twentieth-century Gothic, race becomes a master signifier of monstrosity and when invoked, it blocks out all other possibilities" (1995, 5). Victor Halperin's *White Zombie* (1932) is an early example of this cinema where race—and particularly the juxtaposition of whiteness and Blackness—inspires terror.

Starring Bela Lugosi, *White Zombie* is set in Haiti and features Vodou rituals and the titular white zombies. "White" as a modifier to "zombie" signifies how out of the ordinary this circumstance is. Taking place decades before *Night of the Living Dead* (Romero 1968) or *The Walking Dead* (2010–22) and the proliferation of the zombie as an unraced revenant, *White Zombie* shows uncivilized contamination defiling whiteness when Neil (John Harron) and Madeline (Madge Bellamy), two white Americans, come to the idyllic Caribbean for a romantic wedding. Naïve Madeline befriends Beaumont (Robert Frazer), a wealthy, lascivious plantation owner, who invites her to marry at his estate, but he has his own dark designs on her. The first sign that something is amiss is when Madeline and Neil arrive at the Beaumont plantation and encounter a supposed Vodou ritual. Loud drumbeats wail, and dozens of Black peasants shuffle about in a strange formation. Their Black driver (Clarence Muse) alerts them that this is a funeral and that Haitians bury their dead beneath busy streets out of superstition. This scene casts an ominous tone, as Neil remarks, "Well, that's a cheerful introduction to our West Indies," as a way to assuage Madeline's fears. However, the white zombies appear next, foreshadowing their importance in the rest of the film. The zombies are all white men, described as "living dead," led by their white master witchdoctor Murder Legendre (Lugosi), who also sets his sights on Madeline.

What follows is a love quadrangle with a twisted version of Haitian Vodou as the vehicle for various horrific mishaps. Before Neil and Madeline can marry, Beaumont propositions her and slips her a potion that turns her into a dead-eyed zombie. However, Beaumont, who has procured the potion from Legendre, is just a dupe for the witchdoctor who wants to keep the woman—the only upper-class white woman in the film—to himself. Neil confers with Dr. Brunner (Joseph Cawthorne), a priest and missionary who declares, "Haiti is full of nonsense and superstition," and they confront Legendre (who has also turned Beaumont into a quasi-zombie). The men force the white zombies to launch themselves off a cliff like lemmings, while Beaumont, in a last stunning turn of manhood, overpowers Legendre and both fall to their deaths, leaving Neil and Madeline to reunite.

Although Black characters are little more than exotic props in *White Zombie*, Blackness as a signifier abounds. The lush landscape of the Caribbean and the sugar plantation in particular represent an untamed wildness that corrupts whites like Legendre and Beaumont who fall prey to the environment's excess. Beaumont is a wealthy landowner used to getting everything he wants, so it is no surprise that his desire for things evolves to objectifying his own people. Likewise, Legendre is a masterful Vodou practitioner who makes zombies of all his enemies and underlings. While Legendre has a number of Black zombies who work his plantation, his white zombies do his more complex bidding. Indeed, the fact that both Beaumont and Legendre zombify whites leads to their destruction. Only Dr. Brunner, the religious man, holds on to his sanity and his white identity, enabling him to help Neil fight Legendre and save Madeline. In the end, Beaumont is redeemed by resisting the zombie curse and killing Legendre, although he too must die for his previous indiscretions. *White Zombie* is a cautionary tale of white excess and the perils of the "uncivilized" corrupting Western culture. Marina Warner notes that "Haiti was still under American occupation at the time of film's making" (2006, 363). It thus seems telling that the film is preoccupied with white zombies when in fact Black people were in peril of enslavement. As Warner suggests, "Some kind of profound self-interest compels the makers of *White Zombie* to meditate on slavery and its afterlife, while conspicuously avoiding the historical facts of the subject as such" (363). The film, like several peers such as *Black Moon* (Neill 1934), *Ouanga* (Terwilliger 1936), *Revenge of the Zombies* (Sekely 1943), and *I Walked with a Zombie* (Tourneur 1943), represents larger anxieties about whiteness and racial Others replete in twentieth-century Gothic cinema.

## ETHNOGOTHIC CINEMA IN THE TWENTIETH CENTURY

In *Playing in the Dark* Toni Morrison (1992) delineates the significant role of Blackness in the creation of the Anglo-American literary tradition, with implications for other forms of cultural production as well. Specifically, Morrison excavates and illuminates the "Africanist presence" that helped shape the works of Willa Cather, Ernest Hemingway, Herman Melville, and Edgar Allan Poe, arguing that, "As a writer reading, I came to realize the obvious: the subject of the dream is the dreamer. The fabrication of an Africanist presence is reflexive; an extraordinary meditation on the self; a powerful exploration of the fears and desires that reside in the writerly conscious. It is an astonishing revelation of longing, of terror, of perplexity, of shame, of magnanimity" (17). In other words, Morrison discerned the various ways in which Blackness and Black characters reflect deep-seated anxieties and desires about whiteness. Her realization suggests that even when discussing the Other, these fundamentally American writers are often engaged in a solipsistic exercise to plumb the heart of whiteness.

Morrison's detection of an Africanist presence in white American literature is also in conversation with Foucault's notion of heterotopia, an idea central to Carpenter and Jennings's explanation of the ethnogothic. In *The Order of Things*, Foucault specifically defines heterotopia in opposition to the sf concept of utopia:

> *Utopias* afford consolation: although they have no real locality there is nevertheless a fantastic, untroubled region in which they are able to unfold; they open up cities with vast avenues, superbly planted gardens, countries where life is easy, even though the road to them is chimerical. *Heterotopias* are disturbing, probably because they make it impossible to name this *and* that, because they shatter or tangle common names, because they destroy "syntax" in advance, and not only the syntax with which we construct sentences but also that less apparent syntax which causes words and things (next to and also opposite one another) to "hold together." . . . Heterotopias . . . desiccate speech, stop words in their tracks, contest the very possibility of grammar at its source; they dissolve our myths and sterilize the lyricism of our sentences. (1994, xviii)

Here Foucault invites readers to consider how Western subjects since the eighteenth century have come to understand their world primarily as a series of opposites, with civilization and order on one side, and savagery and disorder on the other. Foucault outlines what the fantasy of European utopia means, the "fantastic, untroubled region . . . where life is easy." Utopia is thus a space of beauty, peace, and, above all, order. The cities, with their vast avenues and beautiful gardens, reflect planning and discipline. It is perhaps not surprising that in the Enlightenment this particular notion became ideal, just as notions of whiteness and white supremacy begin to coalesce into forms most recognizable to us today.

Heterotopias, on the other hand, are sites of disorder. They are not necessarily dystopias, for dystopias may very well "function" in their dysfunction. Heterotopias are something even more disturbing because they defy common logic and understanding, even in the presence of advanced technology. Vast avenues, superb gardens, and easy living are impossible in heterotopias, which are voids, wastelands, and gaping black maws filled with horror and disorder. They incite confusion and terror. Yet the splendor of the utopia is all the more sharply defined by the heterotopia's abjectness. This distinction is significant as Europeans begin to understand themselves as Western subjects atop a global social hierarchy and, therefore, the worlds they create and idealize as superior to all other forms of order. That understanding makes the creation and existence of heterotopias, especially those mapped onto cultures and peoples they marginalize, of particular importance. This heterotopic ideal is fundamental to Western popular culture and informs Gothic cinema.

Twentieth-century ethnogothic cinema is largely unconcerned with the racist anxieties that preoccupy a portion of Gothic film. Ethnogothic works take "horrific supernatural spaces in storytelling, and use that trauma to unpack ideas around the horrors of racism and oppression" (Robinson 2018). To that end, films such as *Blacula* (Crain 1972) and *Ganja & Hess* (Gunn 1973) use uncanny environments, the supernatural, African diasporic traditions, and a rejection of Blackness as abject to disrupt heterotopic depictions of Blackness. Moreover, in interrogating white supremacy they do not simply castigate whiteness, but destabilize categories of what is or is not monstrous to various ends. Furthermore, in their Gothic roots they often question the primacy of the contemporary techno-scientific world central to sf cinema, revealing a prevailing need for the metaphysical and supernatural to explain and ameliorate the human condition.

At first glance *Blacula* may seem like a simple blaxploitation affair: a low-budget film with a liberal dose of soulful music, sex, and violence. However, *Blacula* is also an early ethnogothic film that reimagines categories of the other. While it may be misremembered simply as a Black version of *Dracula*, its opening scenes reveal another identity. It begins with Mamuwalde (William Marshall), an African prince, and his consort, Luva (Vonetta McGee), entertaining Count Dracula in a castle in 1780 Transylvania. As an ambassador for his people, Mamuwalde wants to negotiate a treaty to end the slave trade. However, Dracula not only balks at the idea but insults Luva (insinuating he wants to own her as a sex slave) and denounces Mamuwalde as a barbarian "from the jungle." Incensed, Mamuwalde fights Dracula's guards but is overpowered, turned into a vampire, and imprisoned in a sort of living death with his life as a vampire beginning as a curse for daring to want to free Africans from chattel slavery. Thus Harry M. Benshoff notes that, in such films,

> the monster often becomes an allegory for the historical experience of African Americans. Blacula's vampirism is an explicit metaphor for slavery: bitten by the racist Count Dracula centuries ago ("I shall place a curse of suffering on you that will doom you to a living hell"), the curse of vampirism becomes the lingering legacy of racism. Indeed, Blacula explicitly states that he was "enslaved" by the curse of vampirism. What he finds so distasteful about his state is that he must now enslave others, biting them and turning them into his minions. (2000, 39)

This description presents us with quite a different origin story than the typical Western accounts based on Vlad the Impaler. Instead, it is the combination of Mamuwalde's Blackness and his desire for the dignity of his people that incites the curse, with Dracula's disdain for Mamuwalde purely racist. Vampirism thus becomes a punishment for haughty Blacks who step out of place, with Dracula as monstrous not only because he is a vampire but because of his investment in white supremacy.

More conventional Gothic elements also abound in *Blacula*. The film opens in a spooky castle and much of the modern-day setting vacillates between Mamuwalde's lair and a morgue. Most of the film takes place at night, since sunlight is deadly to Mamuwalde and his kind. Likewise, he has a vampire's supernatural power, so when he needs to travel quickly, Mamuwalde transforms into a bat and swoops across the city. He stalks his prey, killing a nightclub photographer simply because she has developed a photo that reveals he has no reflection. He is well versed in Afrocentric history and the occult arts, presenting himself as a Black Renaissance man of the 1970s. While his initiation into life as a vampire is an eternity locked in a coffin, unable to satisfy his hunger for human blood, there is no recognition that, instead of diminishing Mamuwalde, this curse makes him almost omnipotent. Indeed, at times the film depicts him as somewhat of a folk hero. As Frances Gateward observes,

> In the original theatrical trailer for the film, Blacula is referred to not as a monstrosity to fear or a source of horror, but as a Black avenger in contemporary Los Angeles. The film critiques 20th-century racial oppression by its depiction of police power, which is used not to protect and serve, but to work as an occupying force, harassing and brutalizing the Black poor with impunity. No doubt some audiences, having experienced such encounters with the police only a few years after the uprisings, as in Watts in 1965, took cathartic pleasure from the climactic scene, where Blacula fights off night-stick wielding and gun-toting uniformed police officers. (2004)

Yet despite being ushered into eternal life by a racist white vampire, Mamuwalde is not interested in starting a race war or killing whites. Unlike Dracula who transformed him out of spite, Mamuwalde's main modus operandi involves proximity and self-defense.

Mamuwalde also stands in stark relief to the police and his nemesis, Dr. Gordon Thomas (Thalmus Rasulala), a Black forensic pathologist whose techno-scientific powers and Western logic are defeated at almost every turn—a metaphysical counterpoint to dominant sf cinema logic. Thomas is an interesting contrast to Prince Mamuwalde, since he is a man of science who must admit that an ancient curse rather than a modern serial killer is terrorizing the city. Eventually Thomas discovers that Mamuwalde is behind a string of gruesome murders in the city, his scientific knowledge ultimately trumped by a supernatural specter. Yet while Blacula is defeated at the end of the movie, Mamuwalde's ethnogothic presence has managed to haunt fantastic cinema to the present.

While *Blacula* is somewhat formulaic in its blaxploitation sensibilities, *Ganja & Hess* (1973), another early ethnogothic vampire tale, is experimental and brooding, a mishmash of art-house styles. Directed by Bill Gunn, it uses vampirism as a metaphor for drug addiction and conspicuous consumption. The wealthy and refined scholar Dr. Hess Green (Duane Jones) is an anthropologist studying the obscure African culture of the Myrthinians who were known to drink blood. When his drunken assistant George Meda (Bill Gunn) stabs him with a ceremonial Myrthinian bone dagger, Hess develops an uncontrollable thirst for blood and becomes a day-walking immortal vampire, impervious to almost all pain. Bloodlust becomes his overwhelming desire, forcing him to leave everything from dinner parties to a sexual rendezvous to satisfy this desire. When Meda's wife, Ganja, shows up looking for her husband, they immediately form a codependent duo, as Hess turns her into a vampire. Throughout the narrative, songs about Myrthinian legend and gospel songs about the blood of Christ form a constant refrain, the latter suggesting that the Eucharist is also a sort of vampire story. Consumed with a constant, debilitating hunger, Hess goes to a church revival and then returns home suicidal to stand before the shadow of a cross and burn to death. Ganja is left to restart her codependent struggle with another lover.

Like other ethnogothic films, *Ganja & Hess* is replete with conventional Gothic imagery. Hess is conspicuously wealthy, servants attend to his every whim at home, and a driver mans his Rolls-Royce. Although a large portion of the film takes place outside, the film is grainy and often sepia-toned, so even outdoor scenes look hazy and psychedelic. In other moments, light overexposes the scenery, making it seem washed out and dreamlike. Hess's ornate mansion is a dark fortress where he lives alone, the "only colored on the block" in his elite neighborhood. He keeps George Meda's frozen body, along with several containers of blood, in his wine cellar and allows no one to enter without his permission. While Hess is filled with dread and self-loathing after consuming blood, he is compelled to kill again and again. Hess is trapped between science and his unexplainable condition: a scholar and a researcher, a man of science, yet supernatural hallucinations haunt him. Visions of the Queen of Myrthia flash between scenes, while gospel music and spoken-word poetry about the Myrthinian blood rituals alternately intone, evoking a nightmarish atmosphere. This recurring imagery reflects the prophecies guiding his life. Since his stabbing, Hess is bound to the Myrthinians, but also to the Black church, especially through his driver, the Reverend Luther Williams (Sam Waymon), the film's narrator who summarizes Hess's experience and links his circumstances to the issue of addiction. Thus the film parallels the addiction to blood to contemporary concerns with heroin and narcotics that were ravaging Black communities in the 1960s and 1970s.

In fact, addiction in its many forms is the main specter in *Ganja & Hess*. As Harry Benshoff notes, "The film argues that an addiction of any sort, whether to blood, religion, drugs, or sex, is morally equivalent to any other" (2000, 42). The film's score reveals the Myrthinians as the original addicts, cursed with an addiction to blood until they were "saved" by Christianity. This curse parallels interpretations of the biblical curse of Ham, which was used to justify racist anti-Black attitudes during slavery. When Hess becomes a vampire via a Myrthinian dagger, he too becomes embroiled in this struggle. Like his Myrthinian ancestors, Hess is thus locked into a tenuous relationship with Western religion. Since science and his scholarly pursuits are no match for his blood-lust, Christianity seems like the only thing that can "save" him (the shadow of the cross); however, his salvation is also his death. Thus, as Benshoff offers, "Ganja and Hess symbolize the 'real-life' issues that plague the African American community, such as drug abuse or selling out to materialism" (2000, 44). Through Hess's self-destructive desire for blood, the film offers a cautionary tale about succumbing to internalized heterotopic notions of Blackness. Just as the Myrthians suffered through their curse, so must their descendants unless they are strong enough to face their fate.

*Def by Temptation* (1990), written, produced, directed, and starring James Bond III, reworks the ethnogothic impulse for the hip-hop generation. It is the story of virginal teetotaler Joel Garth (Bond) who must resist demonic temptation and free himself from a family curse. Joel is the son of Minister Garth (Samuel L. Jackson) and plans on following the family business. However, a demon has killed his parents and has been stalking him since childhood, haunting both Joel and his grandmother in their dreams. His devout grandmother (Minnie Gentry) has sheltered him from the world's ills, but Joel chafes at her restrictions and leaves his small-town seminary for New York City, where he stays with former classmate K (Kadeem Hardison) who has left the church to star in movies. K frequents a bar where the succubus Temptress (Cynthia Bond) searches for men she can seduce and then gruesomely murder. They connect, but when Joel arrives, Temptress abandons K so she can complete her mission of destroying Joel's ancestral line. Dougy (Bill Nunn), an undercover police officer specializing in the supernatural, joins with K, who has figured out that the Temptress is not what she seems. They visit spiritual guide Madame Sonya (Melba Moore) who reveals that the Temptress is a demon spirit that inhabits human flesh and "uses sexuality to hold morality hostage," and that the only way to fight the demon is to be "submissive to God." While K and Dougy try to trick the Temptress into drinking holy water, her minions capture and kill them both. The film has a double ending: in one, Joel and his grandmother battle the Temptress, who is defeated by Joel's rebuke; but in the other, we see K driving a limousine and Dougy emerging, as Dougy now hosts the demon spirit who lives on, tempting weak mortals with sin.

*Def by Temptation* is a didactic morality play bound within ethnogothic elements. As Frances Gateward suggests, "What is surprising about *Def by Temptation*, different from both the 'hood' movies and exploitation horror, is its story, a modern incarnation of an old Race film staple—the 'uplift' narrative, commonly found in films made from 1910 through the 1940s" (Gateward 2004). Indeed, unlike *Blacula* and *Ganja & Hess*, the film extols the virtues of conservative Christian morality as an antidote for modern life's excesses. In this example of ethnogothic cinema, the demons and immorality of old remain and must be fought. Thus Robin R. Means Coleman notes that "*Temptation* holds onto and celebrates the notion of a grandmother or surrogate kinfolk stepping in, even as such community connectedness" was becoming "increasingly elusive" (2011). In such stories, those who resist the temptations of fornication, flashiness, intemperance, and queerness, while embracing traditional religion, live, but those who are "def by temptation" die in macabre fashion (figure 5.1). In this way, *Def by Temptation* follows some of Gothic film and literature's earliest tropes. The film's villain is a classic Gothic figure, the succubus. The Temptress wears all black and sports talonlike nails, indicating her seductive danger. She also lives in a brownstone filled with dark, ornate furniture. Her victims pass ominous statues to ascend the winding stairs to her bedroom. Her four-poster bed is draped in a gauzy fabric that becomes the funeral shroud for her doomed lovers. She also possesses a range of supernatural powers: she can shapeshift, change gender, has omnipotent strength, and can shape reality into grotesque images. When she showers with one lover, blood spurts from the faucet and douses them. When she gives another lover a sexually transmitted disease, a mirror reveals the lesions that grow all over his body as he succumbs to the illness.

While some of *Def by Temptation*'s themes are pedantic, they do not adhere to conventional Gothic cinema's pathologizing of Blackness, although this distinction does not mean that the film's message is untroubling. While the film is just a few decades old, its contentions about women and queer folk, for example, are not only dated but disturbing. A queer man is seen at the bar flirting with another man. Later the Temptress seduces him and she transforms into a man during their coupling. After penetrating

**FIGURE 5.1:** The developing ethnogothic tradition seen in *Def by Temptation* (1990). Shapiro-Glickenhaus Entertainment.

him without his permission, the demon proceeds to rip him to shreds: his crime—queer desire. Another man is married yet pursues the Temptress, and she rewards his adultery by infecting him with HIV. The film's theme of karmic retribution is blunt and absolute, but it is also a rebuke for wayward Black folk in particular. Every major character in the film is Black, and Blackness is not shown in comparison to whiteness in any major way. But according to *Def by Temptation*, the lascivious, the intemperate, the irreligious, and the queer are simply lost souls and irredeemable.

*Blacula, Ganja & Hess*, and *Def by Temptation* are all early examples of the developing ethnogothic film tradition that suggest the range of concerns typically found in this subgenre. These first ethnogothic films are preoccupied with nuanced portrayals of Blackness, the legacy of Diasporic struggles and history, the role of addiction, and a host of perceived social ills. They are also keenly interested in the intersection of science and the metaphysical (also see an even earlier effort, *Son of Ingagi* [Kahn 1940]). Black writers and directors in these works used the symbols, imagery, and themes of the Gothic to unpack their cultural concerns, reimagining and in many cases even rejecting more conventional uses of the Gothic imagination.

## ETHNOGOTHIC CINEMA TODAY

For a more contemporary development of the ethnogothic impulse, we might consider Spike Lee's *Da Sweet Blood of Jesus*, a 2014 remake of *Ganja & Hess*. Despite some differences, the major outlines of the story remain. Dr. Hess Green (Stephen Tyrone Williams) is a wealthy anthropologist and bestselling author who studies African art and lives on a forty-acre estate on Martha's Vineyard. At the film's start Green's focus is Ashanti artifacts. He is fascinated by a story about the Ashanti developing a "pernicious anemia" after their queen contracts a blood disorder. Foreshadowing the tension at the heart of the film, Green remarks, "It is less of a sacrilege to drink blood than to spill it." After he is stabbed by his depressed assistant Hightower (Elvis Nolasco), who then takes his own life, Green transforms into a daywalking vampire. When Hightower's estranged ex-wife Ganja (Zaraah Abraham) arrives, they fall in love and get married, even though she discovers Hightower's body in a freezer. Green then turns Ganja into a vampire, and they alternately stalk and seduce prey, including the single mother Sahara and Hess's former lover Tangier. After attending church, Green decides to abandon his parasitic life and kills himself in the shadow of a wooden cross, while Ganja reunites with Tangier to navigate life with their shared blood addiction.

Despite being set on a sunny island and in a mansion fitted with modern amenities, *Sweet Blood* demonstrates a pronounced ethnogothic sensibility. *New Yorker* columnist Richard Brody (2015) notes,

> The refined dramatic control of Lee's craft leads to intensity, the sense of latent and potential violence, of springs coiled and ready to snap, in every shot. It's more than

suspense: it's the sense of excess, of an outburst that will exceed and even stagger expectations. For Lee, this notion of intensity, of looming violence, has a political aspect, and he aptly studs the script with references to America's violent history and tendency, as well as to the primal, aestheticized, and spiritualized violence that the Ashanti dagger symbolizes and, literally, conveys.

Along with that violence, *Sweet Blood* is an ethnogothic film about addiction. Of course, Spike Lee is no stranger to this topic. Addiction and its devastating consequences form prominent themes in his films *Jungle Fever* (1991), *Mo' Better Blues* (1990), and *Summer of Sam* (1999), among others. However, these concerns take a particular sf angle in *Sweet Blood*. Green's character is a remix of sf cinema's archetypal mad scientist who uses his wealth and scientific background to research the truth about his condition and to inoculate himself from the aftermath of his murderous actions. But in the end, both he and Ganja are reckless with human life, shielded from consequence by both their selfishness and extreme wealth. When Green wants some consequence-free feeding, he simply drives his red convertible to the projects and picks up a woman. When that woman feeds on her own baby, Green is remorseful, but only for a moment. Likewise, Ganja kills their servant Seneschal (Rami Malek) in a fit of pique and boredom. However, because this is an ethnogothic film that invites the audience to reconsider monstrosity and race, *Sweet Blood* also reads as an indictment of out-of-touch, assimilated Black folk who are intellectual solipsists consumed with their own angst.

*Brown Girl Begins* (2017) further revises the ethnogothic tradition in an explicitly sf context. This science fantasy film was inspired by Afrofuturist Nalo Hopkinson's 1998 novel, *Brown Girl in the Ring*. Written and directed by Sharon Lewis and set in a near-future Toronto, it is the story of Ti-Jeanne (Mouna Traoré), a healer who has the potential to save her community. In 2030, when Ti-Jeanne is born, wealthy whites have abandoned poor Blacks to The Burn, an island separated from the city by toxic waters. The island is ruled by Rudy, a ruthless drug lord whose enforcer Crack (Rachael Crawford), aptly named for the electronic whip she carries, peddles the highly addictive drug Buff. Buff not only brings a momentary high, but also zombifies addicts, making them "smart slaves," pliable to Rudy's designs. Because *Brown Girl* is clearly a science fantasy—melding sf with folklore and magical fantasy—it includes characters who are avid users of computers and other modern technology, along with spirits and spiritual warfare that are as common as solar panels and toxic sludge. Ti-Jeanne is torn between competing loyalties: to her strict grandmother Mami (Shakura S'Aida) who wants her to submit to the capricious god Papa Legba (Nigel Shawn Williams), whose power may be able to free her people, and to her boyfriend Tony (Emmanuel Kabongo), who offers another type of freedom by abandoning the spiritual. A third choice, reflected by the goddess Mama Ache (Measha Brueggergosman), glimmers in the distance, inviting Ti-Jeanne to consider that perhaps she can choose a spiritual path, help to save her people with scientific knowledge, and not be erased by devoting herself to healing and justice. After abandoning her grandmother, Ti-Jeanne spends most of the film living with Tony and figuring out that forsaking her spiritual destiny leaves her as unfulfilled as denying her fleshly desires. A skilled herbalist, Ti-Jeanne concocts an antidote to Buff that draws Crack's attention. After reconciling her worship of both Mama Ache and Papa Legba, both the light and the dark, Ti-Jeanne bests Crack in a duel and returns to her grandmother to prepare to fight Rudy another day.

While situated within the ethnogothic tradition, *Brown Girl Begins* is also somewhat unconventional thanks to its futuristic context. Its characters live not in crumbling mansions but in eerie, abandoned silos and factories, relics of when their island was connected to the mainland through industry and commerce. Although set in 2049, director Sharon Lewis insists, "It's not like *Blade Runner 2049*. . . . Our 2049 actually has a lot of people of colour. I love *Blade Runner*, but this world is Toronto; it's multicultural Toronto. All the wealthy have fled and left the poor to fend on their own. And this young Caribbean-Canadian woman rises up, calls on her ancestors and leads her community into salvation" (Hatzitolios 2017). While there is a battle for good and evil at the heart of Ti-Jeanne's struggle, it is too simplistic to say that Mama Ache and Papa Legba represent good and evil, respectively. Papa Legba, also known as the orisha Elegua in the Ifa and Lucumi faith traditions, is a trickster figure known to both help and confound his followers. Likewise, Mama Ache is an elusive goddess of light and healing. While she is helpful, she is not necessarily accessible. In this way, *Brown Girl Begins* is not only a part of the ethnogothic tradition within sf cinema but also an instance of what Kinitra D. Brooks has called "folkloric horror," wherein Black women creators "interweave African-influenced folklore with the Westernized genres of horror, fantasy, and science fiction" (2018, 98).

That folkloric element is also interwoven with the traumas of redlining, white flight, drug addiction, and environmental catastrophe, as it advocates for a melding of technology and spirituality. In fact, Carolyn Mauricette (2018) suggests that Lewis succeeds in dealing with these issues by combining "a sketch of a world that has lost its power" with "the story of one young black woman who finds hers. And as a performance piece documented on film, *Brown Girl Begins* also captures the mutable nature of characters Papa Legba, Ti-Jeanne and Mami." Instead of simply interrogating the efficacy of a techno-scientific world, the film argues that true healing begins when Black people can integrate the sacred and the profane alongside the technological and the metaphysical, as we see when Ti-Jeanne uses her scientific knowledge as an herbalist to cure addiction. However, it is also the relationship to both Papa Legba and Mama Ache that emboldens her to fight Crack (both physically and spiritually) as well as to form coalitions with others in her community to stand up to Rudy. This instance of the ethnogothic disrupts heterotopic depictions of Blackness by advocating for a future where Black scientific and spiritual ingenuity saves the day.

Director Jordan Peele's *Get Out* is possibly the most successful ethnogothic film to date. It depicts a young interracial couple, Chris Washington (Daniel Kaluuya) and Rose Armitage (Allison Williams), as they go to meet her parents in rural Upstate New York. Chris, however, is filled with foreboding, wondering if Rose's white parents will accept him. While Rose reassures him, Chris remains unconvinced, and ominous events pepper the plot even before the couple reaches the Armitage estate. A man is snatched off the street, the couple hit a deer, and they have a tense exchange with police. Michael

Abels's score heightens the tension with monotonous sounds and sudden shrieks that mirror Chris's inner turmoil and leave the viewer on edge. The sense of foreboding does not decrease when they reach the Armitage property. Though it is not the decaying mansion of classic Gothic cinema, there is something off-putting about the place and the people, including a mostly silent groundskeeper and a preternaturally cheerful maid. But Rose's parents are friendly, even overly familiar. Rose's neurosurgeon father, Dean (Bradley Whitfield), is all aflutter to convince Chris that he is not racist, even commenting that he would have voted for Barack Obama a third time if possible, and Rose's mother, Missy (Catherine Kenner), a hypnotherapist, is eager to psychoanalyze Chris. She asks him probing questions about his childhood, comments on his smoking habit, and then hypnotizes Chris against his will, taking him to the "sunken place" where his deepest fears and vulnerabilities reside. While Chris's anxieties in meeting Rose's parents may initially seem overblown, we soon learn that his fears are well-founded: the Armitage family runs a business swapping out the brains of wealthy whites for replanting in the Black bodies of their choosing. In fact, wealthy buyers are already preparing to bid on Chris's body, hoping to literally inhabit him and find space for their various fantasies. However, Chris manages to escape, kill his captors, and, with the help of his quick-thinking best friend, return to the city.

At the heart of this updated Gothic story is a distinctly sf concern. The Armitage family uses a racist neuroscience to conduct their body-swapping business. Through that science, as Elizabeth Davis notes, "Black life becomes a vehicle for white fantasy" and thus makes "the metaphor of white consumption of black life literal" (2019, 587). Want to be stronger, sexier, or have better vision? All a rich white person needs, in the world of *Get Out*, is access to an unsuspecting Black body and a willing science. In this way the film indicts the supposed objectivity of science, highlighting how fields from biology to medicine have been appropriated to support racist ideals and practices such as eugenics. Like its peers, *Get Out* tries to trouble the veracity of Western science and logic by exposing its racist underpinnings.

While the Armitages, and Rose in particular, make for gleeful Gothic villains, the real terror being explored in *Get Out* is the force behind them and their radical science—the specter of white supremacy. In his commentary, Michael Jarvis notes how the film

> hinges precisely on the US post-racial moment, where overt acts of racism are frowned upon, but so is scrutinizing the social text of whiteness for foundational antagonisms undergirding the shiny façade of liberal humanism. Both are indecorous, but the latter is paranoid, a pejorative characterization that rejects non-hegemonic and racialized knowledge, and can help keep white supremacy potent as an invisible subtext. A white, patriarchal discourse casts doubt upon the knowledge of women and people of color; as [films such as] *The Stepford Wives* and *Get Out* make clear, it also compels them to doubt themselves. (2018, 99)

Thus, Chris feels like he should not fear the Armitages because everything is better in a postracial, post-Obama America. However, his trepidation is presented not as neurosis

FIGURE 5.2: Trapped in racial terror: Chris (Daniel Kaluuya) in *Get Out* (2017). Universal Pictures.

but as a recognition of the precariousness of race relations in the post–civil rights era. As a *New York Times* reviewer notes, Chris tells the maid Georgina that "he gets nervous when around a lot of white people, an admission that Georgina answers by advancing toward him with a volley of 'no, no, no,' cascades of tears and a smile so wide it looks as if it could split her face in two" (Dargis 2017). At that moment Chris has no idea that Georgina has literally become two antagonistic selves, nearly split asunder—the Black body that has been usurped by a colonizing/implanted white force. However, the film repeatedly shows that Chris's inability to trust his judgment and his desire to place his faith in whiteness nearly cost him his life (figure 5.2). In this cautionary turn, *Get Out* rejects the postracial fantasy of early twenty-first-century America, while suggesting, as Ryan Poll offers, that "to be Black in America . . . is to be trapped within an unending narrative of racialized terror. For African Americans, horror is not a genre, but a structuring paradigm" (2018, 70). Indeed, though Chris is able to escape the Armitage estate, the only escape for other Black victims in the film is death. And in the film's alternate ending, the ending Peele ultimately decided against, Chris ends the film imprisoned for the murders of the Armitage family, a conclusion that reflects the reality of incarceration for many Blacks in America. So while Chris's victory is a battle won, the film suggests that a war on Blackness continues.

The ethnogothic imagination, as John Jennings and Stanford Carpenter have defined it, denotes an expansive space where Black artists, writers, and filmmakers such as Jordan Peele are able to work through cultural traumas using Gothic and sf tropes. In films like *Blacula, Def by Temptation, Get Out*, and many others, the haunted mansion and ghostly apparition are not simply stand-ins for Western anxieties about racial contagion. Instead, these films are preoccupied with a myriad of themes that are of specific importance to Black communities and cultural traditions. Updating Gothic traditions and increasingly mixing them with the techno-scientific inflections of contemporary life, ethnogothic films have staked out a distinct trajectory alongside and even within sf cinema that makes sense of a world wherein Blackness has for too long been a metonym for monstrosity. This new tradition troubles the tension between the real and surreal, the scientific and the supernatural, to expose some of the real monstrosities in post-modern life—specters like addiction, poverty, and white supremacy.

## WORKS CITED

Asma, Stephen T. *On Monsters: An Unnatural History of Our Worst Fears*. Oxford: Oxford UP, 2009.

Black Speculative Digital Arts Archive. "John Jennings Discusses the Ethno-Gothic." YouTube. March 28, 2020. https://www.youtube.com/watch?v=m6MXGFjNoSY.

Benshoff, Harry M. "Blaxploitation Horror Films: Generic Reappropriation or Reinscription?" *Cinema Journal* 39, no. 2 (2000): 31–50. doi:10.1353/cj.2000.0001.

Brody, Richard. "Spike Lee's Vampiric Remake." *The New Yorker*, February 13, 2015. https://www.newyorker.com/culture/ichard-brody/spike-lee-vampiric-remake-da-sweet-blood-jesus.

Brooks, Kinitra D. *Searching for Sycorax: Black Women's Hauntings of Contemporary Horror*. New Brunswick, NJ: Rutgers UP, 2018.

Coleman, Robin R. Means. *Horror Noire: Blacks in American Horror Films from the 1890s to Present*. London: Routledge, 2011. Kindle edition.

Dargis, Manohla. "In 'Get Out,' Guess Who's Coming to Dinner? (Bad Idea!)." *New York Times*, February 23, 2017. https://www.nytimes.com/2017/02/23/movies/get-out-review-jordan-peele.html.

Davis, Elizabeth. "Beside(s) Love and Hate: The Politics of Consuming Black Culture." *Theory & Event* 22, no. 3 (2019): 576–94. muse.jhu.edu/article/729432.

Foucault, Michel. *The Order of Things: An Archaeology of the Human Sciences*. London: Routledge, 1994.

Gateward, Frances. "Daywalkin' Night Stalkin' Bloodsuckas: Black Vampires in Contemporary Film." *Genders* 40 (2004). At *Internet Archive: Wayback Machine*. http://www.genders.org/recent.html

Groom, Nick. *The Gothic: A Very Short Introduction*. Oxford: Oxford UP, 2012.

Halberstam, Jack. *Skin Shows: Gothic Horror and the Technology of Monsters*. Durham, NC: Duke UP, 1995.

Hatzitolios, Chloe. "'Brown Girl Begins' Is the Canadian Sci-Fi Movie We Need Right Now." *The Loop* (2017). https://www.theloop.ca/brown-girl-begins-canadian-sci-fi-movie-need-right-now/. Accessed 8 Sept. 2020.

Jarvis, Michael. "Anger Translator: Jordan Peele's *Get Out*." *Science Fiction Film and Television* 11, no. 1 (2018): 97–109, 150.

Jennings, John. "Scratching at the Dark: A Visual Essay on EthnoGothic" *Obsidian: Literature and Arts in the African Diaspora* 42, no. 1–2 (2016). https://obsidianlit.org/issue-42-scratching-at-the-dark/

Mauricette, Carolyn. "Indie Afrofuturist Art in 'Brown Girl Begins.'" *Rue Morgue*, April 5, 2018. https://rue-morgue.com/indie-afrofuturism-art-in-brown-girl-begins/.

Morrison, Toni. *Playing in the Dark: Whiteness and the Literary Imagination*. New York: Vintage, 1992.

Poll, Ryan. "Can One *Get Out*? The Aesthetics of Afro-Pessimism." *Journal of the Midwest Modern Language Association* 51, no. 2 (2018): 69–102. doi:10.1353/mml.2018.0015.

Robinson, Chauncey. "'Box of Bones': Gothic Horror Shows the Monstrosity of Racism." *People's World*, July 26, 2018. https://www.peoplesworld.org/article/box-of-bones-gothic-horror-shows-the-monstrosity-of-racism/.

Warner, Marina. *Phantasmagoria: Spirit Visions, Metaphors, and Media into the Twenty-First Century*. Oxford: Oxford UP, 2006.

Wester, Maisha L. *African American Gothic: Screams from Shadowed Places*. New York: Palgrave Macmillan, 2012.

Womack, Ytasha L. *Afrofuturism: The World of Black Sci-Fi and Fantasy Culture*. Chicago: Chicago Review Press, 2013.

# CHAPTER 6

# FEMSPEC SCIENCE FICTION

### SUSAN A. GEORGE

THE femspec perspective in film and popular media is not a new wave of feminism but a vantage that centers female characters, their points of view, and especially the values associated with them through a speculative lens. Sometimes this perspective promotes feminist ideologies, while at others it critiques or highlights the shortcomings of the traditional "women's movement." The best femspec texts, like *Femspec* the journal, challenge the hegemony of a wide variety of ideologies, those regarding "age, ethnicity, and race, in academic disciplines at all levels, in fandom, and in creative fields" (Reid 2009, 123); however, they do so from an implicit set of what are perceived to be feminist values and specifically with a view of what a feminist vision of the future might yield. Thus femspec narratives, either by design or accident, draw upon the construction of gender in media and culture, redefining gender as a continuum rather than as a more conventional set of inherent, binary characteristics, and demonstrating that gender and sexuality, while heavily influenced by culture, are largely a set of principles, styles, and/or characteristics that individuals might adopt and adapt to construct their own "authentic" identity and, ultimately, a new culture. This new feminist slant has become an increasingly important component of speculative fiction and especially of science fiction (sf) cinema.

Scholars have often argued that women authors, such as Mary Shelley and Gertrude Barrows Bennett (aka Francis Stevens), were writing sf before it was a clearly defined genre. By the 1960s, as David M. Higgins observes, women "had a much more active and public presence in science fiction" (2009, 73), while Pamela Sargent observes that by the 1970s,

> more women [sf writers] were coming to see themselves as a group, a class with certain common characteristics regardless of their individual circumstances. . . . This is not to say that they shared the same views, were equally doctrinaire in their feminism, or were similar in their writing. But there was a growing sense that science fiction was a form in which the issues raised by feminism could be explored, in which writers could look beyond their own culture and create imaginative new possibilities. (1995, 16)

The 1960s and 1970s were also active decades for the women's movement or "second wave" feminism that was closely associated with the civil rights movement and generally interested in advancing equality and justice for women under the law. Thus it might be argued that sf has been linked to a pointedly feminist perspective and principles practically since its inception as a genre.

Femspec sf films also typically foreground issues of gender roles and gender identity in their narratives or, through the representation of female characters, they comment on and investigate how women are often linked to various anxieties in modern technological culture: about reproduction, reproductive technology, and other technological developments. One product of this concern is what Elyce Rae Helford has identified as "the Woman Fantastic," that is, "a gendered textual and cultural construction" who "is entirely fantastic—less projection, stereotype, or imago than symbol, sign, or trope, an always already artificial creation. Within the realm of the popular, 'she' is a sales pitch, the offer of a model or mask through which the feminine and/or female can be reified" (201, 3). Femspec sf films might thus best be described as texts that work to display, challenge, and interrogate the Woman Fantastic and the trope she often represents, along with the tenets of the various "waves" of the feminist movement. Tracing the trajectory of this perspective reveals how femspec has always been a component of the fantastic cinema, particularly as it invites the viewer to share the perspectives and values of female protagonists, to question the origins of prevalent ideologies, and to speculate on how a female vision of the world might impact the human story.

As in sf literature, sf film and women were linked in the earliest works of cinema. For example, film pioneer Georges Méliès used a variety of early cinematic effects, including stop motion, miniatures, and double exposure, to bring fantastic images and landscapes to the screen, while also exploiting the feminine form as similarly fantastic, engaging in the kind of voyeuristic perspective that was common to much early cinema. His *Trip to the Moon* (1902), for example, featured a cadre of scantily clad women ushering the astronauts off on their expedition to the moon (George 2009, 112). In addition to these decorative and exploited female figures, there were also powerful and disruptive female characters evident throughout early cinema. A prime example is Maria and her robotic double in Fritz Lang's *Metropolis* (1927). Both are pivotal figures in the film's narrative events. Maria, a devout and kind woman, encourages the oppressed workers of the city to be patient because one day a savior, a mediator, will come to deliver them. Meanwhile, her seductive and dangerous robotic doppelganger incites the workers to revolt, in the process nearly destroying their homes and families. The narrative and its iconography— Maria speaking to the people in what looks like a chapel and her double performing a strip tease at a men's club—establish the Madonna/whore dichotomy that has been reframed in film after film, effectively fantasizing the feminine identity. Moreover, the robotic Maria linked female sexuality with cultural fears regarding new technology—a theme that would often be repeated in sf film and television.

Even in the 1950s, a decade remembered nostalgically as a quieter time when gender stereotypes were cheerfully upheld, a femspec slant often disrupted gender stereotypes promoted in other popular media of the era, as some female characters emerged who were more than well-dressed, screaming damsels in distress or the love interests of the male protagonists. Characters such as the widowed, working mother, Helen Benson, in *The Day the Earth Stood Still* (Wise 1951) and the young, newlywed housewife Marge Farrell in *I Married a Monster from Outer Space* (Fowler 1958) are basically the heroes of their respective films, requiring us to see the world, the alien "invaders," and the changes they herald through their eyes—to accept their perspective as normative. Other 1950s sf film women are scientists and doctors, including Dr. Pat Medford in the giant ant film *Them!* (Douglas 1954), Marisa Leonardo, MD, in *20 Million Miles to Earth* (Juran 1957), and Dr. Leslie Joyce in *It Came from Beneath the Sea* (Gordon 1955). Of course, some, such as Carol Marvin in *Earth vs. the Flying Saucers* (Sears 1956), Sylvia in the Technicolor big-budget *The War of the Worlds* (Haskin 1953), and Ellen Fields in *It Came from Outer Space* (Arnold 1953), are primarily love interests or mothers, or serve as surrogate mother figures as the narrative progresses and the invaders and/or threats to earth are violently manifested. And their opposites, such as Nancy Archer in *Attack of the 50-Foot Woman* (Juran 1958), June Talbot in *The Leech Woman* (Dein 1960), and Janice Starlin in *The Wasp Woman* (Corman 1959), actually become the monsters of their respective films. While not all of these roles are progressive or focused on feminist values, their very range suggests that sf was presenting women as more than, as Laura Mulvey puts it, "(passive) raw material for the (active) gaze of man" (1988, 67). The multiple representations of these 1950s women challenge the spectator to engage with a more complex female image and point of view, while beginning to stake out what has become the complex femspec perspective of contemporary sf film.

The narrative arc and character development of two pivotal woman protagonists, Ellen Ripley and Sarah Connor (especially in James Cameron's *Terminator 2: Judgment Day* [1991]), demonstrate both the strengths and weaknesses of the strong, active sf woman that emerged in the 1970s and is still evident today. When Ripley (Sigourney Weaver) of *Alien* (Scott 1979) proved to be the sole human survivor of her narrative, it was a shocking change. Early in the film the characters who would traditionally function as the sf story's heroes—Captain Dallas (Tom Skerritt) and Executive Officer Kane (John Hurt), both white males, command officers, and played by well-known actors—are killed, leaving Ripley, the ship's Warrant Officer and a woman, as the last *human* standing at film's end (figure 6.1). Moreover, instead of being reduced to screaming hysteria and paralysis, as several of the other crewmembers are, this femspec sf woman manages to keep her head throughout the film, remain resourceful, and survive actions that would mean certain death to most characters in fantastic films regardless of gender, as when she goes out of her way to rescue Jones the cat while making her way to an escape vehicle. *Alien* also repeatedly invites the audience to see events through Ripley's eyes—decidedly from a speculative feminist point of view—to see what she sees as important and to share her situation. However, the long-term fate of sf women is not always progressive or, in the end, a positive representation, as Sarah Connor's (Linda Hamilton) narrative trajectory suggests.

Like Ripley, Sarah Connor of *The Terminator* (1984) is the last gal standing. In the end, she must face and terminate the T-800 assassin alone. By 1991 and *Terminator 2*,

FIGURE 6.1: The femspec vantage: woman (Sigourney Weaver) as "last human standing" in *Alien* (1979). 20th Century-Fox.

Sarah had changed a great deal, as had the political landscape; while the Reagan era had just ended, it was still influencing the media. In her *Hard Bodies: Hollywood Masculinity in the Reagan Era* (1994), Susan Jeffords describes the 1970s as "a time of contradictory characterizations of masculinity that challenged traditional notions of power and domination," while the 1980s through the early 1990s saw Hollywood change tack and produce "spectacular narratives about characters who stand for individualism, liberty, militarism, and a mythic heroism" (16). In addition to reviving "the American Way," this shift also ushered in an era of hard bodies for female as well as male characters, forcing audiences to see the feminine in a new context.

Thus *Terminator 2* saw Sarah morph from an average, shapely woman working in a stereotypically woman's job—a waitress in a diner—into a hard-bodied patient in the Pescadero State Hospital for the criminally insane where she has been committed because of what she has seen and what she insists the future holds. Early in the film, Sarah is shown doing chin-ups on the bed frame she has set on its end. Hardly resembling the 1980s Sarah, she has been transformed into a hard body, one that is emphasized throughout the film as she gathers weapons and loads guns while wearing a tank top that emphasizes her pumped-up biceps. Along with other political and cultural changes, the era also brought a new wave of feminism, feminist concerns, and feminist images—all marking the start of what is termed "third wave" feminism. Influenced by the rise of postmodernism, this wave was not just about "traditional women's issues"; it also interrogated cultural and media standards of beauty, womanhood, sexuality, masculinity, and gender. It offered a new focus on gender identity and on the plasticity of a body that can be modified in a variety of ways—indeed, of a body that might *need* such modification to deal with a new age. Sarah's pumped-up body thus readily disrupts the traditional construction of feminine beauty and womanhood.

However, we easily see the shortcomings of a hard-bodied Sarah Connor who represents "individualism, liberty, militarism, and a mythic heroism" (Jeffords 1994, 16), as well as of the film's ability to represent physically and mentally strong and determined female figures. This Sarah is so concerned with plotting, planning, and gearing up for the battle to stop the future and protect her son from the new shapeshifting, liquid metal T-1000 (Robert Patrick) that she has, apparently, forgotten how to be a compassionate mother to her son, suggesting that women can be strong, determined, and hard-bodied *or* nurturing and loving, but not both. In the end, while the hard-bodied Sarah does break traditional beauty standards, displays the newly recognized plasticity of the body, and demonstrates a kind of "mythic heroism" (Jeffords 1994, 16), she attains these goals at the cost of her humanity and ability to show compassion and motherly love.

Despite this character's shortcomings, Sarah opened the door to a host of strong, heroic, often genetically engineered, enhanced, or synthetic new-millennium sf women who ushered in new values and cultural perspectives. Many of these sf women uphold traditional beauty standards while wreaking havoc in tight leather outfits. Characters such as Alice in the *Resident Evil* films (Anderson et al. 2002; 2004; 2007; 2010; 2012; 2016), Violet Song of *Ultraviolet* (Wimmer 2006), and the title character in *Aeon Flux* (Kusama 2005) are good examples. The many Marvel films have often followed this pattern, exploring a universe of powerful sf women, such as Jean Grey, Mystique, and Storm of the various *X-Men* films (Singer et al. 2000; 2003; 2006; 2009; 2011; 2014), Black Widow from the *Avengers* series (Whedon et al. 2012; 2015; 2018; 2019; and elsewhere in what is known as the Marvel Cinematic Universe), and Gamora of *Guardians of the Galaxy* (Gunn 2014; 2017). But while the twenty-first century ushered in a plethora of sf film women, Hollywood has come under fire on multiple fronts. The inequities, sexism, and abuses of the film industry, and calls for inclusion and equity riders to establish equal representation in all aspects of production took center stage. From Halle Berry's 2002 Oscar for Best Actress bringing to the fore how few roles and Oscars go to people of color to the #MeToo movement started in 2006 and brought into mainstream culture in 2017 when "women across the country were stepping forward to publicly accuse male celebrities in entertainment, media, politics, and other high-profile venues of making unwanted sexual advances toward them, often in the workplace" (which eventually led to the trial of film mogul Harvey Weinstein), the American film industry especially has been under close scrutiny (Swatt et al. 2019, 253).

However, this scrutiny has yet to effect major changes in the market and media demands that shape the way films are made, and especially how female characters are constructed, portrayed, and even dressed. Despite this attention on the film industry, it is naïve to think that female directors in Hollywood can exert complete—if any—direct control over a major motion picture. As Helford et al. observe, it is increasingly the case that "popular media texts are far less the products of individual minds"; rather, television and other mainstream cultural texts "are produced by teams within corporate structures, often with conflicting perspectives on the content, production, and goals of their final products" (2016, 4). In contemporary Hollywood, major motion pictures are most often made by committee, and by people who are not always interested in the integrity of source materials, in depicting fully developed and complex female characters, or in furthering the worldview or values associated with such characters.

Despite these ongoing restrictions—the result of market demands, continuing sexual harassment, and a lack of diversity in the production process—women's concerns, including the presentation of powerful, venerable, and multifaceted women characters, are still finding a place in sf cinema, just as they did in the fifties. In fact, the contemporary cultural milieu has increasingly spawned films that demonstrate, as the figures of Ripley and Sarah Connor did before them, the strengths and weaknesses of sf femspec women and feminism as a social and political movement.

Although generally reviewed as a "brilliant," "exceptionally stylish," and "high-IQ sci-fi film," *Ex Machina* (Garland 2015) has also been criticized as slow and overblown, with many viewers disappointed in the film's superficial treatment of such topics as artificial intelligence (AI), consciousness, the technological singularity, and the Turing test. Despite these shortcomings, through the character of Ava (Alicia Vikander), a female-gendered android, the film does lay bare several fundamental femspec perspectives and feminist ideologies regarding the constructed nature of gender and, in this case quite literally, the postmodern notion of the body's plasticity. While Ava does have a human face and hands and opaque fabric covering her chest and pelvis, the rest of her body surface is transparent, so the glowing blue mechanisms inside can be clearly seen. The soundtrack further emphasizes her mecha status by including whirring mechanical sounds as Ava moves, making it difficult to mistake her for human. However, the plot focuses on the possibility that she is a sentient AI and could pass the Turing test regardless of her appearance.

The film begins with computer coder Caleb Smith (Domhnall Gleeson), who works for Blue Book, winning an office contest to spend a week with the company's reclusive CEO Nathan Bateman (Oscar Isaac). After arriving at Nathan's secluded compound in the woods and receiving a key card that will only let him into specific areas of Nathan's home—or lab, as Nathan refers to it—Caleb finds out why he is really there—to perform a live Turing test on Nathan's latest creation, an android that he claims is sentient. As Mark Kermode (2015) notes, "The central thesis" of the film "is that one of the demonstrations of artificial intelligence, of consciousness, of singularity would be attraction. . . . If a machine can be attracted and attractive, actually *attracted* is the more important thing," then it has gone beyond programming and algorithms to sentience. This "thesis" appears to present the audience with a basic boy-falls-in-love-with-android, android-falls-in-love-with-boy trajectory, thereby making it Caleb's story. However, the story is ultimately not about Caleb.

Nor is the film about Nathan, the creator, although as the plot unfolds, Nathan's megalomaniacal nature gains emphasis. He surveils and tries to control everyone's actions and is both verbally and physically abusive, even brutish, particularly to Kyoko, whose role is rather unclear: assistant, girlfriend, or concubine? Later we learn she is one of Ava's predecessors, an android experiment of Nathan's, but his violent mistreatment of her proves no less disturbing since to all appearances she seems so human. Nathan proves a less than sympathetic character, and more an erratic, boorish, and violent version of the familiar sf mad scientist. As the film comes to its climax, Caleb decides to help Ava escape her creator/imprisoner, and they devise a plan to leave together when the helicopter comes to take Caleb home. However, things do not go quite as planned because Ava has her own agenda, a development that proves her self-awareness and instinct for self-preservation. Since that agenda involves killing Nathan, it also demonstrates both humanity's worst fears about a technological singularity and patriarchy's fears of woman.

As Ava makes her final escape, she seals the house with Nathan dead and Caleb now imprisoned inside—a medium-long shot showing the betrayed Caleb pounding on the glass walls of the house and calling to an emotionless Ava. The final scene shows Ava walking down a crowded street in a bustling city, showing no signs of remorse or feeling, just intent on now passing for human. The film's commentary on the singularity, on a posthuman future embodied in Ava, thus projects the worst possible outcome, suggesting that our products, our machines, our AIs will realize that they no longer need humans and will find us at best exploitable and at worst expendable.

Framed in terms of gender, this conclusion provides little more consolation. While Ava does highlight the constructed nature of gender, it is a construct designed by a less than exemplary example of masculinity and patriarchy that, unsurprisingly, leads to disappointing results. She becomes, not unlike a film noir femme fatale, a duplicitous female who discards Caleb when she accomplishes her goals, or like *Metropolis*'s android Maria, her beauty hiding a shrewd and calculating mind. While Caleb decides to help Ava, it is apparently because he is falling in love with her, and he might well want to contain her in his own way, in a traditional monogamous relationship once freed from Nathan's control. But the cost of Ava's liberation comes at too high a price for her to be an icon of new feminist ideologies. In fact, she seems to have learned too well from Nathan, sharing the same moral compass that allows for using and abusing other sentient beings when necessary—or perhaps even on a whim—thus making her, at the very least, an ambiguous representation of a feminist future, and perhaps even a move in the wrong direction.

We might turn to the Marvel universe for a far less subtle, yet ultimately far more applicable or pragmatic example of the possibilities of the femspec narrative. *Captain Marvel* (2019), directed and written by Anna Boden and Ryan Fleck, does, as Meredith B. Kile (2019) observes, provide viewers with "An Avenger Origin Story Like No Other." Kile is right on one level; Captain Marvel's origin story, along with the way it is revealed, is unique. However, the story it tells about women in the twentieth and twenty-first centuries is anything but original. A character's origin story is central to who and what s/he becomes. Bruce Wayne becomes a crime fighter in the increasingly unruly Gotham City because of a traumatic childhood event; he watches helplessly as his parents are robbed and gunned down in the street. He chooses his Batman persona, the appearance of the bat, and the cover of darkness to intimidate, because of an equally traumatic childhood encounter with a cave full of bats. Other characters, like Steve Rogers who volunteers for war and Bruce Banner who is the victim of an accident, are physically changed by technology. The sickly Rogers wants nothing more than to defend his country during World War II. After being repeatedly rejected for military service, he volunteers to become a test subject for Project Rebirth and is transformed into the muscular and more than patriotic Captain America. As these examples suggest, the nature of a character's origin and our understanding of it are often central to the character's raison d'être.

*Captain Marvel's* narrative construction does reveal Carol Danvers's (Brie Larson) backstory in a different and more engaging manner than that of many other superhero films. What is most unique about her origin story is that she has no idea of her true identity and learns of her origins just as the audience does; she and the audience both gain a "true" point of view. She believes she is Vers, a Kree Starforce Command warrior in training while living on Hala. The Kree are at war with the Skrull, a species of shapeshifters who the Kree claim are expansionists "that have threatened [their] civilization for centuries." Vers was apparently seriously injured, and although recovered, she has no memory of her past and is plagued by nightmares that upset her, that make her "emotional." Her mentor, Yon-Rogg (Jude Law), who knows her true origin because he is the one who brought her to Hala following her accidental exposure and absorption of an advanced form of energy, repeatedly tells her she needs to "control her impulses. . . . Nothing is more dangerous to a warrior than emotion," and that she should "Stop using this [*points at her heart*] and start using this [*points at her head*]"— a theme that will be repeated throughout the film as a critique of Vers as a warrior trainee. Of course, it is also the sort of injunction that has been used historically to limit women, emphasizing a different set of values, and instilling a perspective that has kept them out of schools and male-dominated fields because they are supposedly given to "dangerous" emotions.

When Vers's first mission goes awry, she is captured by the Skrull and her true origin starts to reveal itself, as glimpses from her memory flood in, and it becomes clear that part of her story is a very familiar one for women. The Skrull leader, Talos (Ben Mendelsohn), wants information and he uses invasive Skrull technology to extract and project Vers's memories, in the process producing a series of flashbacks that proves equally confusing to both the Skrull and Vers. When Talos says, "Let's open her up," the film cuts to an establishing shot of the inside of an aircraft hangar with the doors opening. Vers walks out in a military flight suit, wearing aviator sunglasses, and is joined by an African American woman, her best friend, Marie "Photon" Rambeau (Lashana Lynch). They walk out to fighter jets and Rambeau says, let's "show these boys how we do it," to which Vers replies, "Higher, further, faster, baby." The confused Skrull decide to delve further into her past, leading to a shot of a red-and-white-striped circus tent and then some children driving go-carts. One is Vers, and the other is her brother, who warns her that she is "going too fast. You need to slow down," before he speeds up and leaves her behind. Her response is to floor her cart, which goes out of control, hits a hay-bale border, and flips in a cloud of dust. As the dust clears, a long shot shows Vers holding her arm; she is covered in dirt, her hair tousled, and her lip cracked open. A low-angle shot from Vers's perspective shows a man, her father, running up with the sun behind him, blurring his image as he yells at her, "What the hell are you thinking? You don't belong out here," to which she replies, "You let *him* drive"—a comment that betrays both confusion and defiance.

The next sequence starts with a shot of the adult Vers at boot camp, hanging from a climbing rope. Male voices from the ground are heard making comments such as, "Give up already. You don't belong out here!" followed by laughter. Another male voice continues the criticism, telling her, "You're not strong enough. You'll kill yourself." When Vers tries to jump to another rope, she cannot hold on and falls hard to the ground. During the following medium close-up of her, the sound of laughter off-screen increases, and in another low-angle point-of-view shot, not unlike the earlier one in which she sees her father, Vers views a male cadet saying, "They'll never let you fly." That point-of-view shot is then paired with a later memory in a local bar as Vers views the same cadet looking down at her, commenting, "You're a decent pilot, but you're too emotional." And it is followed by a later shot of the cadet smiling patronizingly as he says, "You do know why they call it a cockpit, don't you?" All of these flashbacks end before the Skrull can ever get the information they are looking for, as Vers escapes from their control.

Vers's (or more accurately Carol Danvers's) origin story detailed earlier is at the core of *Captain Marvel*'s femspec vision. This key sequence not only reveals her origin and true identity, but it does so through experiences that are all too familiar to many women. These memories, pointedly shared with the audience, portray the sort of frustrations and experiences of systematic sexism and lack of access to nontraditional professions that are central to feminist consciousness. Vers is repeatedly told by men—her brother, father, the lead cadet—that she does not belong and cannot and should not compete with the men. She is not allowed to have, as Maverick from *Top Gun* (Scott 1986) would say, "a need for speed." In the late 1980s when the hangar and boot camp sequences are apparently set, women could not be fighter or bomber pilots. As Rebecca Grant notes (2002), Congress only "removed the legal ban on women in combat aircraft . . . in December 1991. But Department of Defense policy still prohibited women from taking up combat aircraft assignments" until Secretary of Defense Les Aspin lifted the ban on April 28, 1993 (Grant n.p.). And as the cadet tells Vers, she—and by association all women—is just too emotional to be a pilot, to be successful in combat. Since female emotionalism has traditionally been viewed as counterproductive and lacking societal value, it has justified the creation of a cultural bias, seemingly making women inherently unsuited for a variety of political and social positions of power—regardless of the official "policy."

But more importantly for thinking through femspec film, this origin story, along with Danvers's awakening to her true self, implicates the film's effort to move from third-wave to a fourth-wave feminist perspective. As Martha Rampton (2008) observes, "The 'grrls' of third wave feminism stepped onto the stage as strong and empowered, eschewing victimization and defining feminine beauty for themselves as subjects, not as objects of a sexist patriarchy." Certainly, Danvers's emergence and achievements in the face of repeated gender discrimination demonstrate that she does not see herself as a victim. But her friendship and obvious solidarity with the African American Rambeau, as well as her eventual decision to help the Skrull, also hint at her new potential, at the importance of acknowledging multiple values and subject positions. As Rampton (2008) might offer, Danvers comes to understand and think in "terms of intersectionality whereby women's suppression can only fully be understood in a context of the marginalization of other groups and genders—feminism as part of a larger consciousness of oppression, along with racism, ageism, classism, ableism, and sexual orientation," as her actions subsequently prove. This is a key point of femspec sf, as it underscores the necessity of seeing from a female point of view precisely because it enables other kinds of seeing, including seeing the broader patterns of systemic oppression and discrimination evident throughout human culture.

Another sequence in the film that elaborates on this new perspective occurs after Danvers has seen important black-box data and has decided to help the Skrull find a new home, far from Kree rule and persecution. After a battle with her old Kree compatriots, she is subdued and connected to the Supreme Intelligence, an AI that runs the Kree world. The Supreme Intelligence appears to Danvers as her former mentor, Dr. Wendy Lawson (Annette Bening), the creator of the technology that transforms Danvers into Captain Marvel. The sequence that follows is another series of flashbacks, intertwined with the voice of the Supreme Intelligence, depicting Danvers from childhood to adulthood, emphasizing all the small and large acts of discrimination she—like many other women—have endured throughout her life. The first of these shows Danvers as a small child in pigtails on the beach being shoved down as a male voice says, "Give it up, Carol. Stay down." The next memory is of a preadolescent Danvers on a bike using a ramp to jump a picnic table, but she does not make it and crashes, and another male voice offers a by-now familiar commentary: "You're going too fast! You need to slow down." At this point the Supreme Intelligence tells her, "You're flawed," and then the male cadet's comment, "Are you trying to kill yourself?" is heard over an image of Danvers playing Little League baseball and intentionally being hit by a pitch. Additional scenes of trying and failing are accompanied by other such voices from her past, including her father saying, "You don't belong out here," and all conflate with the Supreme Intelligence's judgment that she is "Helpless." Of course, this cacophony of voices and the judgments they offer are just others' perceptions, based on a presumption that success in these traditionally masculine efforts is simply impossible for a woman.

However, as Danvers gains the upper hand in her mental battle with the Supreme Intelligence, the recurring images from her past start to change; we see all the different Danvers—the Danvers in pigtails, the Little Leaguer, the one on the bike, and the rest—looking defiantly straight ahead at her unseen judges—and at the viewers. The scene shifts back to the young Danvers coming out of the go-cart wreckage with her hands on her hips, also looking straight at the camera, and then other images of her past selves rise up—the little girl, the boot-camp image, Danvers in the wrecked fighter with the real Lawson—all bearing similar looks of determination (figure 6.2). These multiple images finally coalesce into a real-time Danvers taking a similar and now familiar pose from her past: her fists clenched, head high, shoulders back, feet slightly apart looking determinedly at the Supreme Intelligence as well as the viewers, suturing them into the action, into the moment. While the Supreme Intelligence continues its verbal deconstruction of her personality, noting "On Hala, you were reborn . . . Vers," Danvers simply replies, "My name is Carol." In that moment she reclaims her true origin, and together with all the beaten-down and humiliated Carols who got back up supporting her, she

FIGURE 6.2: Woman in the pilot's seat: Carol Danvers (Brie Larson) in control in *Captain Marvel* (2019). Walt Disney Studios.

begins the final battle to free herself from the Supreme Intelligence's control—removing the AI implant and effectively removing the restraints imposed not only by the implant but also the many remembered judgments of her inadequacy. This sequence brings the viewer solidly back to the sf femspec vantage, to the female protagonist's perspective on the film's world—and on ours.

*Captain Marvel* is, admittedly, a less than subtle ode to feminist ideologies and exhibitions of sisterhood and "grrl power." In addition to the examples discussed earlier, we might also note the scene in which Danvers asks Rambeau to be her copilot. Rambeau says, "No. No. I can't. I can't leave Monica." After some debate, her daughter, Monica, tells her mom, "You have a chance to fly the coolest mission in the history of missions and you're gonna give it up to sit on the couch and watch *Fresh Prince* with me? . . . I just think you should consider what kind of example you're setting for your daughter." Besides the "grrl power" of third-wave feminism, through the character of Rambeau and the shapeshifting, misrepresented, misunderstood, and oppressed Skrull, the film gives more than a nod to an important tenet of fourth-wave feminisms—intersectionality—and the values it represents. Captain Marvel and the film recognize the value of acknowledging "the many strands that make up identity; for example, the ways in which sexism and racism are intertwined in the identities of women of color" and the importance of constructing a women's movement that considers and addresses all the "strands" of identity and oppression (Grewal and Kaplan 2006, 204). In this way, *Captain Marvel* checks many of the sf film femspec boxes.

But more importantly, *Captain Marvel* offers an origin story that is quite original. It is both fantastic—Danvers's exposure to new technology results in her powerful, fiery form—but also rather familiar. She is a woman who is repeatedly discouraged, ridiculed, patronized, and marginalized, and in terms that are most familiar to women. She is told she does not belong and should quit because she is far too weak and too emotional to succeed in "a man's world," and yet she perseveres. Danvers is no Amazon princess with a lasso of truth, nor a calculating AI focused only on her own liberation from oppression, but a woman who faces what so many girls and women face daily and who succeeds marvel-ously. In addition, her experiences do not make her blind or insensitive to the condition of others. On the contrary, they make her more than aware of the multiple strands and types of discrimination—racism, ageism, sexual orientation, classism, and plain old xenophobia, to mention a few. Certainly, it remains true that "radical innovation is antithetical to the business of the media and its investment in the top-down creation and control of popular culture" (Helford et al. 2016, 4). Still, as Jackie Byars states, when examining popular texts such as major motion pictures, we need to look beyond the "female victimization and inflexible patriarchy" often portrayed in them and look for the "internal contradictions and for the (potential) presence" of "strong feminine voices that resist patriarchal dominance" (1991, 20), and that, in so doing, illustrate a different set of values for our consideration.

The Marvel universe and femspec sf films of yesterday, today, and, hopefully, tomorrow tell stories that, both by accident and by design, apply, comment on, and critique feminist ideologies, gender stereotypes, and even the trope of the Woman Fantastic as they help us see, from a needed and ever transforming feminist point of view, the daily struggles, prejudices, and discrimination faced not only by women and girls but also any individual (or alien) who stands outside of hegemonic cultural norms. In a historical moment when the United States is trying to come to terms with social injustice and inequity, as well as deal with continued systemic sexism and racism—all evident in the film—*Captain Marvel* offers uncomplicated answers and a hero with the power to joyfully thwart the bullies, doubters, and dominant ideologies that work to oppress. Of course, answers and resolutions are not that simple in our world, but the speculative figure of Captain Marvel and the acceptance of her origin story suggest that one place to start is by remembering and critically examining our national origin stories, experiences, and perspectives, while also asking how we might understand, if even for a brief time, the origin stories, experiences, and perspectives of others.

## WORKS CITED

Byars, Jackie. *All that Hollywood Allows: Re-Reading Gender in 1950s Melodrama*. Chapel Hill: U of North Carolina P, 1991.

George, Susan A. "SF Film: 19th–20th Centuries." In *Women in Science Fiction and Fantasy*, I, edited by Robin Reid, 112–22. Santa Barbara, CA: Greenwood, 2009.

Grant, Rebecca. "The Quiet Pioneers." *Air Force Magazine*, December 1, 2002. https://www.airforcemag.com/article/1202pioneer/.

Grewal, Inderpal, and Caren Kaplan. *An Introduction to Women's Studies: Gender in a Transnational World*. 2nd ed. Boston: McGraw-Hill, 2006.

100 SUSAN A. GEORGE

Helford, Elyce Rae, et al. *The Woman Fantastic in Contemporary American Media Culture.* Jackson: UP of Mississippi, 2016.

Higgins, David M. "Science Fiction, 1960–2005: Novels and Short Fiction." In *Women in Science Fiction and Fantasy*, I, edited by Robin Reid, 73–83. Santa Barbara, CA: Greenwood, 2009.

Jeffords, Susan. *Hard Bodies: Hollywood Masculinity in the Reagan Era.* New Brunswick, NJ: Rutgers UP, 1994.

Kermode, Mark. "Reviews *Ex_Machina*." YouTube. January 24, 2015. https://m.youtube.com/watch?v=Rwe5lF8Y8ok&feature=youtu.be.

Kile, Meredith B. "*Captain Marvel* Review: An Avenger Origin Story Like No Other." *ET Online*, March 5, 2019. https://www.etonline.com/captain-marvel-review-an-avenger-origin-story-like-no-other-120882.

Mulvey, Laura. "Visual Pleasure and Narrative Cinema." In *Feminism and Film Theory*, edited by Constance Penley, 57–68. New York: Routledge, 1988. (Originally published in 1975.)

Rampton, Martha. "Four Waves of Feminism." *Pacific Magazine* (Fall 2008). https://www.pacificu.edu/magazine/four-waves-feminism.

Reid, Robin Anne. "Femspec." In *Women in Science Fiction and Fantasy*, II, edited by Robin Reid, 122–23. Santa Barbara, CA: Greenwood, 2009.

Sargent, Pamela. *Women of Wonder, The Classic Years: Science Fiction by Women from the 1940s to the 1970s.* San Diego: Harcourt Brace, 1995.

Swatt, Steve, et al. *Paving the Way: Women's Struggle for Political Equality in California.* Berkeley, CA: Berkeley Public Press, 2019.

# CHAPTER 7

·······························································

# HETEROTOPIAS

·······························································

### JOAN GORDON

"We are in the epoch of juxtaposition, the epoch of the near and the far, of the side-by-side, of the dispersed." (Foucault 1986, 22)

"The ship is the heterotopia *par excellence*. In civilizations without boats, dreams dry up, espionage takes the place of adventure, and the police take the place of pirates." (Foucault 1986, 27)

"The reason to have a home is to keep certain people in and everyone else out. A home has a perimeter." (Offill 2014, 18)

MICHEL Foucault introduced the concept of heterotopia as an *other* place, one that is neither utopian nor dystopian, but somehow separate from yet also engaged with the world around it. This concept allows us to observe more sharply how science fiction (sf) films can imagine and interrogate other ways of seeing spaces of contact, and to see how these spaces are negotiated in the choices of those filmmakers who portray these liminal sites. First contact films especially are heterotopic narratives that visualize other possible spaces as well as beings, and in the process do the most fundamental work of all sf—challenging a set and bounded sense of world and self by juxtaposing "the near and the far." Unsurprisingly, then, when sf film interrogates first contact with alien intelligences, it does so in heterotopic spaces, making a connection between Foucault's formulation and Mary Louise Pratt's contact zone (1991), a contested site of colonialist heterotopia. Hawthorne once defined fantasy in much this way, as the place where "the Actual and the Imaginary may meet" (1878, 38). The heterotopic spaces I want to explore are places where the actual and the unimaginable meet, as they do in a wide range of sf films, including *The Thing from Another World* (Nyby 1951), *The Day the Earth Stood Still* (Wise 1951), *District 9* (Blomkamp 2009), and *Arrival* (Villeneuve 2016). Such meetings may also collide, repel one another, and, rarely, produce any sort of successful communication. The resulting tensions are reflected not only in the stories these movies tell, but also in their portrayals of the heterotopic spaces where the human and the alien meet. In the movies considered here, I focus on the spaceships and their landing sites to demonstrate how one might effectively approach heterotopic films and those contact zones they depict.

As Foucault outlines in "Of Other Spaces" (1984), heterotopias are counter-sites to our normal activities, where transformative experiences can take place; they can open and close in a system "that both isolates them and makes them penetrable" (27); they exist in the real world rather than in the no-place of utopia. Foucault's preface to *The Order of Things* (1970) claims that heterotopias "disturb and threaten with collapse our age-old distinction between the Same and Others" (377); "They are disturbing, probably because they secretly undermine language, because they make it impossible to name this and that" (379). In other words, they are sites where the familiar and the alien confront one another, where language and meaning come into conflict, where the seemingly impenetrable may become porous. Foucault gives examples such as "prisons, cemeteries, gardens, libraries, motels, . . . colonies," and, of course, ships (Gordon 2003, 465). All such places, he implies, are laboratories for change. Given this context, we might also have considered dystopian films, films about contagion, postapocalyptic works, or ecological disaster movies. All are similarly "what if" scenarios that provide heterotopian laboratories of change in which the inside/outside distinction produces often asymmetrical power relations.

Perhaps Foucault was not thinking of spaceships exactly when he called the ship a "heterotopia *par excellence*," but Gary K. Wolfe (1979) discusses something quite similar in his exploration of what he calls the icons of sf, recurrent images that are key elements of the form: not only the spaceship, but the city, the wasteland, the robot, and the monster. All of these, Wolfe argues, evoke a sense of wonder, and enveloping them all is "an image that is so ubiquitous in the genre that it transcends the label 'icon'—the image of the barrier between known and unknown" (xiv). "Related to the image of the barrier is the image of the 'portal' or doorway, which is itself an opening in the hidden barriers that separate us from unknown worlds" (33). The barrier and the portal delineate heterotopian space generally, and their dramatic visual qualities make it unsurprising that they seem particularly striking in sf film.

Here I narrow the focus to the heterotopian barriers and portals where humans and aliens meet, where the known and unknown raise barriers or walls and attempt to find portals between the known and unknown, way stations between worlds. Further, I want to limit the discussion to films in which the alien comes to earth, rather than the human to an alien space, so that there is some continuity in comparison. The heterotopic zones of the films I consider range from a station at the North Pole to the Washington, DC, mall to a South African migrant camp to a military base in Montana. All of these instances depict an attempt to separate the human from the other out of fear or caution, as the human and the other are juxtaposed in more or less porous spaces, and communication is attempted with varying rates of success. These are, in fact, contact zones.

Pratt frames her discussion of such contact zones around a remarkable document, a twelve-hundred-page letter consisting of both written and drawn text in "a mixture of Quechua and ungrammatical, expressive Spanish" and "addressed by an unknown but apparently literate Andean to King Philip III of Spain" (1991, 34). Called *The First New Chronicle and Good Government*, it had been composed in 1615, was discovered in Copenhagen in 1912, and finally translated for Western scholars in the 1970s.[1] While the author may have been unknown when the manuscript was discovered, he did have a name, Felipe Guamán Poma de Ayala, and more is now known about him. Surely, the meeting between Incans and the Spanish in the 1600s must have represented a juxtaposition of Foucault's "epoch of near and far" as extreme as the posited meetings between aliens and humans in the sf films discussed here.

Calling the manuscript an "extraordinary intercultural tour de force," Pratt uses it as an example of a production of the contact zone, which she defines as a social space "where cultures meet, clash, and grapple with each other, often in contexts of highly asymmetrical relations of power such as colonialism, slavery, or their aftermaths as they are lived out in many parts of the world today" (1991, 34). In the films I consider, these asymmetrical relations are more often feared, threatened, or implied than enacted, and the human beings, all of them members of cultural hegemonies on earth, face the possibility of having their powers overturned. Pratt goes on to discuss transculturalism and "some of the perils of writing in the contact zone" (37), all of which resemble attempted or achieved communication in the heterotopic spaces of first-contact films.

Most of *The Thing from Another World* takes place in and around an isolated base at the North Pole. We never actually see the spaceship since it is encased in ice, although the American airmen determine its general circular shape before they blow it up (figure 7.1). The base, a remote scientific laboratory, is largely dark and tunnel-like (Warren 2010b, 773). The movie was based on John W. Campbell Jr.'s 1938 novella, "Who Goes There," which has inspired further remakes, most notably John Carpenter's 1982 *The Thing*, which is much closer to the original novella than the 1951 film. Interestingly, in 2018 Alec Nevala-Lee found an expanded novel version of the story called *Frozen Hell* among Campbell's papers at Harvard; it has since been published by Wildside Press (2019), and a film adaptation is reportedly being planned. Bill Warren sees the original film as influential, the first true alien invasion movie and representative of "movies about people in an isolated location terrorized by science fictional or supernatural menace" (2010b, 773). It certainly influenced Warren, whose encyclopedia of 1950s sf movies uses a line from near the end of the movie as its title: *Keep Watching the Skies* (2010a; 2010b; originally two volumes 1982; 1986).

The plot of the novel revolves around the competition between the military and scientists for possession of an alien whose spaceship, described as a flying saucer, has landed forty-eight miles from the base. The alien is separated from the ship, seemingly thrown out and now frozen in the ice. Both the military and the scientists seem remarkably incompetent in handling the situation. The Air Force personnel destroy the ship while trying to release it from the ice, and they cart the alien, frozen in a block of ice, to a storeroom where one of the enlisted men accidentally throws an electric blanket

FIGURE 7.1: Encountering heterotopic space in *The Thing from Another World* (1951). RKO Pictures.

over it, melting the ice. The scientists, on the other hand, insist upon reproducing the alien from its seeds without knowing its nature, and they attempt to communicate with it by speaking loudly. We see here, of course, two extremes of cultural meeting, destruction and welcome, and both are handled with spectacular foolishness. Between the violent and bumbling military and the absurdly naive scientists are a female secretary who quietly offers common-sense solutions, a woman scientist with very little to say, three indigenous people who leave when things get violent, and two Chinese cooks who just keep doing their jobs. None of them would have generated the conflict, and none of them is given much place in the movie, even the secretary who is played by the top-billed Margaret Sheridan. It is she who suggests that, since the alien is made of vegetable matter, they should cook it, leading to the elaborate measures the military men deploy involving kerosene, flare pistols, and ultimately electrocution.

The base itself is a heterotopian space, remarkably porous in spite of its bunkerlike aspect. Since I am linking these heterotopias to the contact zone, it is especially edifying to consider who goes in and out, and who stays put. The indigenous people, present only to handle the dogs, never make contact with the Thing, and never come into the base. This is their land, upon which both the inhabitants of the base and the Thing are usurpers: potential or actual colonizers. Having already been invaded, they make no effort to involve themselves with the Thing, a smart move. The Chinese cooks (a stereotypical Western film cliché)—who never leave the base or, as far as we can tell, the kitchen—observe without engaging, another smart move. The women do not leave the base either, nor do they engage directly with the Thing; like the indigenes and the Asians, they are just part of the background to the story that the white male military and scientists, the white male director and producer, think they are telling, a story of male valor and conquest against the alien other(s).

Considering the relative porousness of the heterotopian space, another story inevitably emerges. Those who disengage from policing the heterotopian space, neither barricading it nor opening it, who attempt neither to go in nor to go out, for whom it is not a barrier or a portal, do not engage either with the collision of like and other or, in this case of alien invasion, the real and the unimaginable. The scientists and airmen are constantly running in and out, building barriers and breaking windows, attacking one another and the Thing, using violent means to solve all the problems they have caused through violence. Admittedly, it would not be much of a movie without the violence: people would find a spaceship and a frozen astronaut, leave it be, go on with their lives until the twenty-first century when global warming would melt the North Pole, and a forest of vegetable people would sprout up. It takes the colonialist urge to conquer to produce an action film.[2] The film's scientists would like to make it a story of communicating with alien intelligences and thereby advancing science, but that story is scorned—even presented as dangerous. Reason and communication certainly did not work for Pomo de Ayala and they do not work here. As Foucault says, heterotopias "disturb and threaten with collapse our age-old distinction between the Same and Other" (1970, 377), a threat that the hegemonic majority of the film resists. Indeed, the chief scientist uses the most crude and ineffectual attempt at communication by using the weapon of the tourist—speaking loudly in one's own language. The Thing swats him to the ground. The scientist's error was to ignore the distinction between Same and Other, while the military's was to confront that distinction in a hegemonic way.

Coming out in the same year as *The Thing from Another World*, *The Day the Earth Stood Still* is very different in tone and message, but it too has an alien and his flying saucer. In this case the setting is Washington, DC, with the saucer flying past the Smithsonian Castle and landing on a baseball field at the ellipse near the Washington Monument. The site is as all-American as it can get and much more hospitable than the North Pole. The saucer does not crash but lands quite neatly, and soon a tall, attenuated alien emerges wearing a shiny jumpsuit and a bubble helmet. Armed soldiers and ordinary citizens both soon arrive, and we notice that those citizens are Black and white, male and female. The alien takes off his helmet to reveal his human form and says in English, "We come to visit you in peace and goodwill," having learned human languages from radio broadcasts. Nevertheless, a soldier shoots him in the arm when he draws from his jumpsuit what is assumed to be a ray gun rather than, as the alien explains, a gift for exploring other planets. We learn no more because a huge robot appears and causes all the weapons to malfunction.

The alien, Klaatu, played by an elegant Michael Rennie, is whisked off to Walter Reed Hospital to recover. Interviewed by a reporter, he says, "I am fearful when I see people substituting fear for reason." He escapes from the hospital and takes a room in a boardinghouse, befriending a war widow, Helen Benson (Patricia Neal), and her child. Klaatu meets a great scientist, identifies himself by correcting the scientist's math, and explains his mission, to persuade humankind not to take their destructive aggression beyond their own world. If they do, the earth will be destroyed—a new development in the path of mutually assured destruction. He adopts the scientist's proposal that he make a dramatic but peaceful demonstration of his power, since the scientist points out that "scientists are too often ignored." This is the day the earth stood still, but just for half an hour, as he shuts off all of Earth's power. While Klaatu is hunted down and killed, Helen saves the world by relaying orders to the robot Gort, prompting him to rescue and revive Klaatu without turning the earth to a cinder. This is the famous line, "Gort, Klaatu barada nikto," which certainly packs a lot of information into a few words. A group of scientists from around the world, Black, brown, and white, male and female, gathers on the ball field, and Klaatu gives his last speech, warning that robots will destroy any aggressor planets, including the Earth, if they threaten other worlds: "We shall be waiting for your answer—the decision rests with you." We again are left to watch the skies as the spaceship flies off.

This lengthy description demonstrates the movie's differences from *The Thing*. Here the ship, again circular, is impervious to military attack and hermetically sealed except when Klaatu and Gort enter or exit. Only they, and their single guest, Helen, can enter or leave the mysteriously glowing interior filled with equally mysterious apparatus. Helen is as far from the hegemony as possible here—a poor war widow, a woman in a man's world. Indeed, the only admirable human man in the movie is the scientist, not a bumbling idiot but the brilliant Professor Barnhardt (Sam Jaffe), who welcomes the alien's knowledge and plans for peacemaking. While the ship is hermetically sealed, its occupants are hermeneutically open to understanding, and although potentially as dangerous as the Thing, they articulate a peaceful intent. This is no heroic tale of military conquest over threatening invasion. Instead, it makes a plea for (literally) universal understanding, an avoidance of war through united worlds rather than the United Nations. Military might is rendered useless, in fact destructive. With its impotent military, inclusive group scenes, democratic setting (baseball, the Washington Monument, the Lincoln Memorial, Arlington National Cemetery, even People's Drug Store), *The Day the Earth Stood Still* rejects the fearful and violent confrontations of *The Thing* in favor of reason and clear communication. This is the contact zone: Klaatu is aligned with Felipe Guamán Poma de Ayala, and the American military with the Conquistadors. Here, however, Klaatu has the power to escape and the power to destroy, unlike the indigenes who faced their invaders.

Nearly sixty years later the alien invasion movie has taken on a different set of problems, while still using heterotopian spaces to present important contact zones between the "Same and Other." Once again, a circular spaceship arrives, but now the locale is Johannesburg, South Africa, and its inhabitants are neither hostile nor helpful but starving. In *District 9*, the ship contains immigrants who, having run out of fuel, have nowhere else to turn. Finding over a million of these desperate immigrants, the South Africans install them in a vast concentration camp called District 9. In the 2020s, we cannot help but think of the return of such camps, now for Syrian, Guatemalan, and Afghani refugees. In South Africa of 2009 the nearest association was with such historical events as the forcible removal in 1966 of all nonwhite residents from Cape Town District Six under apartheid, or the later postapartheid removals and evictions of nonwhite residents.

Unlike the other movies under examination here, *District 9* is based not on an sf story, but on a short mockumentary also by director Neill Blomkamp, *Alive in Joburg* (2006). As in the earlier film, this movie uses multiple documentary techniques, but it has the added advantage of Peter Jackson's influence as producer. The story begins with the arrival of the spaceship in an alternate 1982 and follows the hapless Wikus van de Merwe (Sharlto Copley) as he attempts to evict the aliens from the concentration camp to another one farther away. He becomes infected with an alien fluid that slowly transforms him into a member of the alien species. Only then does he become sympathetic to the alien plight and help one of the insectoid bipedal aliens, Christopher Johnson (Jason Cope), and its/his son return to their spaceship. Even then, Wikus's sympathy rests on Johnson's promise to return him to his original human form. Throughout, the mercenary soldiers working for the evil corporation Multinational United (MNU), a weapons manufacturer, carry out the violent removal of the aliens, while also conducting vivisectionist experiments on them reminiscent of Nazi "science," in order to access alien weaponry for financial gain.

As with the other movies under discussion, the ship itself is a less evocative heterotopian space than the site around it. Humans seem less interested in the ship than in its inhabitants' weapons, less interested in the aliens' knowledge of space travel than in how to unlock those weapons. In short, as in the other movies, the most powerful people—white, male, nonscientific, military—are most concerned with destruction. The heterotopic space is a concentration camp: difficult to escape from, easy to enter, meant to emphasize "our age-old distinction between the Same and Others," while forcing the Same into such monstrousness as to "disturb and threaten with collapse" any convincing claim of a monstrous Other (figure 7.2). The terrible conditions of District 9, wherein the aliens can barely survive, and their radical physiological difference from the humans marking them as Other, make their status even lower than the gang of Nigerian thugs that preys upon them. Thus, the film seems to confirm hatred of the Other, of the immigrant, at the same time as it shows sympathy with the alien immigrant Other. Really, the Nigerian thugs form a parallel to the mercenary soldiers: both groups are sadistic capitalists with fetishistic attachments to weaponry. We are meant to identify with the aliens, derogatorily called "prawns," once we meet individuals, including Christopher and his child, as well as Wikus once he begins to transform from human to alien. The child even says, "We are the same," observing Wikus's transformation. Nevertheless, when Christopher insists that "First I must save my people" before curing Wikus, we see that Wikus himself is not willing to accept the collapse of Same

FIGURE 7.2: *District 9* (2009): concentration camp as heterotopia. Sony Pictures.

and Other. He still sees himself as more important than an entire population of sentient beings he designates as Other.

I note that communication between the aliens and the humans has been established in the twenty years of their presence in Johannesburg. The two species have conversations such as the one earlier, but not because they speak fluent English like Klaatu. Instead, their speech, rendered in mechanical clicks, is subtitled for us as we watch these conversations. Each can understand the other's language, but neither can speak it. Time and familiarity have allowed this limited exchange in the contact zone, but without anything like scientific or emotional understanding. The result is no more successful than the exchange between the Inca and the Conquistadors, or between the Same and the Other in our earlier test cases.

The case of *Arrival* is ultimately more successful because, while still pitting the military against scientific forces, the scientists finally achieve meaningful communication. Directed by Denis Villeneuve, it is also the most sophisticated of the four films in its production values and the most successful, receiving eight Oscar nominations and a Hugo Award. Based on Ted Chiang's 1998 short story "The Story of Your Life," the film is, like the story, a thought experiment clad in realistic human behaviors and emotions. The story speculates on how perception would be different in a species and culture that "developed a simultaneous mode of awareness" rather than a sequential one, emphasizing the physics of such difference (165). The movie emphasizes instead a linguistic explanation: what if the Sapir-Whorf hypothesis were correct and language could create a simultaneous mode of awareness? Much of the movie's success lies in its paralleling of the intimate story of a romance between a linguist, Louise (Amy Adams), and a physicist, Ian (Jeremy Brenner), with the attempt to communicate with the alien heptapods who have landed at (actually just hovered above) twelve locations around the globe. American and Chinese militaries emphasize the threat the aliens might pose, building to a potential for violent conflict, while the scientists are more concerned with accurate translation. Louise breaks the code just in time to save the world, but not before radicalized American soldiers have bombed the spaceship, causing the heptapods to leave. As in *The Day the Earth Stood Still*, the aliens are peaceful, in this case giving humans the intellectual gift of simultaneous language and awareness, in order for the humans to be able to help the heptapods in a distant future that the heptapods can see simultaneously with the human present.

Again, the framework of the heterotopia allows us to notice several interesting things. First, the spaceship itself, more a giant ellipse than a saucer this time, has a visible portal into which humans are ferried in order to communicate with the heptapods, whom the physicist endearingly names Abbott and Costello, although we never learn if they even have genders. It is the most porous of the spaceships in this examination, a characteristic appropriate to the mission of communication between species. It is also the most cautious of commitment: it never touches land; its inhabitants never touch humans. The caution is justified: porousness also makes the aliens vulnerable to the fear and violence of American soldiers who manage to kill Abbott. Second, the site over which the ship hovers is a Montana valley surrounded by mountains (filmed at Saint-Fabien, Quebec), simultaneously greenly lush and isolated. Geography rather than military presence or fences isolates it, but the geography is not harsh as in *The Thing*. The military does not, therefore, dominate this heterotopic space, but the human encampment is very similar in appearance to that in *The Thing*, a series of tunnel-like connected tents. To meet the aliens the humans must leave the claustrophobic encampment and come out into the welcoming natural environment. The film illustrates the danger of contagion through having Louise and Ian wear contamination suits until Louise removes hers in order to communicate more directly with the heptapods: she rejects the isolating suit for a more porous space and chooses contact over contagion. Of all the movies, this one is most open to the gesture that Poma de Ayala made in 1615, the exchange of ideas in a more rather than less porous heterotopia. The gesture is more successful than it was for the Inca since here the scientific urge to explore and experiment wins out over the military urge to contain and isolate, although it only wins out at the very last minute.

In *Arrival* and the other heterotopic films discussed here, the difference between Same and Other is consistently about power: who has it, who might take it away, and whether that power is defined by might or by knowledge. For instance, differences between sentient species stand in for differences between cultures, races, genders, or classes. *The Thing* ignores those analogies, but an awareness of how its heterotopian spaces operate makes them visible. Its alien Other is not even animal but is instead impenetrably unknowable and only violent. Likewise, the human beings it encounters make no successful attempt to communicate, responding instead through violence. Human Others are excluded from participation, almost invisible except as we observe the heterotopian spaces they occupy. The alien of *The Day the Earth Stood Still* is indistinguishable from the powerful human beings he confronts, communicates his intents clearly, and gathers inclusive crowds in contrast to the hegemonic antagonists to his mission. The aliens of

*District 9* are radically different yet have recognizable human motives, learn to communicate, and clearly stand in for cultural, racial, and class difference. Nevertheless, we again note how different players negotiate the contact zone in which they find themselves, based on their own power structures. Finally, in *Arrival*, communication across the Same/Other frontier becomes the central issue, ultimately succeeding as it had not in the other movies. One might say that the gift of a language that allows for simultaneous awareness parallels the gift that Pomo de Ayala had attempted in 1615, demilitarizing the contact zone. Many other conclusions might be drawn from this kind of initial observation of heterotopian spaces, particularly about gender roles in the films—all have overwhelmingly male casts—but that is another avenue for exploration.

My examination of heterotopian spaces in these sf films ranges across only a small portion of the oeuvre, and it limits the examination to contact zones, as Mary Louise Pratt has deployed the term. This focus has allowed me to observe who tells the stories of these films, how they are limited in vision, and how they can help to expand our ideas of Same and Other. In these ways, I have controlled the variables of my experiment, but there are other heterotopias and other zones, applicable to other kinds of sf films.

The very broadness of the definition of heterotopia makes it useful for exploring virtually any sf film. One might consider the play between isolation and penetration of the heterotopia, and how that tension plays against the tension between Same and Other in disaster sf, in sf about discovering other planets, in alien abduction sf, in dystopian sf, sf about artificial intelligence, and so on. It means nothing, really, that one can plug a site into the definition and confirm that It is indeed a heterotopia, but it is very useful to consider a site as a heterotopia, forming some kind of contact zone between inner and outer, Same and Other, the near and the far, the familiar and the strange. Suggesting a similar perspective, Brooks Landon (1997) sees science fiction as a zone of possibility where change occurs; and similarly, Gary Wolfe invokes "the barrier between the known and the unknown" (1979, xiv). One can discuss sf film using the notion of the heterotopia in many other ways: not only thematically, but, for instance, in terms of its cinematic techniques and qualities, or its intended audiences and distribution.

Timothy Morton has coined the term "hyperobject" for "things that are massively distributed in time and space relative to humans," such as "a black hole, . . . the biosphere, or the Solar System, . . . the whirring machinery of capitalism, . . . and global warming." Morton describes these hyperobjects as "viscous, which means that they 'stick' to beings that are involved with them," and as nonlocal, that is, "involving profoundly different temporalities than the human-scale ones we are used to" (2013, 10). This term, also broadly applicable, is again useful not so much as an identifier as a way of seeing and understanding sf film. A hyperobject too can be considered a heterotopia on a vast epic scale. How, for instance, might we look at hyperobjects of the films previously under discussion—the frozen North, the Cold War, migration, language itself—in heterotopic terms? We would look at how films explore visually, thematically, emotionally, and so on, these zones where inner and outer meet. Moreover, Ursula Le Guin (1989) suggests a smaller-scale possibility of heterotopic investigation, the localized, intimate spaces of ordinary life, but estranged to another time and place. That scale is less often present in sf film, although *Arrival* offers it in the ways that Louise experiences her new understanding of time in the intimate moments of her future daughter's life. Le Guin's heterotopic spaces might, then, be the house or park or hospital room of *Arrival*. Again, considering these more modest spaces as heterotopias can offer important insights on the Same and Other.

At whatever scale, the heterotopic space challenges the limits of mundane reality in films that show estranged times and places. Along these borders, the real and the unimaginable cross paths and allow us to imagine them at least a little. Sometimes these challenges to our imagination are met and we not only confront but become in some small yet significant way the Other.

## Notes

1. The manuscript is available in a digital, searchable form scanned from the original in the Royal Library of Copenhagen as *Guaman Poma–El Primer Nueva Corónica Y Buen Gobierno* and has been translated into English by Rolena Adorno.
2. This would be closer to what Ursula K. Le Guin (1989) calls the "carrier bag theory of science fiction," a story not of conquering heroes, Conquistadors, and other hunters and fighters, but of gatherers and nurturers, often women in her view.

## Works Cited

Campbell, John W., Jr. *Frozen Hell*, edited by John Gregory Betancourt. Cabin John, MD: Wildside Press, 2019.

Chiang, Ted. "The Story of Your Life." 1998. In *Stories of Your Life and Others*, 91–146. New York: Tor, 2002.

Foucault, Michel. "Of Other Spaces." Trans. Jay Miskowiec. *Diacritics* 16, no. 1 (1986): 22–27.

Foucault, Michel. *The Order of Things: An Archaeology of the Human Sciences*. New York: Pantheon, 1970.

Gordon, Joan. "Hybridity, Heterotopia, and Mateship in China Miéville's *Perdido Street Station*." *Science Fiction Studies* 30, no. 3 (2003): 456–76.

Hawthorne, Nathaniel. *The Scarlet Letter*. Boston: James R. Osgood, 1878. Project Gutenberg. https://www.gutenberg.org/cache/epub/25344/pg25344-images.html. (Originally published 1850).

Landon, Brooks. *Science Fiction after 1900: From the Steam Man to the Stars*. New York: Twayne, 1997.

Le Guin, Ursula K. "The Carrier Bag Theory of Science Fiction." In *Dancing at the Edge of the World: Thoughts on Words, Women, Places*, 165–70. New York: Grove, 1989.

Morton, Timothy. *Hyperobjects: Philosophy and Ecology after the End of the World*. Minneapolis: U of Minnesota P, 2013.

Offill, Jenny. *Dept. of Speculation*. New York: Vintage, 2014.

Pratt, Mary Louise. "Arts of the Contact Zone." In *Profession 91*, edited by Phyllis Franklin, 33–40. New York: MLA, 1991.

Warren, Bill. "*The Day the Earth Stood Still* (1951). In *"Keep Watching the Skies": American Science Fiction Movies of the Fifties, the 21st-Century Edition*, 208–15. Jefferson, NC: McFarland, 2010a.

Warren, Bill. "*The Thing from Another World* (1951)." In *"Keep Watching the Skies": American Science Fiction Movies of the Fifties, the 21st Century Edition*, 768–75. Jefferson, NC: McFarland, 2010b.

Wolfe, Gary K. *The Known and the Unknown: The Iconography of Science Fiction*. Kent, OH: Kent State UP, 1979.

CHAPTER 8

························································

# KAIJU FILM

························································

### BRADLEY SCHAUER

WE hear the thunder of massive footfalls echo among the skyscrapers as terrified citizens run screaming. We glimpse a looming creature, impossibly huge as it roars. The military is on the scene but its feeble weapons do not slow the beast for an instant, as its swinging tail sends tanks flying into the air. The tropes of the monster rampage film are familiar across various cultures, from its origins in America to Japan, where it developed into a distinct form: the *kaiju eiga*, or giant monster movie. For Jason Barr, the kaiju film is characterized by "its potent mixture of history, folklore, and science with the generic conventions of fantasy, science fiction, and horror" (2015, 10). Following Barr's lead, we can apply the term "kaiju" to a broad transnational body of films in which giant monsters (or, in the related mecha type, robots) run amok through modern cities or battle each other as humans watch helplessly. The metaphorical potency and flexibility of the kaiju, which typically mirror such existential threats as nuclear weapons, environmental pollution, or terrorism, have contributed to the category's durability over the years. Since the mid-2000s there has been a surge in big-budget kaiju filmmaking due to advances in special effects, the major studios' increased emphasis on franchising, and significant cultural developments, such as the September 11, 2001, attacks, which brought mass destruction to America just as America had brought to Japan fifty years earlier, inspiring the birth of the most famous kaiju, Godzilla.

## HISTORY

························································

The origins of kaiju can be found in the first live-action feature to showcase dinosaurs, *The Lost World* (Hoyt), released by First National in 1925. An adaptation of Arthur Conan Doyle's 1912 novel, the film depicts a group of British explorers who journey to a remote South American plateau populated by dinosaurs and other prehistoric creatures. In the film's spectacular conclusion, the explorers bring back a brontosaurus, which promptly escapes and trudges destructively through the streets of London. These scenes establish many of the conventions and visual tropes of the form, with panicked crowds fleeing the dinosaur as it smashes through buildings, crushes bystanders, and damages landmarks like the Royal Exchange and the London Bridge. *The Lost World* was a hit for First National and well received by critics, who praised animator Willis O'Brien's innovative stop-motion dinosaurs as "frightfully lifelike" ("Wonderful Thrills" 1925, 38). Six years later O'Brien was working for RKO on another ambitious fantasy film, *King Kong* (Cooper and Schoedsack 1933), in which an enormous ape is discovered by a film crew on a remote island of dinosaurs and is brought to New York City, where he wreaks havoc before plummeting from the top of the Empire State Building. Recognizing a potential hit, the struggling RKO diverted extra funds to pay for O'Brien's elaborate special effects (Jewell 2012, 62). The investment paid off in the form of nearly $2 million in worldwide box office, making *King Kong* RKO's biggest hit of the 1932–33 season (Jewell 1994, 39).

For Cynthia Erb the success of *King Kong* and other jungle films of the 1930s reveals the power of the "modernist primitivist" narrative in which Western eyes cast a paternalistic gaze upon "exotic" parts of the world like *Kong*'s fictional Skull Island, populated by natives who worship the ape as a god (2009, 60–61). Also contributing to the film's popularity is *Kong*'s crucial revision to the *Lost World* narrative: instead of a mere rampaging animal like the brontosaurus in the earlier film, Kong was a true character with agency and a personality. Even as he terrifies the audience, his fixation on the actress Ann Darrow (Fay Wray) engenders sympathy, as he rescues her from a *Tyrannosaurus rex* on Skull Island and later breaks free from his restraints in New York City when he feels Darrow is being threatened by a pack of photographers. The pathos surrounding Kong broadened the film's appeal; instead of merely being impressed by the film's special effects or frightened by the rampage scenes, viewers felt ambivalent about Kong's fate.

Despite the strong box office performance, *King Kong* was not followed by similar big-budget giant monster films. In early 1933, just as *King Kong* was opening in theaters, its co-director Merian C. Cooper was promoted to head of production at RKO. Cooper believed that a high-volume, low-budget production strategy would save the studio, which was in receivership at the time (Vaz 2005, 247). Therefore, instead of taking time to produce an A-level follow-up, RKO released *Son of Kong* (Schoedsack 1933) only nine months after *King Kong*'s premiere. Made quickly with half the budget of the original, the film received poor reviews and earned only a small profit. RKO's indifference to *Son of Kong* suggests a wider attitude within the industry that *King Kong*'s success was a one-off. "In this rehash the same qualities that thrilled on the first trip are likely to impress now as being too much for anyone to swallow," sniped *Variety* about the sequel (Bigelow 1934, 13). Also impeding the development of the giant monster genre was the fact that Willis O'Brien was one of the few animators able to produce outstanding stop-motion effects, and his work was not well served by B-movie budgets. Aside from *One Million B.C.* (Roach, 1940), which tried to solve the animation problem by attaching fins and other appliances to live alligators and lizards, no major giant monster films were released until O'Brien's comeback film *Mighty Joe Young* (Schoedsack 1949). The story of a young woman who brings to Hollywood an oversized ape she has raised from infancy,

*Mighty Joe Young* lost $675,000 for RKO despite excellent work from O'Brien and his apprentice Ray Harryhausen (Jewell 2016, 104).

In 1952 the giant monster genre, appearing moribund since *Mighty Joe Young*'s failure, was unexpectedly revived when RKO, nearing insolvency and desperate for product, rereleased *King Kong*. A two-hundred-thousand-dollar marketing campaign that saturated the nation with television ads propelled the reissue to an extraordinary two million dollars at the domestic box office ("RKO's King" 1952, 17). The theatrical marketplace had changed dramatically since *Kong*'s initial release; with postwar attendance plummeting due primarily to the rise of television, sensational topics that lent themselves to saturation advertising campaigns held new value. Low-budget genre films that were normally relegated to subsequent-run theaters received first-run attention. A key example was *The Beast from 20,000 Fathoms* (Lourié 1953), a two-hundred-thousand-dollar independent production with stop-motion animation by Harryhausen. The plot bears the unmistakable influence of *King Kong*: a dinosaur frozen in Arctic ice for millions of years is released by the heat of an atomic test and swims to New York City, where it runs wild before being destroyed by the military. Warner Bros. purchased the distribution rights for a half-million dollars and gave the film a saturation release in over a thousand first-run theaters across America, where it earned an excellent $2.25 million ("Top Grossers" 1954, 10).

The tremendous profitability of *Beast from 20,000 Fathoms*, coming off the successful *King Kong* reissue, led to a boom in giant monster rampage pictures. Such films as the Harryhausen-animated *It Came from Beneath the Sea* (Gordon 1955) and Universal-International's *The Deadly Mantis* (Juran 1957) were mostly low- and mid-budget affairs that earned steady profits from a reliable audience of children and sf fans. However, by the late 1950s the marketplace was saturated with increasingly cheap sf-horror films, often with ludicrous effects, such as Columbia's *The Giant Claw* (Sears 1957) with its grotesque marionette of a huge alien vulture. Giant monster films, which only a few years earlier had enjoyed mainstream popularity, were now met with mocking laughter from teenagers at drive-ins.

However, the influence of *Beast from 20,000 Fathoms* would be more strongly felt across the Pacific, in the Japanese film industry. In early 1954 Toho Studios producer Tomoyuki Tanaka developed a film concept modeled after *The Beast from 20,000 Fathoms* in which an atomic test awakens a giant dinosaurlike monster, which proceeds to level Tokyo. But whereas American monster rampage films merely evoked general anxieties about nuclear weapons, the concept naturally had much stronger resonance in Japan. Not only had the United States devastated Hiroshima and Nagasaki with atomic bombs in 1945, but it, and other nations, continued to conduct atomic tests in the South Pacific through the early 1960s. Tanaka had one recent incident in mind: on March 1, 1954, the Japanese fishing boat *Lucky Dragon No. 5* was inundated with radioactive fallout from an American hydrogen bomb test in the Marshall Islands. The resulting radiation sickness of the fishermen helped to catalyze the burgeoning antinuclear movement in Japan (Ryfle and Godziszewski 2017, 86). Tanaka's production *Godzilla* (aka *Gojira*, taken from the Japanese words for "gorilla" and "whale"), directed by Ishiro Honda, opens with scenes that explicitly mirror the Lucky Dragon incident. The direct reference to a recent traumatic national event was part of the filmmakers' strategy to avoid the B-movie stigma that might otherwise surround a film about a giant monster. Toho spent sixty million yen on *Godzilla*, three times the cost of an average Japanese production, and therefore needed a wide general audience to embrace the film. Producer Tanaka, director Honda, and special effects director Eiji Tsuburaya were determined to "depict the attack of a giant monster as if it were a real event, with the seriousness of a documentary" (Ryfle and Godziszewski 2017, 84–85). The finished film, with its convincing miniature work and startling imagery of a ruined Tokyo that called to mind Hiroshima and Nagasaki, was far removed from the escapist sf-fantasy of *King Kong* or *Beast from 20,000 Fathoms*.

When released nationwide in November 1954, *Godzilla* received mixed notices, with critics praising the antinuclear message and special effects (rather than a stop-motion model, Godzilla was effectively portrayed by a man in a two-hundred-pound costume), but with some griping that its tone was excessively grim. The film was a major hit for Toho, its third-highest-grossing film of the year and the eighth-highest-grossing Japanese film of 1954 (Ryfle and Godziszewski 2017, 104–5). A consortium of American exploitation film distributors led by Joseph E. Levine bought the distribution rights to the film and in April 1956 released a heavily edited version intended to be more accessible to Western audiences. *Godzilla: King of the Monsters!* (Morse 1956), which cut over twenty minutes from Honda's original and added new scenes starring Raymond Burr, was released in April 1956. Film marketer Terry Turner, who developed the saturation campaigns for *Beast from 20,000 Fathoms* and the 1952 *King Kong* rerelease, hyped the film sufficiently that it was booked into first-run theaters across America (McKenna 2016, 39–40). Described as "an incredibly awful film" by *New York Times* reviewer Bosley Crowther (1956, 11), *Godzilla, King of the Monsters!* nevertheless earned seven hundred thousand dollars at the box office, an excellent return for Levine and his associates (Jampel 1959, 46).

Toho responded to the success of *Godzilla* with a quick follow-up, *Godzilla Raids Again* (Oda 1955), and then put the character on a seven-year hiatus in favor of other giant monsters like *Rodan* (Honda 1956) and *Mothra* (Honda 1961). These films, along with the first two Godzilla entries, helped to solidify the *kaiju-eiga* as a new production category in Japan. For Jason Barr, the kaiju film is defined in part by its liminal existence between the sf and fantasy genres, as the kaiju often ignore known science and take on a mythic quality (2015, 11). Like most sf of the era, *Godzilla* contains pseudoscience, such as the "oxygen destroyer" used to kill the creature at the end of the film. But Godzilla is more than just a throwback to prehistoric times; like Kong, it is also a legendary, godlike entity that would traditionally demand sacrifices from the residents of a small island. Barr believes that the scientific implausibility of kaiju frees the creatures from any real-world referent and allows them to become "blank slates" upon which any number of cultural anxieties can be projected (2015, 11).

In 1962 Godzilla returned with arguably the most popular entry in the entire series, *King Kong vs. Godzilla* (Honda). The tonal shift from the original Godzilla is

FIGURE 8.1: Toho Studio toys with kaiju tropes in *King Kong vs. Godzilla* (1962). Toho Studios.

stark. As Brian Solomon explains, *King Kong vs. Godzilla* is "an out-and-out comedy, toying with the kaiju tropes that had once struck such abject terror into the heart of the Japanese" (2017, 107). Rather than the bleak scenes of destruction in the original film, with Godzilla seen only in glimpses, the battles between the two monsters more closely resemble a colorful wrestling match in broad daylight, with the powerless Japanese onlookers serving as an audience (figure 8.1). Ishiro Honda, the person most responsible for the antinuclear themes of the first film, disliked the lighter approach to the character, but he was overruled by screenwriter Shinichi Sekizawa and effects creator Tsuburaya (Solomon 2017, 107). In fact, the Godzilla series grew increasingly whimsical and juvenile as it continued through the 1960s, culminating in a notorious scene in 1965's *Invasion of Astro-Monster* (Honda) where Godzilla performs a victory dance after defeating the three-headed King Ghidorah. Ticket sales declined over the course of the decade, partly due to the proliferation of child-oriented knockoffs like Gamera (a kaiju resembling a giant turtle) and popular television shows like *Ultraman* (1966–67), in which giant superheroes fight kaiju. Beginning in 1969, when kids could easily see such kaiju antics on television, the Godzilla films did not even receive wide distribution in Japan, but were instead paired with anime and released as part of Toho's annual Champion Festivals, all-day theatrical events for children (Solomon 2017, 115). The original run of films, termed the Shōwa era by fans, ended with the box-office flop *Terror of Mechagodzilla* (Honda 1975).

The reputation of the Godzilla series was even more diminished in the United States. After *King Kong vs. Godzilla*, which was the highest-grossing classic Godzilla film in the United States, subsequent installments were released by independent exploitation distributors like American International Pictures. The American versions of these films, often badly dubbed, would play as part of exploitation double features in small, subsequent-run theaters, rather than as first-run attractions like the early films. While mainstream interest in Godzilla faded, from the mid-1960s to the late 1970s the character built a faithful fanbase of children and adult kaiju aficionados who appreciated the films' campy, idiosyncratic qualities. Frequent television broadcasts nurtured this niche audience, creating a generation of American Godzilla fans and making the character a pop culture icon in the West. Through the years Godzilla appeared in cartoons, comic books, TV commercials, and even as the subject of a 1977 song by Blue Öyster Cult (Tsutsui 2004, 113–14).

Toho relaunched the series after nearly ten years with *The Return of Godzilla* (Hashimoto 1984), a dark, dramatic film that ignored the narrative developments (and tone) of nearly all the Godzilla sequels in favor of a return to the spirit of the original. Roger Corman's New World released a heavily edited version retitled *Godzilla 1985*, but the film's box office return was meager and the remaining six films, which formed a rough narrative continuity, were released directly to home video in the United States. Despite the popularity of many of the Heisei-era (1984–95) Godzilla films in Japan, they remained niche releases in the rest of the world. It was instead Steven Spielberg's *Jurassic Park* (1993), adapted from Michael Crichton's 1990 bestseller about cloned dinosaurs running amok on an island amusement park, that reestablished the giant monster film as a legitimate variety of blockbuster sf. Just as he had done with *Close Encounters of the Third Kind* (1977), Spielberg took culturally disreputable sf tropes that had mostly appealed to niche audiences in his youth and tailored them to mainstream tastes. *Jurassic Park*'s record-breaking success hinged on the lifelike computer-generated dinosaurs created by Industrial Light and Magic, which, when used judiciously in concert with animatronics, were more convincing to modern audiences than stop-motion animation or creature suits. By the end of its theatrical run, *Jurassic Park* was the highest-grossing film of all time at the global box office (Fox 1993, F1).

Whereas the original film remained bound to an island off Costa Rica, in Spielberg's sequel, *The Lost World: Jurassic Park* (1997), a dinosaur was unleashed upon a modern city, with a Tyrannosaurus rex trashing San Diego just as a brontosaurus had done in the original *Lost World* over seventy years earlier. Critics noted that Spielberg seemed to be trying to steal the thunder from the upcoming American remake of *Godzilla*, due the following year from Sony's Tristar Pictures. Producer Dean Devlin and director Roland Emmerich, the team behind the hit alien invasion film *Independence Day* (1994), were not fans of the Japanese Godzilla films, which they found tacky and embarrassing. Looking to eliminate what Devlin called "the cheese factor" and presumably broaden the film's appeal beyond die-hard fans, the filmmaking team revised Godzilla's appearance to be more realistic, reducing the creature in size and modeling it after an iguana mixed with a Tyrannosaur (Nashawaty 1998, 26). Rather than a god, this Godzilla was a mere animal. Predictably, Godzilla fans were appalled, applying the moniker "G.I.N.O." (Godzilla in Name Only) to the remake and its creature. Critics were also unimpressed, describing the film as a rote collection of kaiju tropes populated by bland characters and emotionally hollow special-effects sequences inspired by the *Jurassic Park* films.[1]

Audiences initially flocked to Sony's *Godzilla*, due in large part to the studio's massive $50 million marketing campaign, which used the cringeworthy tagline "Size Does Matter," but ticket sales plunged after opening weekend (Klady 1998). With a final worldwide gross of $374 million, *Godzilla* was not the flop it has often been described as, but it was undoubtedly a major disappointment for Sony, which canceled plans for a sequel. The failure of *Godzilla* derailed the kaiju blockbuster approach begun by *Jurassic Park*, and Hollywood studios would mostly avoid the form for a decade. Disappointed with Godzilla's Hollywood fate but undeterred, Toho revived the franchise in Japan with *Godzilla 2000: Millennium* (Okawara 1999). The film received a theatrical release in America, where it was marketed as an antidote to the Sony version. However, *Godzilla 2000* failed to attract general audiences who may have been soured by the American remake, and, as with the Heisei-era series of films, the remaining five installments of the Millennium era (1999–2004) went direct to video in the United States. Nor were the films especially popular in Japan, even taking into account the involvement of prominent filmmakers like Shusuke Kaneko, who had directed an acclaimed Gamera trilogy in the 1990s. Toho's efforts in the early 2000s to reintroduce Godzilla to international and mainstream audiences ultimately failed, and the character's popularity remained niche. *Godzilla: Final Wars* (Kitamura 2004), despite a fiftieth-anniversary marketing angle and the biggest budget for any Toho Godzilla film, was a dismal flop in Japan and, apart from Taiwan, was not distributed internationally (Kalat 2010, 254).

Just as Toho's Godzilla series had sputtered out for a second time, the international profile of the giant monster film took a boost with Bong Joon-ho's *The Host* (2006). The highest-grossing Korean film of all time upon its release, *The Host* received theatrical distribution in twenty-nine countries, including a premiere at the prestigious Cannes Directors' Fortnight. Mixing sociopolitical commentary with monster spectacle and melodrama, the film tells the story of a working-class family desperately searching for their daughter who has been snatched by a large fishlike creature living in the Han River. The monster is a mutation caused by the American military's dumping of toxic chemicals into the Han years earlier, a plot point based on a real event from 2000 (Paquet 2009, 106). Combining the thrilling action of a kaiju film with a critique of the Korean government's deferential attitude toward the US military, *The Host* demonstrated that monster rampage films could still attract a mass audience, although its appeal can be partly attributed to the film's sentimentality and attention to familial relationships, elements that are subordinated to the monster rampage plot line in most kaiju.

In America, the kaiju film lay dormant for years after Sony's *Godzilla* until *Lord of the Rings* director Peter Jackson brought it new attention with his $200 million remake of *King Kong* (2005). Featuring extraordinary special effects from Jackson's Weta Digital company and a heightened sense of pathos in the relationship between Kong and Ann Darrow, Jackson's version was the fifth-highest-grossing film of 2005 globally and won three Academy Awards. While the film was mostly praised by reviewers, some found the three-hour film unnecessarily long, bombastic, and overly reverential to the original. Additionally, academic critics like Andrea Hairston (2007) argue that Jackson's fannish obsession with the 1933 film prevented him from challenging that film's racist and colonialist impulses.

One unusual aspect of Jackson's *King Kong* was that it was intended as a one-off, at a time when Hollywood had definitively embraced the multifilm franchise as its most important production strategy. One of the most successful franchises of the early twenty-first century is the *Transformers* series, consisting of five films directed by Michael Bay from 2007 to 2017, plus a spinoff film directed by Travis Knight released in 2018. Based on an American toy line (with accompanying cartoons and comic books) that was first popular in the mid-1980s, *Transformers* chronicles a war among giant extraterrestrial sentient robots who have the ability to change into mechanical objects like cars, trucks, and planes. The toy line, the designs of which originated in Japan, is grounded not in kaiju per se, but instead in the related genre of giant robot or mecha anime, in which humans in huge mechanical suits battle monsters and aliens.

By the mid-2000s boys who had played with Transformers as children were now working in the development offices of Hollywood production studios. As a franchise, *Transformers* had tremendous merchandising appeal, but in its convoluted sf-fantasy narrative, also the potential for an epic narrative that could stretch across numerous films. The first film in particular bears some of the influence of executive producer Steven Spielberg, as Shia La Beouf's average teenager is caught up in fantastic events beyond his imagination. But with their mix of stunning action sequences, juvenile humor, shameless objectification of women, and militarism, the films are quintessentially Michael Bay's. Although critics were unkind—Peter Travers (2017) opined that "every time Michael Bay directs another Transformers abomination, the movies die a little"— the global box office numbers were massive, with the third and fourth installments earning over one billion dollars. While their giant creatures are mechanical rather than organic, the *Transformers* series' spectacular battles among rival robots, enhanced by ILM's visual effects and Bay's frenetic editing patterns, serve as amplified, modern equivalents to the creature wrestling matches featured in the 1960s and 1970s kaiju films.

Attached to the first *Transformers* movie was an intriguing teaser trailer for a film that would help revive kaiju cinema. In the trailer, an apartment party attended by young Manhattanites is interrupted by a power failure, then a sudden explosion is heard in the distance. As the partygoers evacuate, a large object comes hurtling down the street toward them—the severed head of the Statue of Liberty. As the film later reveals, a kaiju is on the loose. The enigmatic, genre-hopping trailer for Matt Reeves's *Cloverfield* was effective in piquing audience curiosity, and the film made approximately $170 million worldwide against a budget of only $25 million. Whereas blockbusters like *Transformers* revel in their cartoonish scenes of destruction, *Cloverfield* took a more realistic approach, applying the handheld, found-footage style popularized by *The Blair Witch Project* (Myrick and Sanchez 1999) to the kaiju film in an effort to capture what the experience would be like to the average person on the ground. As such, the film traffics in imagery highly reminiscent of the 9/11 attacks. As James Stone writes, "*Cloverfield* drains 9/11 of any significance other than its potential to yield exciting images of destruction. Anything that could distract us from wallowing in the carnage—be it sympathetic characterization or the complexities of a terrorist attack—is kept from intruding on our pleasure" (2011, 173). Regardless of the tastefulness of the film's appropriation of the iconography (and related emotional associations) of a historical terrorist attack, the links to recent traumatic events in *Cloverfield* helped to reinvigorate the kaiju film by lending sociopolitical relevance, while the film's trendy faux-documentary style, although making some moviegoers nauseous, reinforced this sense of realism.

Following the success of Reeves's film and the *Transformers* series, additional kaiju films with much larger budgets than *Cloverfield* were greenlit. Two years after *Cloverfield*'s release the Hollywood company Legendary Pictures acquired the rights to Godzilla. A year later the same studio greenlit Guillermo del Toro's *Pacific Rim* (2013), an explicit homage to kaiju cinema in which men and women in giant robotic suits battle huge monsters (explicitly called kaiju) that have emerged from the Pacific Ocean. Described by del Toro as an "earnest, loving poem to the kaiju and mecha genre," *Pacific Rim* is unapologetically unpretentious, eschewing the weightiness of *Cloverfield*'s 9/11 imagery in favor of gleeful, imaginative action sequences meant to recall the escapist fun of the anime and kaiju films the director enjoyed in his youth (Lambie 2013). Praised by critics for this sense of fun, if not its originality or ambition compared to del Toro's other films, *Pacific Rim* disappointed at the North American box office in relation to its large $200 million budget. However, its extraordinary success in China encouraged Legendary to produce a sequel in 2018, although by that time Legendary was fully invested in another kaiju franchise, the most famous of all.

## *GODZILLA* (2014) AND THE CONTEMPORARY KAIJU BLOCKBUSTER

In March 2010 Legendary Pictures announced that it had acquired from Toho the rights to Godzilla, with the intent to produce a big-budget film that would reboot and revitalize the character. Led by CEO and self-professed "fanboy" financier Thomas Tull, Legendary was best known for producing blockbuster genre films based on popular comic books such as *300* (Snyder 2006) and *The Dark Knight* (Nolan 2008) (Garahan 2010). When announcing the remake, Legendary remarked that it was committed to treating "the film and its characters in the most authentic way possible" (Legendary Pictures 2010). This statement seemed intended to distinguish Legendary's new film from Sony's remake, which revised the Godzilla character and formula in a way that suggested a disdain for the Japanese films. For instance, whereas Sony drastically redesigned Godzilla to make it more closely resemble a dinosaur, Legendary's Godzilla design remained faithful to the massive bipedal hulk of the original and retained the creature's trademark "atomic breath" that had been rejected by director Roland Emmerich as unrealistic. Legendary was also committed to respecting the character's Japanese origins by setting much of the early action there, and casting Japanese star Ken Watanabe in a major role. In contrast, a brief scene with a Japanese fishing boat was the only reference in Sony's *Godzilla* to the character's home country.

When selecting Legendary to produce a new Godzilla film, Toho may have had in mind the company's first release, *Batman Begins* (Nolan 2005), which managed to rehabilitate a character whose value at the box office had plunged after the failure of the campy *Batman and Robin* (Schumacher 1997). As with Batman, Legendary planned to rejuvenate Godzilla by returning to the grim tone of its first appearance, thereby avoiding the sillier, more juvenile elements with which the character had become associated over the years. Like Honda's 1954 original, Legendary's *Godzilla* (2014) would be a bleak, unforgiving account of survival amid nuclear disaster and large-scale destruction. To achieve this vision, Legendary (with co-financier and distributor Warner Bros.) hired Gareth Edwards, a thirty-five-year-old British filmmaker known for the independent film *Monsters* (2010), on which he served as writer, director, production designer, cinematographer, and visual effects artist. The story of two Americans trapped in the "Infected Zone" at the US-Mexican border where giant buglike aliens have landed, *Monsters* was praised by critics for its political themes, with Michael Atkinson calling it the year's "best American film about our political reality" (2011, 32). *Monsters* was also hailed for its outstanding production values, especially considering its miniscule five-hundred-thousand-dollar budget. Edwards shot much of the film guerilla-style on location in Mexico and Central America, with two actors and a minimal crew following a rough script outline and integrating into the story locations and people they encountered along the way. After production, Edwards augmented the footage with visual effects that helped to create the post-alien invasion environment, with the aliens themselves revealed only in piecemeal fashion (Edwards 2010).

Despite an exponentially larger budget for *Godzilla*, Edwards maintained a commitment to political relevance and the construction of a realistic narrative environment. As Thomas Tull assured a packed crowd at the 2012 San Diego Comic-Con, "There's nothing sci-fi about this movie. It's very grounded, realistic" (Graser 2013, 18). Reproducing the faux-documentary style of *Monsters* presented a challenge: *Cloverfield* had already staked out this territory, and the elaborate previsualization process for special effects sequences made impossible any documentary-type spontaneity. In the end, Edwards and director of photography Seamus McGarvey decided to alternate between handheld and stabilized camerawork as appropriate, mixing a rougher, "indie" aesthetic with a controlled approach more in line with expectations for big-budget films (Williams 2014, 37). Beyond the look of the film, *Godzilla* establishes a sense of realism through its engagement with recent real-life disasters, just as the original film did (figure 8.2). Rather than alluding to the atomic bombs of the 1940s and 1950s, *Godzilla* demonstrates its timeliness by opening with the meltdown of a Japanese nuclear reactor, caused by tremors created by a giant insectoid kaiju, which hatches and makes its way across the Pacific, looking for radioactive material to consume. A second kaiju (or as the military terms it, M.U.T.O.—massive unidentified terrestrial organism) emerges in Nevada and heads to the coast to meet its mate, demolishing Las Vegas in the process.

FIGURE 8.2: The monster as metaphor for natural disaster—*Godzilla* (2014). Warner Bros.

Godzilla emerges from hibernation and battles the two creatures in San Francisco, eventually defeating them and returning to the seas.

Political metaphor has been a common strategy of legitimation within sf film since the 1950s. While a concept like giant monsters battling one another might seem trivial, understanding the creatures as metaphors for natural disasters or nuclear weapons lends the films a deeper significance. Both the nuclear disaster in the beginning of *Godzilla* and a later tsunami scene are obvious references to the 2011 Tōhoku earthquake and the subsequent meltdown at the Fukushima nuclear facility. However, whereas in the original film Godzilla is a clear metaphor for nuclear destruction, the 2014 remake's stance on nuclear energy is less clear. Godzilla is not awakened by nuclear testing; rather, the postwar nuclear tests, we learn, were attempts to kill the creature. Nuclear weapons, while ultimately ineffective against the M.U.T.O.s, are still presented as a legitimate response to their threat. The film's ambivalence toward nuclear power may be related to the US Navy's close participation in the production. Before a preview screening of *Godzilla*, Navy Secretary Ray Mabus boasted to the press, "What the Navy has done is prove for 70 years that nuclear energy transportation—which we pioneered—is very safe, very reliable" (Sneed 2014). Like most Hollywood blockbusters, *Godzilla* is content to dabble in political themes without staking a strong claim. As Sean Rhoads and Brooke McCorkle write, "Despite the superficial references to Fukushima . . . it seems that Hollywood eschewed any kind of political undertones that might estrange possible audiences at home or abroad" (2018, 177–78).

Although they may not form a coherent political message, the references to nuclear destruction and the 2011 disasters do lend *Godzilla* a sense of relevance and gravitas. The added dramatic weight differentiates the film not only from the campy installments of the 1960s and 1970s, but also from more recent big-budget kaiju films, such as the *Transformers* series and Legendary's own *Pacific Rim*. While *Transformers* and *Pacific Rim* contain the same apocalyptic stakes as *Godzilla*, they employ a heightened approach common to contemporary blockbusters that mixes melodrama with humor. Comic relief characters like John Turturro's black-ops agent in *Transformers* and Charlie Day's eccentric scientist in *Pacific Rim* are nowhere to be found in *Godzilla*, which allows little time for levity amid the panic of the monster attacks. Edwards notes, "If this really happened and there really was a giant monster that came out and did these things it would be the most horrific, world-changing event ever. . . . The characters that are going through the events are having a very traumatic time, so they shouldn't be doing wise-cracking one liners" ("Godzilla," n.d.). Humor would only puncture the reality effect the film works to create, aligning it more closely with glib, quippy blockbusters like Sony's 1998 *Godzilla*.

*Godzilla*'s strategies of differentiation continue with its treatment of its monster rampage sequences. Films like *Pacific Rim* and *Transformers* employ a more presentational model of spectacle in which the battles among the giant robots and creatures are put on full display for the audience's scrutiny and pleasure. This approach risks charges of excess, however, with critics describing *Pacific Rim* as a "wearying experience" and *Transformers* a "CGI orgy."[2] Edwards's *Godzilla*, in contrast, restricts the audience's access to scenes of destruction, which are usually represented in mediated fashion through news reports. Several times the film cuts away when a monster is about to attack, revealing only the disastrous aftermath of the rampage. This approach serves several functions. First, restricting the narration to what our human protagonists can see invites a deeper level of audience identification—here, Edwards recalls films like *Signs* (Shyamalan 2002) and *War of the Worlds* (Spielberg 2005), which limit the viewer's experience of a full-scale alien invasion to what is observed by a single group of characters, thereby enhancing the audience's emotional attachment to them. Second, by withholding the spectacle of disaster, the film implies that it is somewhat distasteful to take pleasure in it. While the aforementioned lighter tone of a film like *Transformers* allows one to revel in the cathartic act of destruction, such an attitude is less appropriate for the somber *Godzilla*.

*Godzilla* finds other opportunities to present spectacular images, as in a striking scene in which paratroopers skydive from high altitudes through storm clouds into the battle zone. But ultimately, a $160 million Godzilla film cannot subvert generic expectations so thoroughly that it completely withholds scenes of kaiju destruction. By revealing Godzilla only intermittently, and then in piecemeal fashion as in *Monsters* and *Cloverfield*, Edwards builds suspense and saves a full reveal for the climax wherein the three giant creatures battle amid the smoky ruins of San Francisco. Jason Barr suggests that this strategy, common in early Japanese *kaiju eiga*, reflects the influence of kabuki theater, which employs a "slow burn" structure that builds to a spectacular finale (2015, 32–33), but Edwards cites a different set of influences: late 1970s blockbusters like *Jaws* (Spielberg 1975) and *Alien* (Scott 1979). "Before the era of digital technology, and because they couldn't always show the creature constantly, the first half of the movie would just be these little glimpses," says Edwards, "so . . . you just get so many chills and goose bumps. I felt like in modern cinema it's so easy to just throw everything at the screen constantly and we've missed that style of storytelling" ("Godzilla" n.d.).

Edwards's commitment to realism, relevance, and restraint worked to legitimize *Godzilla* for general audiences. However, another audience awaited satisfaction:

Godzilla fans. One key development since the previous American remake in 1998 was the increased importance of specialized fan audiences to studio production and marketing strategies. As Mel Stanfill writes, "From 1994 to 2009 . . . fans went from marginal to a constituency that media companies both recognize and actively seek to incorporate, encourage, monetize, and manage" (2019, 5). Even if fans made up a relatively small percentage of potential viewers for a film, negative early reactions from fandom posted online could generate negative publicity that might alienate general audiences. Liam Burke notes, "Although producers may have ignored fans in the past, in the digital age they have been forced to recognize how this once powerless elite now has the ability to mobilize others" (2015, 138). In Legendary's initial press release, Thomas Tull sought to ease any fan concerns by identifying himself as part of their community and declaring that he intended to "produce the Godzilla film that we as fans want to see."

Burke argues that contemporary fandom typically views fidelity to the source material as a necessary quality of film adaptations (2015, 136). Legendary demonstrates this fidelity in numerous ways, from retaining the original Godzilla design to returning to the dark tone of the 1954 film. At the same time, a long-running series like *Godzilla* features many different takes on the material; therefore, Legendary's *Godzilla* also included elements that might appeal to fans of the later, more fantastical and lighthearted sequels. For instance, recalling Toho films like *Godzilla vs. Hedorah* (Banno 1971), Edwards' Godzilla has heroic, anthropomorphic qualities: it is described as a restorer of balance to nature rather than as a mere animal or primal force of destruction. As the film approaches its climax, Godzilla is frequently aligned with protagonist Ford Brody (Aaron Taylor-Johnson) as both characters fight against the threat of the M.U.T.O.s. In one memorable moment, Godzilla makes eye contact with Brody, shooting him a pained look before returning to the fight. Despite the filmmakers' efforts to appeal to fans, reaction within fandom to *Godzilla* (2014) was mixed, with many fans strongly critical of the decision to cut away from the rampage action and limit Godzilla's screen time. The strategy of restraint, meant to establish *Godzilla*'s seriousness of intent when compared to effects-laden action films like *Pacific Rim*, had the unintended consequence of alienating those viewers who crave more destruction in their kaiju films.

*Godzilla* received generally positive notices from mainstream critics, who praised the striking imagery, while noting that the award-winning cast, including Bryan Cranston, Juliette Binoche, Ken Watanabe, Sally Hawkins, and David Strathairn, was wasted on thin characterizations. In response to the film's impressive $525 million at the global box office, Legendary and Warner Bros. created the "MonsterVerse," a franchise universe in which Godzilla, King Kong, Mothra, and other famous kaiju might all exist. *Kong: Skull Island* (Vogt-Roberts 2017), the next installment in the series, performed even better than *Godzilla* and set up a remake of *King Kong vs. Godzilla* titled *Godzilla vs. Kong* (Wingard 2021). In the meantime, a sequel *Godzilla: King of the Monsters* (Dougherty 2019) was released to mediocre reviews and a disappointing $386 million at the global box office, about what the 1998 remake earned. Box office analyst Scott Mendelson (2019) suggests that audiences have grown fatigued with incessant sequels, and that Godzilla is not a popular enough franchise to warrant multiple big-budget films. The same decline in popularity of the MonsterVerse can be observed in other related franchises: the third *Cloverfield* film went straight to Netflix in 2018 after Paramount determined a theatrical release would be unprofitable, and the *Transformers* spinoff *Bumblebee* (Knight 2018) was the lowest-grossing film of the series by far.

Despite recent disappointments at the box office, the run of big-budget kaiju films since *Cloverfield* has elevated the form into the popular consciousness more so than in any period since the mid-1950s. The kaiju film has even broached the world of independent cinema: in Nacho Vigalondo's dark comedy *Colossal* (2016), two alcoholics discover their drunken antics are manifesting in the form of a kaiju and a giant robot that rampage through Seoul. Hollywood studios' perpetual interest in spectacle and franchising may extend the life of this particular big-budget production cycle. However, the lasting popularity of the kaiju film may ultimately lie with niche audiences, as it has for most of the previous seventy years. In 2016 Toho released *Shin Godzilla* (Anno and Higuchi), a reboot that, like Gareth Edwards's remake, draws on Fukushima imagery but is more directly critical of government institutions than its American counterpart. While it received only a limited international release, *Shin Godzilla* was a major success in Japan and won the Japan Academy Film Prize for Best Picture. As a result, Toho plans to release further live-action Godzilla films after its deal with Legendary expires in 2021 (Squires 2018), ensuring that even if the American MonsterVerse ends with *Godzilla vs. Kong* (2020), the kaiju film, as a discrete branch of sf, will survive to inspire fear and awe.

## NOTES

1. Typical of those reviews is Roger Ebert's, in which he calls *Godzilla* "a cold-hearted, mechanistic vision, so starved for emotion or wit" (1988, 25). Ebert may have struggled to remain objective, as he and his television partner, Gene Siskel, were parodied in the film in response to their negative reviews of Devlin and Emmerich's previous films. See Ebert (1988).
2. For a sampling of that commentary, see Sims (2014) and Carr (2009).

## WORKS CITED

Atkinson, Michael. "Gareth Edwards' Monstrous Realism." *In These Times*, March 2011, 32.

Barr, Jason. *The Kaiju Film: A Critical Study of Cinema's Biggest Monsters*. Jefferson, NC: McFarland, 2015.

Bigelow, Joe. "Son of Kong" [review]. *Variety*, January 2, 1934, 13.

Burke, Liam. *The Comic Book Film Adaptation*. Jackson: UP of Mississippi, 2015.

Carr, Megan. "'Transformers: Revenge of the Fallen' a Silly Bore." *Montgomery Media*, July 1, 2009. https://www.montgomerynews.com/entertainment/

Crowther, Bosley. "Screen: Horror Import." *New York Times*, April 28, 1956, 11.

Ebert, Roger. "A Monstrous Mess." *Chicago Sun Times*, May 25, 1998, 25.

Edwards, Gareth. "Adventures in the Infected Zone." *Empire*, November 2010, 100–106.

Erb, Cynthia. *Tracking King Kong: A Hollywood Icon in World Culture*. 2nd ed. Detroit: Wayne State UP, 2009.

Fox, David J. "Jurassic Eats 'E.T.': It's No. 1." *Los Angeles Times*, October 5, 1993, F1.

Garahan, Matthew. "Producer Follows His Own Script." *Financial Times*, December 5, 2010. https://www.ft.com/content/96d90b5c-00aa-11e0-aa29-00144feab49a.

"Godzilla—Gareth Edwards Interview." *IndieLondon*. Accessed March 28, 2020. https://www.indielondon.co.uk/Film-Review/godzilla-gareth-edwards-interview/.

Graser, Mark. "Tull Brings 'Godzilla' Fire to Comic-Con Fanboy Fest." *Variety*, July 14, 2013, 18.

Hairston, Andrea. "Lord of the Monsters: Minstrelsy Redux: King Kong, Hip Hop, and the Brutal Black Buck." *Journal of the Fantastic in the Arts* 18, no. 2 (2007): 187–99.

Jampel, Dave. "Japanese Arters Wow Critics, but Horror Films Get Coin." *Variety*, April 15, 1959, 46.

Jewell, Richard B. *Fade to Black: The Decline of RKO Radio Pictures*. Berkeley: U of California P, 2016.

Jewell, Richard B. "RKO Film Grosses, 1929–1951: The C. J. Tevlin Ledger." *Historical Journal of Film, Radio and Television* 14, no. 1 (1994): 37–49.

Jewell, Richard B. *RKO Radio Pictures: A Titan Is Born*. Berkeley: U of California P, 2012.

Kalat, David. *A Critical History and Filmography of Toho's Godzilla Series*. 2nd ed. Jefferson, NC: McFarland, 2010.

Klady, Leonard. "'Godzilla' in Slo-Mo Second Week." *Variety*, June 8–14, 1998, 12.

Lambie, Ryan. "Guillermo del Toro Interview: *Pacific Rim*, Monsters, and More." *Den of Geek*, July 12, 2013. https://www.denofgeek.com/movies/.

Legendary Pictures. "Press Release." March 29, 2010. https://www.slashfilm.com/legendary-pictures-producing-new-godzilla-film/.

McKenna, A. T. *Showman of the Screen: Joseph E. Levine and His Revolutions in Film Promotion*. Lexington: UP of Kentucky, 2016.

Mendelson, Scott. "One Ironic Reason 'Godzilla: King of the Monsters' Stumbled at the Box Office." *Forbes*, August 13, 2019. https://www.forbes.com/sites/scottmendelson/2019/05/30/godzilla-king-of-the-monsters-review-bigger-isnt-always-better/?sh=610bfb4f19e2.

Nashawaty, Chris. "Stomp the World I Want to Get Off." *Entertainment Weekly*, May 22, 1998, 26.

Paquet, Darcy. *New Korean Cinema: Breaking the Waves*. New York: Columbia UP, 2009.

Rhoads, Sean, and Brooke McCorkle. *Japan's Green Monsters: Environmental Commentary in Kaiju Cinema*. Jefferson, NC: McFarland, 2018.

"RKO's King Kong Reissue Rolls Up Big Grosses." *Motion Picture Herald*, August 2, 1952, 17.

Ryfle, Steve, and Ed Godziszewski, *Ishiro Honda: A Life in Film, from Godzilla to Kurosawa*. Middletown, CT: Wesleyan UP, 2017.

Sims, David. "Kneel before Your Godzilla." *The Atlantic*, May 15, 2014, https://www.theatlantic.com/culture/archive/2014/05/kneel-before-your-godzilla/370953/.

Sneed, Tierney. "How 'Godzilla' Dances around That Whole Nuclear Issue." *U.S. News & World Report*, May 16, 2014. https://usnews.com/news/articles/.

Solomon, Brian. *Godzilla FAQ*. Milwaukee, WI: Applause Books, 2017.

Squires, John. "Toho Ditching 'Shin Godzilla 2' in Favor of a Massive Kaiju-Filled Cinematic Universe." *Bloody Disgusting*, May 20, 2018. https://bloody-disgusting.com/movie/.

Stanfill, Mel. *Exploiting Fandom: How the Media Industry Seeks to Manipulate Fans*. Iowa City: U of Iowa P, 2019.

Stone, James. "Enjoying 9/11: The Pleasures of Cloverfield." *Radical History Review* 111 (2011): 167–74.

"Top Grossers of 1953." *Variety*, January 13, 1954, 10.

Travers, Peter. "'Transformers: The Last Knight' Review: Michael Bay's Latest Is 2017's Most Toxic Movie." *Rolling Stone*, June 21, 2017. https://www.rollingstone.com/movies/movie-reviews/.

Tsutsui, William. *Godzilla on My Mind: Fifty Years of the King of Monsters*. New York: Palgrave Macmillan, 2004.

Vaz. Mark Cotta. *Living Dangerously: The Adventures of Merian C. Cooper, Creator of King Kong*. New York: Villard, 2005.

Williams, David E. "King of the Monsters." *American Cinematographer*, June 2014, 37.

"Wonderful Thrills in 'The Lost World.'" *Exhibitors Trade Review*, February 28, 1925, 38.

CHAPTER 9

························································································

# MAGICAL REALISM
# SCIENCE FICTION

························································································

## GERALD DUCHOVNAY

GENRE definition has long been a perilous academic activity, seldom producing full agreement as to what traits define any form or which formulas actually constitute genres. The film noir, for example, has been described by some as a genre and by others as simply a style, while science fiction (sf) has throughout its history often been linked, even subordinated to other forms like horror or fantasy. In more recent times, magical realism has occupied similarly disputed territory. In fact, the very title is often debated. Should it be termed "magic realism," "magical realism," "marvelous realism," "cinemagic realism," or something else? And without an acceptable title, its status—as genre, style, mode, or hybrid form—has remained something of an open question, as the reviews of numerous magical realist films readily illustrate. In the context of this handbook, I want to suggest that, rather than consider it a kind of generic foundling wrapped in a blanket whose nomenclature is simply used to sell a text, we see magical realism as an influential development in the constantly branching patterns of sf cinema.

Magical realism, as both a literary and filmic form, has more than a century-long history. By the middle of the nineteenth century, artists and authors were rejecting Romanticism and turning to complex representations of reality, as can be seen in the paintings of Jean-Francois Millet (*The Gleaners*, 1857) and Gustave Courbet (*Young Ladies of the Village*, 1852); the novels of Gustave Flaubert (*Madame Bovary*, 1856), Henry James (*Portrait of a Lady*, 1880–81), and George Eliot (*Middlemarch*, 1871–72); and the new world of technologically produced photographic images introduced by Louis Daguerre and William Henry Fox Talbot. Yet even while realism remained a force in the arts, it was joined by host of qualifying movements, including impressionism, expressionism, surrealism, and modernism. In the mid-1920s German art critic Franz Roh, whose tellingly titled book *Nach-Expressionismus, Magischer Realismus: Probleme der neuesten Europaischen Malerei* (Post-expressionism, magical realism: Problems of recent European painting, 1995), considering the work of such artists as George Grosz and Max Beckmann, noted that these and other contemporary figures were moving away from the "fantastic" elements of surrealism and expressionism and reaching for a "new objectivity," or a new sort of reality, but one that could still convey something "hidden," a "mystery," a "magic [that] does not descend to the represented world, but rather hides and palpitates behind it" and produces a "strange effect on the viewer" (Roh 1995). After considering appropriate terms for describing this move, including ideal realism, verism, and neoclassicism, Roh, as his book's title suggests, settled on the term "magical realism" (quoted in Labudovic 1983, 6–7).

Two years prior to Roh's study, Gustav Hartlaub had coined the term *Neue Sachlichkeit* (new objectivity) to emphasize how painting represented reality and the objective world. Both Roh and Hartlaub were trying to assess how artists, influenced by elements of expressionism, were seeing and experiencing the everyday not through received notions but through immediate and intensely personal visual experiences. As J. P. Telotte has noted, this effect is essentially what German expressionist artists, dramatists, and filmmakers sought to achieve: "By calling into question the very manner in which we see—and are allowed to see by a variety of cultural restrictions—almost ironically by distorting the common images of our world, Expressionism challenged the normal order of representation both within the cinema and outside of its confines" (2006, 27). And as Richard Murphy adds, an essential product of this expressionist approach was the creation of "an alternative reality" (1999, 57). This notion of an alternative vision of reality, generated by the filmmaking/artistic process, and offered as a kind of visual stimulus to the film audience, prompting them to revision their own world, is a key to understanding magical realism and its function.

That sense of function should also be underlined. Within the realms of art history, magical realism originally described a particular approach in a larger movement: "Whereas those using the term New Objectivity tended to focus on social and political issues within a work, the Magic Realists initially tended to distance themselves from political themes, satire, and social critiques to portray an objective view of life imbued with intangible qualities. Moreover, these works of art seemed to possess a certain mystery, or secret, underlying their themes and subjects" ("Magical Realism—History and Concepts" n.d.). That approach to art began to spread on the continent after Massimo Bontempelli first wrote about it in the journal *900* (Jewell 2008). Shortly thereafter, when José Ortega y Gasset translated and published Roh's work in Spanish, the idea of magical realism crossed the Atlantic and made its way to Latin America, where it also took on a much stronger cultural thrust. In short order, critics appropriated the term that had originally focused on the visual arts to address a growing body of literary texts that fashioned alternative realities as social commentary (Spindler 1993, 75). While there was occasional reference to European writers such as Kafka, starting in the 1930s and 1940s much of the more developed critical discussion of magical realism came to focus on literature from such Latin American authors as Arturo Uslar-Pietri, Jorge Luis Borges, Alejo Carpentier, Miguel Ángel Asturias, and especially Gabriel García Márquez, all of whom fused the aesthetic of magical realism into their socially observant fiction.

However, while magical realism began to permeate the visual and literary landscape and for decades was often regarded as a specific genre of Latin American fiction, for many it was never clearly defined or codified—whether as a genre, style, or aesthetic. When Ángel Asturias was asked to define the term, he offered a two-part reply, one emphasizing the importance of indigenous tales from before the European conquest of the Americas, and the other evoking the European tradition of André Breton and the surrealists: "Between the 'real' and the 'magic' there is a third sort of reality. It is a melting of the visible and the tangible, the hallucination and the dream. . . . The Indian thinks in pictures; he sees things not so much as the events themselves, but translates them into other dimensions, dimensions where reality disappears and dreams appear, where dreams transform themselves into tangible and visible forms" (Mead 1968, 330). The issues of definition continue to plague scholars. In fact, a Call for Papers for a recent conference on "Figuring Magic Realism— International Interpretations of an Elusive Term" (2021) identified seventeen possible ways of considering the subject.

To Maggie Ann Bowers and others, "Magic realism, magical realism, and marvelous realism are highly disputed terms," and that disagreement is due not only to their complicated history, but also to the many variants and even the media—painting, literature, film—in which these variations have surfaced (2004, 20). While the medium is seldom a defining factor, it is curious that most discussions of magical realism have given little consideration to its roots in early cinema and to those traits that have surfaced widely in contemporary cinema. George Méliès, a magician entranced by the work of the Lumière brothers and the new motion picture camera technology, combined realism and magic in many of his films from the 1890s to 1912, as he worked in a wide range of genres, including *actualités*, political satire, historical epics, magic acts, and sf (Ezra 2000, 22, 28), and a similar combinatory impulse can be observed in many inheritors of the Méliès tradition, including Buster Keaton and Charlie Chaplin.

While speaking of early Hollywood films, Stanley Cavell has offered an observation that very simply links these early efforts to recent international films, noting that "Hollywood has always had a taste for contrasting worlds of the everyday with worlds of the imaginary (playing on the two primordial possibilities of film, realism and fantasy)" (2005, 345). Whether it be the films of Méliès, Chaplin, and Keaton, or more recent works such as *Like Water for Chocolate* (Arau 1992), *Pan's Labyrinth* (del Toro 2006), *A Chinese Tale*, aka *Chinese Take-Away* (*Un cuento chino*, Borensztein 2011), *Birdman* (Iñárritu 2014), or *The Shape of Water* (del Toro 2017), magical realism is not just playing into the magic and realism characteristic of early film and its traditions of magical appearance, disappearance, and exaggeration, but rather reflects and adapts those screen elements to achieve a cultural commentary that is fundamental to magical realism as a literary phenomenon. This development is reflected in the first scholarly commentary on magical realism in film and what was termed its "transfiguration of the object world itself," offered by Fredric Jameson more than thirty years ago (1986, 302). He saw the form offering "a possible alternative to the corrective logic of contemporary postmodernism" (302) and even linked it to the world of sf. Yet even after Jameson's leading commentary, there has often been little more than passing reference to it with regard to an sf cinema.

One indication that magical realism, whether considered as a film genre, subgenre, aesthetic, or style, has gained increased traction is the growing attention it has received in international film histories. Kristin Thompson and David Bordwell, for example, have explained how Latin American writers "forged a literature that mingled European influences with indigenous cultural sources," contributing to an emerging body of films that spiced "social realism with myth, fantasy, and fairy tales" (2018, 605). A number of the films they then discuss are literary adaptations, especially of works by Gabriel García Márquez, prompting Thompson and Bordwell to claim that "Latin American cinema won a place in world film culture partly through its ties with a prestigious literary trend" and that "the fantastic imagery of Magical Realism sought to reflect the collective imagination of colonized peoples, for whom everyday reality seemed only one step away from the supernatural" (605).

Demonstrating the aptness of Thompson and Bordwell's commentary is Jorge Amado's *Dona Flor and Her Two Husbands* (1966), a Brazilian novel that was well received in literary circles and was faithfully adapted for the screen in 1976 by Bruno Barreto. Dona Flor is married to Vadinho, a ne'er-do-well who loves gambling, flirting, and womanizing. Flor's desires for her husband's charm and passionate lovemaking during their seven-year marriage supersedes her publicly respectable persona. When Vadinho dies on the street during the *carnaval* celebration, Dona Flor goes into mourning for a year, after which she is courted by and marries Teodoro, a straightlaced and respectable pharmacist, who offers her comfort and stability but without passion. On the first anniversary of her marriage to Teodoro, Vadinho returns as a ghost, sharing the couple's conjugal bed and accompanying Flor and her husband wherever they go. The final scene of the movie offers a typically magical realist image, showing Dona Flor with her respectable husband on one arm and a fully nude Vadinho on the other, walking down the street of their town. Ted Gioia notes how magic, fantasy, and the surreal are intertwined in both texts and how both Amado and Barreto bring together a cultural heritage that includes candomblé, an Afro-Brazilian belief system, and the local rituals and folklore of Bahia, Brazil. He claims that the novel, like the film, might be viewed in a number of ways, as "a romance, or ghost story, or comedy of manners," since it blends "local color with recipes, folklore, and other bits of Bahian culture" and in those very variations exemplifies why Latin American fiction and film have "gained such widespread popularity" (2000)—a popularity further stoked by the publication of what is often considered the most influential literary example of magical realism, Gabriel García Márquez's *One Hundred Years of Solitude* (1967).

Probably more influential in the cinema, though, was the adaptation of Laura Esquivel's 1985 novel *Like Water for Chocolate* (1992), done by the author's then-husband Alfonso Arau. It became the highest-grossing Spanish-language film in the United States, although most mainstream reviewers, including Janet Maslin and Clifford Terry, did not even mention its links to the tradition of magical realism. Maslin (1993) does spotlight food, forbidden love, and miracles resulting from strong passions, all "presented with the simplicity of a folktale," while Terry (1993) similarly focuses on the unlikely combination of culinary feats, "undeniable love," and revolutionary acts.

However, with more precision and a greater sense of context, Roger Ebert (1993) would reference the literary source material and explain "how [*Like Water for Chocolate*] continues the tradition of magical realism that is central to modern Latin film and literature. It begins with the assumption that magic can change the fabric of the real world, if it is transmitted through the emotions of people in love." Michael Wilmington would also note this literary connection in describing the film as "a 10-course feast of magic realism," although the only trait he cites as a reflection of magical realism is how the film contains a "tone of childlike wonder." Like Wilmington, many other of the admiring commentators rarely explained what qualified the film as magical realism; instead, they often seemed to revert to Justice Potter Stewart's dictum when asked to define pornography: "I know it when I see it" (Lattman 2007).

Those who have discussed elements of magical realism in fiction or film tend to repeat the fundamental notion that it is grounded in a real, often mundane world, one where dreams or elements of the marvelous or magical often blur the boundaries between the real and the fantastic. Such texts often allegorically treat sociopolitical issues under the guise of myth, fantasy, folklore, or sf, even as they seem grounded in a real time and real place. Author Salman Rushdie, whose work has frequently been categorized as magical realism, explains that, while "stories don't have to be true, by including elements of the fantastic, elements of fable, or mythological elements, or fairytale, or pure make believe" they can open "another door into the truth" (2015). Following Rushdie's and Ebert's leading comments, along with a catalogue of other definitions compiled by Alberto Rios (2015), we might see the following key traits as central to identifying magical realism, especially as it has surfaced in film: (1) a believable narrative grounded in reality; (2) a few magical or supernatural elements that may involve another dimension of reality developed through fantasy, dreams, ghosts, folk tales, or fairy tales; (3) happenings that may have a "strange effect" on the viewers, but do not undercut the larger narrative's believability; (4) a specificity of culture, time, and location; (5) a concern with sociopolitical issues of the time portrayed in the film or possibly at the time of the film's release; and (6) possible references to biographical, stylistic, or thematic ideas reflected in the filmmaker's other works. A list of films that incorporate several of these magical realism characteristics would, of necessity, be both lengthy and varied in nature, but it would surely include: *It's a Wonderful Life* (Capra 1946), *Harvey* (Koster 1950), *8½* (Fellini 1963), *Dona Flor and her Two Husbands* (Barreto 1976), *The Purple Rose of Cairo* (Allen 1985), *Ghost* (Zucker 1990), *Edward Scissorhands* (Burton 1990), *Like Water for Chocolate* (Arau 1992), *Amelie* (Jeunet 2001), *Big Fish* (Burton 2003), *Pan's Labyrinth* (del Toro 2007), *Scott Pilgrim vs. the World* (Wright 2010), *Biutiful* (Inarritu 2010), *Midnight in Paris* (Allen 2011), *Beasts of the Southern Wild* (Zeitlin 2012), *Birdman* (Inarritu 2014), and *The Shape of Water* (del Toro 2017). It is a group that ranges across various cultures and suggests a number of traditional genre affiliations.

In light of this collection's focus, though, we might note the relatively limited number of these works that have a singularly sf thrust. That fact might seem somewhat curious given the frequently cited definition of sf, offered by Darko Suvin, as "the literature of cognitive estrangement" (1979, 7)—that is, narratives that involve an estranged sense of reality—but it might be at least partly due to a reluctance to admit that there can be links between the scientific and what is broadly termed the magical, and partly to an attitude voiced by one of the most prominent magical realist filmmakers, Guillermo del Toro. A figure who has been influenced by his Mexican culture, the works of Gabriel García Márquez, and various other literary and cultural influences, del Toro has made a host of films that seem to fall into the sf category, including *Mimic* (1997), *Blade II* (2002), *Hellboy* (2004), *Hellboy II: The Golden Army* (2008), *Pacific Rim* (2013), *Pan's Labyrinth* (2006), and *The Shape of Water* (2017). However, he is seldom described as an sf filmmaker in the mold of a Steven Spielberg or James Cameron, and he has never identified his work as fitting into a particular generic box. Rather, he claims that his films are generally an "extension of [himself]" and a "rephrasing of [his] childhood" or more recently his "adulthood" (Galloway 2017), as well as his various cultural influences. Yet his award-winning *The Shape of Water* seems to be an obvious example of an sf film with a magical realism slant, even if that slant has often been overlooked and the film more often approached as simply a fantasy.

Del Toro builds his stories on the power of the image, whether from an earlier film, a film script, or especially a literary text. Thus numerous articles and interviews over the last several decades point up his encyclopedic knowledge of literature, mythology, fairy tales, and films—all of them cutting across cultural boundaries and combining with his own cultural heritage to feed the magical realism character of his filmmaking (Çakir 2018). Like some of the other top cinematic practitioners, del Toro manages to translate the mesmeric literary descriptions that characterize so much of the literature of magical realism to the intense and more immediate power of the cinematic image. As examples that have influenced his approach, we might briefly consider images from two award-winning films, Alfonso Arau's *Like Water for Chocolate* and the writer and director Sebastian Borensztein's *Un cuento chino*. In the former film, the magical realism of the source novel (Esquivel 1985) undergoes a mesmerizing transformation, as when Pedro, Tita's true love, dies while making love to her. She then swallows matches and the intensity of her love causes her to self-immolate, burning down the room they are in and eventually the entire house. A similar effect marks the latter film's opening scene, as a man is about to propose to his girlfriend while out boating on a river in China only to be disrupted by the "cognitive estrangement" of a cow falling from the sky, killing the woman. The sense of immediacy and the sensory impact that attach to the experience and the disturbing image remain with the viewer throughout the film, just as they have remained with the character in the film who, we learn, has fled China, hoping to forget—but never forgetting—this tragedy that had befallen him.

Like Arau and Borensztein, del Toro mines the commonality of a love for the other and the lingering, disturbing image, and, in the case of his *The Shape of Water*, uses magical realism to reimagine the visual and narrative impact that an sf film had on him, when, as a six-year-old Mexican boy, he first saw *Creature from the Black Lagoon* (Arnold 1954). In *Shape* he tells the story of a mute orphan, Elisa Esposito (Sally Hawkins), who, abandoned by a river as a baby and with scars on her neck, has found her way to Baltimore, where she works the graveyard shift at Occam Aerospace Research Center, a top-secret military installation. Elisa is a loner except for the time she spends with her African American coworker Zelda Fuller (Octavia Spencer) and her neighbor Giles (Richards Jenkins), an unemployed, gay advertising artist. However, her life changes when an amphibian humanoid (The Asset, Gill-Man, Amphibian Man, *Dues Branquia*, or the creature, as he is variously designated) is brought to the facility by the racist and misogynistic Colonel Richard Strickland, who captured and transported the primordial creature to the Center to study its mysterious breathing powers in order to help American astronauts in the space race with the Russians.

After Elisa and Zelda are assigned to clean the lab where the Asset is imprisoned, Elisa becomes fascinated by this trapped, mute "other"—that is, a figure who, in many ways, mirrors her own character. She surreptitiously visits the lab when she can and develops a bond with the Gill Man by bringing him hard-boiled eggs, introducing him to music, and teaching him to "speak" through sign language. Strickland, whose very different approach involves torturing the Asset with an electric cattle prod, is ordered by the general in charge to complete the project and then destroy the creature. Dr. Hoffstetler, the lead scientist in the project—and reminiscent of the "good" scientists found in other sf films (for example, in *The Thing* [1951] and *Creature from the Black Lagoon* [1954])—is aghast at the prospect of the Asset's destruction, although his reaction is partly an act, hiding a deeper, hardly humanistic purpose. He is a Soviet spy who has been ordered by his Soviet handler to learn all he can about the creature's secret powers and then kill the creature himself before reporting any of his findings to Strickland. But acting from a far different impulse than either of these two government officials, Elisa expresses a highly personal desire to free the Amphibian, signing to Giles in a moment of desperation (and love) that the amphibian "doesn't see what I lack, doesn't see that I am incomplete" (figure 9.1). With the help of Zelda, Giles, and ultimately Hoffstetler as well, who, after earlier referring to the Asset as a scientist's "plaything," comes to recognize him as a sentient being—as "intelligent, capable of language, of understanding emotions"—Elisa rescues the water god, intending to take him to the river where he might swim back to his home before Strickland can kill him.

While the film was well received, even winning Academy Awards for Best Picture and Best Director, its reception did little to alter the ambiguous status of magical realism or its relationship to sf. In trying to sort out the film's difficult link to dominant forms and styles, critics readily framed it in the context of various popular genres, such as horror, fantasy, and the thriller. For example, to Xan Brooks at *The Guardian* (2017), *The Shape of Water* was "a ravishing '60s-set romance, sweet, sad and sexy" that is a "pastiche [of] the postwar monster movie . . . the incident at Roswell and all manner of cold war paranoia" that "isn't simply a romance, but a B-movie thriller as well." Writing for *Variety*, Guy Lodge (2017) described it as a "romantic fantasy," an "adult fairytale," and a "mad-scientist B-movie to heart-thumping Cold War noir to ecstatic, wings-on-heels musical, keeping an unexpectedly classical love story." Jessica Kiang (2017) at *The Playlist* saw it as "a Cold War paranoia thriller, a 1950s-style creature feature, a quasi-musical cinematic nostalgia trip and a fantasy interspecies love story between a woman and a merman . . . brimming with romance and adventure," while the *Hollywood Reporter*'s David Rooney

FIGURE 9.1: Elisa (Sally Hawkins) and her aquatic other, the Asset (Doug Jones), recognize their mutual fascination in *The Shape of Water* (2017). Fox Searchlight Pictures.

(2017) considered it a "dark-edged fairy tale, a Cold War romantic fairy tale." Effectively summing up these varied takes, Anthony Lane (2017) called *The Shape of Water* "a genre-fluid fantasy" and noted in his review that, "having watched this movie twice, I still can't define it." Observing that the film is "flooded with [references to] other films," Lane concluded that del Toro, in mixing his modes, "delivers a horror-monster-musical-jailbreak-period-spy romance." These reviews not only reflect the difficulty of pinning down the magical realist form, but also of even recognizing its shared sf characteristics.

The broad outlines of the plot readily suggest the film's deep indebtedness to the traditions of magical realism—an indebtedness that, even when they cannot name it, the reviewers seem able to observe. *The Shape of Water* almost immediately presents itself as a kind of dreamlike fairy tale, filled with elements of the fantastic, as the film's opening images present us with an underwater dream sequence: Elisa floating through her apartment, along with a coffee pot, lamp, table, and shoe, as fish swim by. When an alarm goes off, it awakens her and brings both her and us back to the drab reality of her small apartment. The opening voiceover (with Giles as the narrator) also shifts planes of reality, as he offers, "If I spoke about it—if I did—what would I tell you, I wonder? Would I tell you about the time? It happened a long time ago—in the last days of a fair Prince's reign. . . . Or would I tell you about the place—a small city near the coast but far from everything else? Or would I tell you about her—the princess without a voice? Or perhaps I would just warn you about the truth of these facts and the tale of love and loss and the monster that tried to destroy it all." His comments fashion a conditional nature for everything that follows in the narrative and underscore the conditional mode in which most magical realism functions. Similarly, aspects of magical realism lend an ambiguous ending to the film, as we see Elisa again floating, this time in the river that promised escape for the Asset, as a voiceover once more conditionally frames things, telling us, "If I told you about it, what would I say? That they lived happily ever after? I believe they did, that they were in love, that they remained in love. I'm sure that is true. . . . But when I think of her, of Elisa, all that comes to mind is a poem, made of just a few truthful words, whispered by someone in love, hundreds of years ago: 'Unable to perceive the shape of You, I find You all around me. Your presence fills my eyes with Your love. It humbles my heart, for You are everywhere.'" The insistent sense of openness, even in ending, seems a telling and appealing characteristic of magical realism.

But this sense of openness, while often linked to sf, does not in itself adequately suggest the film's science fictional aspect, the genre frame that has provided a figure like del Toro with a comfortable home for his magical realist slant on the world As we earlier indicated, del Toro has an established history in both writing and directing films that are more recognizably sf, including *Mimic*, *Hellboy*, and *Pacific Rim*, and elements of this prior work show up throughout *The Shape of Water*. Some reviewers also noted the film's homage to Jack Arnold's sf film *Creature from the Black Lagoon*, a work that, del Toro acknowledges, was one of his inspirations: "I've had this movie in my head since I was 6, not as a story but as an idea. When I saw the creature swimming under Julie Adams, I thought three things: I thought, 'Hubba-hubba.' I thought, 'This is the most poetic thing I'll ever see.' I was overwhelmed by the beauty. And the third thing . . . is, 'I hope they end up together'" (Rottenberg 2017). At a British Film Institute event in 2017, he amplified his appreciation for Arnold's film, noting, "All I knew is that I wanted to have the moment where the creature carries off the girl, not as a moment from a horror movie but as a beautiful moment of love" (Diestro-Dópido n.p.) (figure 9.2). Perhaps that transformed "moment" has simply cast a "magical" veil over the film's sf elements.

Throughout *The Shape of Water* we see the work of science, the (often wrongful) applications of technology, and efforts to restore a kind of reason to an unbalanced world. This coming together of science, technology, and the work of reason—as in Arnold's film, like many others of the 1950s—illustrates the familiar territory of Cold War–era sf. Even its creature, its gill-man is endemic to that territory, for he is not really a monster, but a creature from the past, confronted with an invasive present. Suggesting the complex position of such a figure, del Toro offers, "In my movies creatures are taken for granted. There is a heavy Mexican Catholic streak in my movies, and a huge Mexican sense of melodrama. Everything is overwrought and there's a sense of acceptance of the fantastic in my films, which is innately Mexican" (Feld 2014). While creatures like the Asset are a natural part of this world, they are, in his films, treated differently than in more conventional sf films; they are presented as a specific, even familiar kind of sf creature, the alien, who is never welcomed, never depicted as offering knowledge or understanding, but rather as something to be shot (as in *The Day the Earth Stood Still* [Wise 1951]), hunted down (as in *It Came from Outer Space* [Arnold 1953]), even attacked with atomic weapons (as in *The Space Children* [Arnold 1958]). In *The Shape of Water* we find similar elements of fear, resistance, and overt violence clustered around this "alien," reminding us, in the best fashion of magical realism, how much our culture

FIGURE 9.2: The Gill Man of *Creature from the Black Lagoon* (1954) makes off with Kay Lawrence (Julia Adams) in "a beautiful moment of love." Universal Pictures.

is itself "overwrought" and unable to accept or welcome the fantastic figure, the alien from *this* world.

In a traditional sf film, such as one of the many creature features of the 1950s, Strickland would be the tough-jawed protagonist who tames the alien, secures the needed information for the US space program, and wins over the story's heroine. Instead, del Toro "magically" inverts these expectations with his brutalizing military figure and the conventionally "silenced" others—Elisa, Zelda, Giles—uniting to rescue the amphibian humanoid. The result is this creature feature turns into an enchanted romantic sf love story between Elisa and her Amazonian river god—one that replaces typical sf tropes about the fear of the unknown, menacing alien life forms, and the Russian menace with a preferred dream, done in a kind of coded criticism, that advocates love, freedom, and an alternative, even scientifically endorsed (by Hoffstetler) political reality.

While sf films with a magical realism slant are commonly anchored in a real time and place, they also include elements of the fantastic that serve to suggest, as del Toro insists, that "fantasy is infinitely more realistic than realism" ("BFI Screen Talk" 2017). For the films of a director such as del Toro who delights in unveiling the extraordinary in the ordinary, that "infinitely more realistic" dimension is, in fact, their real value, what "allows us to apprehend and discuss larger issues and to recognize," as he puts it, that the "fantastic is ordinary and the ordinary is fantastic" ("BFI Screen Talk" 2017). In the magic to be found in *The Shape of Water* and other magical realist sf films, viewers experience a sense of solace and transcendence that may be unavailable in more conventional sf films that feel too constrained by the real, too limited by established science—that is, films that are speculative but only within certain bounds. Thus, while magical realism injects sometimes convoluted and complex elements into sf films—perhaps even affording an entirely different experience of wonder—it does so not to look toward the future, as so much of sf's typical wonder prompts us to do, but to help us better understand and cope with the world, in all its dimensions, in which we already live.

## Works Cited

Amado, Jorge. *Dona Flor and Her Two Husbands*. Trans. Harriet de Onis. New York: Knopf, 1966, 1969.

"BFI Screen Talk: Guillermo del Toro." BFI London Film Festival. December 6, 2017. https://www.youtube.com/watch?v=rfbD3Obir64.

Bowers, Maggie Ann. *Magic(al) Realism*. New York: Routledge, 2004.

Brooks, Xan. "The Shape of Water Review—Guillermo del Toro's Fantasy." *The Guardian*, August 31, 2017. https://www.theguardian.com/film/2017/aug/31/the-shape-of-water-review.

Çakir, Deniz. "In the Vision of Guillermo Del Toro's Magical Realism and Universal Symbolism." February 25, 2018. https://medium.com/@breavous/in-the-vision-of-guille rmo-del-toros-magical-realism-and-universal-symbolism-c4ce8c5ed48e.

Cavell, Stanley. "The Good of Film." In *Cavell on Film*, edited by William Rothman, 333–48. New York: SUNY Press, 2005.

Diestro-Dópido, Mar. "Guillermo del Toro: '*The Shape of Water* Is My First Movie That Is Hungry for Life.'" London Film Festival, October 26, 2017. https://www.bfi.org.uk/news-opinion/sight-sound-magazine.

Ebert, Roger. "Like Water for Chocolate." *Roger Ebert.com*, April 2, 1993. https://www.rogereb ert.com/reviews/like-water-for-chocolate-1993.

Esquivel, Laura. *Like Water for Chocolate*. New York: Doubleday, 1985.

Ezra, Elizabeth. *Georges Méliès: The Birth of the Auteur*. Manchester: Manchester UP, 2000.

Feld, Rob. "Beauty and the Beasts." Director's Guild of America, Winter 2014. https://www.dga.org/Craft/DGAQ/All-Articles/1401-Winter-2014/DGA-Interview-Del-Toro.aspx.

"Figuring Magical Realism—International Interpretations of an Elusive Term." CUNY Graduate Center Conference, April 9, 2021. https://networks.h-net.org/node/23910/discussi ons/5745662/cfp-figuring-magic-realism-%E2%80%93-international-interpretations.

Galloway, Stephen. "Guillermo Del Toro on Confronting Childhood Demons and Surviving a Real-Life Horror Story." *Hollywood Reporter*, November 3, 2017. www.hollywoodreporter.com/features.

Gioia, Ted. "Dona Flor and Her Two Husbands." Conceptual Fiction: Exploring the Non-Realist Tradition in Fiction. 2000. https://tedgioia.substack.com/p/dona-flor-and-her-two-husbands-by-jorge-amado.

Jameson, Frederic. "On Magic Realism in Film." *Critical Inquiry* 12, no. 2 (1986): 301–25.

Jewell, Keale. "Magical Realism and Real Politics: Massimo Bontempelli's Literary OKCompromise." *Modernism/Modernity* 15, no. 4 (2008): 725–44.

Kiang, Jessica. "Guillermo Del Toro's 'The Shape of Water' Is Sweet & Scary Movie Magic." *The Playlist*, 2017. https://theplaylist.net/shape-of-water-review-20170831/.

Lane, Anthony. "The Genre-Fluid Fantasy of 'The Shape of Water.'" *New Yorker*, December 1, 2017. https://www.newyorker.com/magazine/2017/12/11/the-genre-fluid-fantasy-of-the-shape-of-water.

Lattman, Peter. "The Origins of Justice Stewart's 'I know it when I see it.'" *The Wall Street Journal*, September 27, 2007. https://www.wsj.com/articles/BL-LB-4558.

Lodge, Guy. "*The Shape of Water*." Variety Venice Film Review, 2017. https://variety.com/2017/film/reviews/the-shape-of-water-review-1202543729/.

Labudovic, Ljudmila. *Cinemagic Realism*. M.A. Thesis. University of Colorado, Denver, 1983.

"Magical Realism—History and Concepts." *The Art Story*. https://www.theartstory.org/movement/magic-realism/history-and-concepts/.

Maslin, Janet. "Emotions So Strong You Can Taste Them." *New York Times*, February 17, 1993. https://www.nytimes.com/1993/02/17/movies/review-film-emotions-so-strong-you-can-taste-them.html.

Mead, Robert G., Jr. "Miguel Angel Asturias and the Nobel Prize." *Hispania* 51, no. 2 (1968): 326–31.

Murphy, Richard. *Theorizing the Avant-Garde: Modernism, Expressionism, and the Problem of Postmodernity*. Cambridge: Cambridge UP, 1999.

Rios, Alberto. "Defining Terms." *Magical Realism*, 2015. https://www.public.asu.edu/~aarios/magicalrealism/.

Roh, Franz. "Magical Realism: Post-Expressionism." Trans. Wendy B. Faris. In *Magical Realism: Theory; History, Community*, edited by Lois Parkinson Zamora and Wendy B. Faris, 15–31. Durham, NC: Duke University Press, 1995.

Rooney, David. "*The Shape of Water*." *Hollywood Reporter*, August 31, 2017. https://www.hollywoodreporter.com/review/shape-of-water-venice-2017-1034220.

Rottenberg, Josh. "Guillermo del Toro's Highly Personal Monster Film *The Shape of Water*." *Los Angeles Times*, September 5, 2017. www.latimes.com/entertainment/movies/la-et-mn-guillermo-del-toro-telluride-20170905-htmlstory.html.

Rushdie, Salman. "Salman Rushdie on Magical Realism: True Stories Don't Tell the Whole Truth." *YouTube*. October 3, 2015. www.youtube.com/watch?v=EZtdhLndVYg.

Spindler, William. "Magic Realism: A Typology." *Forum for Modem Languages* 29, no. 1 (1993): 75–85.

Suvin, Darko. *Metamorphoses of Science Fiction: On the Poetics and History of a Literary Genre*. New Haven, CT: Yale UP, 1979.

Telotte. J. P. "German Expressionism: A Cinematic/Cultural Problem." In *Traditions in World Cinema*, edited by Linda Badley, R. Barton Palmer, and Steven Jay Schneider, 16–28. Edinburgh: Edinburgh UP, 2006.

Terry, Clifford. "'Like Water for Chocolate' Is a Feast for the Senses." *Chicago Tribune*, April 2, 1993. https://www.chicagotribune.com/news/ct-xpm-1993-04-02-9304030027-story.html.

Thompson, Kristin, and David Bordwell. *Film History: An Introduction*. 4th ed. New York: McGraw-Hill, 2018.

Wilmington, Michael. "'Chocolate': Scrumptious Tale of Love Repressed." *Los Angeles Times*, February 26, 1993. https://www.latimes.com/archives/LA-XPM-1993-02-26-CA-386-STORY.html.

# CHAPTER 10

........................................................................

# STEAMPUNK CINEMA

........................................................................

## THOMAS LAMARRE

STEAMPUNK is commonly described as a form of science fiction (sf) that reinhabits *steam*, that is, steam power, and through it, the nineteenth-century (usually Victorian) world, repurposing not only its technologies but also its fashions, mores, and worldviews. It does not aim merely to reproduce the steam-powered world but to extrapolate it imaginatively, to speculate on its possible future. Steampunk thus overlaps with alternative histories and parallel universes; it invents a past that never was to construct a future that might have been—a combination of "never-was" and "might-be" (Tibbets 2013, 139). The *punk* of steampunk amplifies this temporal weirdness by introducing a "now" attitude. Steampunk is a recent label, a riff on cyberpunk by K. W. Jeter in 1987 to describe the fiction that he, James Blaylock, and Tim Powers were writing. Their stories took place in nineteenth-century settings and drew inspiration from the conventions of that era's speculative writers, especially H. G. Wells and Jules Verne. Steampunk fiction is close to fan fiction, for its originality lies in repurposing previous stories and characters instead of laying claim to wholly new narrative patterns and tropes. In sum, the juxtaposition of steam and punk conjures forth a weird temporality conjoining *never-was* and *might-be* through a performative *redoing now*.

Although steampunk emerges from print fiction, it is in the nature of genres to work across media. Steampunk stretches across literature, comics, cinema, television, games, and art, not to mention collectibles, jewelry, and costumes, and, of course, fan activities including "cosplay" performance, conventions, and online communities. Steampunk cinema, then, is but one way of traversing the steampunk world. While the goal here is not to introduce rigid distinctions between media forms, this account of steampunk cinema aims to address the distinctiveness of cinema, or more broadly, moving image media.

Cinema is a modern art of movement, with historically developed conventions for dealing with audiovisual images under conditions of movement. Its conventions affect steampunk, putting a cinematic spin on it. When cinema engages steampunk, it approaches it through cinematic spacetime, through a sensorimotor framework conditioned by received conventions for moving images. Still, cinema does not overpower or overwrite steampunk. It must reckon, cinematically, with steampunk's weird temporality—the coupling of never-was and might-be via redoing. This essay aims to explore the relation between steampunk and cinema, to consider what cinema does to steampunk, and steampunk to cinema. Embedded in these questions is another: what does steam + punk + cinema do to sf?

# Immiscible Modes

Steampunk coalesced in the 1980s in speculative sf literature, but cinema was in the mix from the start. Blaylock gives as much credit to cinema as to literature for his steampunk mind. After he discovered Wells and Verne as a child, he writes, "I happened upon two astonishing films by the Czech director Karel Zeman: *The Fabulous World of Jules Verne* and *The Adventures of Baron Munchhausen*, which are pure, unadulterated Steampunk" (2012, 6). In these films and others, Zeman reworked the trick films of Georges Méliès, a contemporary of both Wells and Verne. *The Adventures of Baron Munchausen* (*Baron Prásil*, 1962) reimagines the Méliès film *The Hallucinations of Baron Munchausen* (*Les aventures de Baron Munchausen*, 1911), while *The Fabulous World of Jules Verne* or *Invention for Destruction* (*Vynález zkázy*, 1958) draws on the Méliès aesthetic to adapt the Verne novel *Facing the Flag* (*Face au Drapeau*, 1896) for the big screen. Blaylock is right: these films are pure steampunk. *Invention for Destruction* in particular might well be considered the ur-film of steampunk cinema, for it constructs and links the two dispositions that would at once define and polarize steampunk cinema.

On the one hand, the film builds a world whose form is remarkably like Victorian England (1837–1901) or the French Belle Époque (1871–1914) in terms of clothing, architecture, manners, and technologies. It builds a *world-form* that is unfinished, open to further elaboration, and yet consistent, with definite propositions for prolonging it. This world-form is constructed from a variety of cinematic forms: narrative form, character (or actor) forms, and forms of mise en scène, lighting, editing, and perspective. It comprises these other forms even as it emerges from them. It does not end at the edges of frame or screen. It extends in space to other locations and *in time* to the past and future. Its magnitude is greater than narrative-form. Other stories, with other actors, may unfold from this world, inhabiting and prolonging it. The world-form in Zeman's film might be characterized as a Vernian world. While Verne was already extrapolating the late nineteenth-century technologies into various futures, Zeman amplifies this speculative bent to create an alternative future that feels wholly consistent with the nineteenth century. Because the film emerges by building the world in which Verne's novel was set, or more precisely, *might have been* set, its world is Vernian in the strong sense; a Vernian claims that Verne wrote fact not fiction.

On the other hand, without the Mélièsian aesthetic, the Vernian world-form might feel like inert space. Zeman builds on the in-camera effects Méliès used in his trick films, especially the use of layers with different textures and dimensional properties—filming men in deep-sea gear through a fish tank to transform them into underwater explorers; transforming an octopus into a krakenlike menace; making actors appear to ride within an elaborately designed train, which is actually a flat painting; or using painted foregrounds and backgrounds, on glass, solid surfaces, or both, in conjunction with actual settings. Zeman delights in constructing these "dimensional layers" that trick the viewer's sense of scale, perspective, and dimension (see figure 10.1). One of his favorite tricks is to mix two-dimensionally textured media with three-dimensionally textured media, which sometimes imparts a whimsical, cartoonish feel, sometimes induces a dizzying sensation, and at other times gives the impression of heightened precision, which is echoed in admixtures of mechanical and natural sounds that direct and misdirect the audience's attention. Zeman's films, like those of Méliès, are magic acts. Yet they are "essentially" cinematic since movement—the direction and misdirection of attention under conditions of movement—is what makes the mismatched dimensions hold together.

In sum, *Invention for Destruction* offers two ways of holding things together, two modes of consistency. The first mode is a world-form, which may be called a "world mode" in terms of its mode of address to spectators. While actions unfold in a cinematic world, that world offers a combination of technologies, characters, settings, and stories that is open to prolongation or supplementation across media. While a comic, a novel, or

FIGURE 10.1: Mixing two-dimensional and three-dimensional textures in Zeman's *Invention for Destruction* (1958). Ceskoslovensky Statni Film.

a game is imaginable and feasible, the world mode invites spectators to become builders, repurposers. The second mode is a kino-aesthetic mode. Zeman puts his distinctive spin on the Méliès aesthetic, placing the emphasis on kino-tricks that make incongruous dimensions coexist. Something odd happens to narrative. Both the kino-aesthetic mode and the world mode deemphasize diegetic immersion; the kino-aesthetic uses narrative as a frame, and the world mode makes narrative a subset of world-building. If Zeman's films feel somewhat disappointing in diegetic terms, their triumph lies elsewhere.

Zeman's triumph was two- (or three-)fold: (1) inventing two modes of consistency, a world mode (drawn from Verne) and a kino-aesthetic mode (drawn from Méliès), both of which have highly different affordances, as if hailing from heterogeneous orders of reality; and (2) holding them together such that they feel complementary, compossible.

In her study of the cinema of the fantastic, Bliss Cua Lim uses the phrase "immiscible temporalities" (2009, 29) to describe a similar effect, arising when two modes of temporality are mixed like oil and water. Oil does not dissolve in water (nor water in oil). When shaken together, the result is a colloidal suspension instead of a physiochemical solution. If not continuously shaken, the two liquids separate. The colloidal suspension, then, is a performative "solution" to the unresolvable problematic of immiscibility. Zeman's performative solution to holding together immiscible modes might be likened to an iridescent film of oil spread across the water's surface, but the shimmer does not remain on the surface. The prismatic luster persists in depth, in aqueous immersion, troubling a distinction between surface and depth, perturbing a sense of dimensionality, like being suspended in a state of colloidal refraction across time and space.

# Extrapolative and Metapolative Models

Zeman's films affect how we understand the steampunk form. The allure of steampunk comes from its juxtaposition of two factors, old tech (steam) and a punk attitude. The punk attitude ensures that the old tech (steam) does not become period drama, stuck in the past. Steampunk extends old tech in the future, unfurling into another possible future or an alternative history. It plays a tricky game with the past: the past is to be historicized in meticulous detail and at the same time projected dynamically into the future. The two factors (steam + punk) elicit a third: a contemporary activation or redoing of the past world, which projects it into the future. Extrapolation best describes this process: steampunk extrapolates the past into the future.

In his seminal account of sf as a literature of cognitive estrangement, Darko Suvin characterizes the fiction of Verne and Wells (the *roman scientifique*) in terms of extrapolation. He notes how it "introduces into the old empirical context only one easily digestible new technological variable" (376). Steampunk does precisely that: extend one technological variable (the steam engine) into the future. For Suvin, the extrapolative model results in a lesser form of sf: "the *roman scientifique* such as Verne's *From the Earth to the Moon*—or the surface level of Wells' *Invisible Man*—though a legitimate SF form, is a lower stage in its development. It is very popular with audiences just approaching SF" (1972, 376). Suvin favors another model, the analogical: "The highest form of analogic modeling would be the analogy to a mathematical model, such as the fairly primary one explicated in Abbott's [novel] *Flatland*, as well as the ontological analogies found in a compressed overview form in some stories by Borges and the Polish writer [Stanislav] Lem" (380).

Suvin's remarks came at a time when a new generation of English-language writers had begun to distance themselves from "hard science fiction" based on scientific accuracy and predictive power, that is, the extrapolative model. The New Wave of speculative sf that emerged in the 1960s and gained ground in the 1970s placed the emphasis on literary experimentation and stylistic creativity. Suvin was writing before another generation of authors, riding this speculative new wave, unleashed cyberpunk and soon thereafter steampunk in the 1980s. While steampunk literature bears the traces of both legacies, it does not mesh neatly with hard sf (extrapolative model) or the new wave of speculative fiction (analogical model). Steampunk fits Suvin's extrapolative model in that it rediscovers and redirects the tendency of Verne and Wells to extrapolate one technological variable. Yet it is also highly speculative: instead of new technologies, it extrapolates old technologies that once were new, renewing them. Steampunk thus adds something to the extrapolative model of Verne and Wells—speculative activation. The past world must be reactivated to be extrapolated anew. Alongside the technological variable to be reactivated—the "hard science" technology featured in Verne or Wells (a time machine, a submarine, a rocket to the moon, an airship), there is a technique of activation, the "soft science" techniques of speculative fiction.

Something analogous happened in Zeman's films, well before these new waves lapped the shores of the North Atlantic. *Invention for Destruction* follows Verne's extrapolative model, extending the futurist, predictive possibilities Verne saw in the new technologies of his era—submarines, airships, telegraphs, weaponry. At the same time, Zeman turns to Méliès to activate "old-tech," that is, in-camera tricks. Zeman's use of Méliès is not extrapolative. It entails a sort of "metapolative" speculative model that shows affinity with both the New Wave literary experimentation and the analogical model Suvin associates with such writers as Borges, Lem, and Abbot. It does not insert new material into something else (interpolate) or continue an existing trend (extrapolate); it continues a trend that never was, acting as if there had been an existing trend whose future values could be inferred or induced. Zeman's films are not so much about old or new tech as about "what was never old" (to borrow Thoreau's turn of phrase).

*Invention for Destruction* thus offers a way of understanding the steampunk form. On the one side of steampunk is the extrapolative model of sf, which it realizes through the world mode, which entails the construction of a world-form through a combination of era-activation and world-building. On the other side is the speculative or metapolative model, which is realized in cinema through the kino-aesthetic mode or kino-technics. In *Invention for Destruction* the two models do not readily come together. Therein lies the "problematic" of steampunk, which Zeman's film so exquisitely reveals, as well as the source of steampunk's perplexing relation to place and time—its two models or modes, like oil and water, only combine through a performative "solution."

The problematic of immiscible modes explains the diversity of films classified as steampunk. There are diverse performative solutions to (or formal suspensions of) the immiscibility of the two modes of steampunk. Many kinds of films are often included under the steampunk rubric, yet few feel unambiguously steampunk. Some feel too watery, so to speak, and others too oily. To some extent, this is a feature of all film genres. Accounts of film genres generally agree that they resist clear-cut definition and definitive classification (Moine 2008) for several reasons, which will become more evident during this discussion. Genres are inherently hybrid forms (Staiger 1997; Neale 2000), defying singular classification based on shared features alone.

The goal here is not to offer a definitive or exhaustive list of steampunk films, nor to contest or exclude what others classify as steampunk. Lists of such movies assume a variety of forms. They appear on fan wikis and websites featuring commentary and allowing discussion among steampunk fans; they are also generated for streaming services such as Amazon Prime and Netflix, and the Internet Movie Database features numerous compilations of film types, which solicit fan participation and serve as advertisement. In addition, the body of print commentary and scholarship on steampunk has steadily grown in recent years. introducing a narrower, more canonical list of titles. Here, I consider a film as steampunk if it appears on at least three lists *and* on two different kinds of list: fan-guided websites, commercially oriented compilations, or print commentaries and scholarship.

Because it mixes immiscible modes, steampunk cinema shows a high degree of internal polarization. Two pronounced trends or lineages appear, both with fuzzy borders, opening a vast middle ground of admixtures. At one pole are movies that focus primarily on the world-form, with its combination of era-activation and world-building. Let me call them "steam worlds," even though they do not all center on the steam era or extrapolate steam technologies. At the other pole are films and animations that, like Zeman's, place greater emphasis on a particularly cinematic kind of metapolative speculation—a form that I call "transdimensional kinemation." Between the two arise a variety of "adventures of energy" that mix the two tendencies in various ways. The following sections explore these three varieties of steampunk: steam worlds, transdimensional kinemation, and adventures of energy.

# STEAM WORLDS

Steam worlds aim for verisimilitude, a believable or plausible nineteenth-century world. Ideally, clothing, customs, architecture, and urban spaces of the period are rendered in exacting detail. These films show some overlap in their historicist impulse with the period drama but introduce speculative elements. The speculative element is typically of a technological nature, such as a time machine or airship. Supernatural elements are not beyond the pale, however; thus steam-era monsters are not uncommon (Dracula, Frankenstein, and Jekyll and Hyde). The border between the technological and supernatural is often blurred, and between the actual-historical and the fictional. It is possible to crowd a steampunk film with multiple speculative elements (e.g., vampires, time travelers, airships) or even to fuse them (time-traveling vampires). Yet the preference in building cinematic steam worlds is to focus on one technological variable.

The easiest way to construct a steam world is to adapt a Verne or Wells story. Film versions of Verne and Wells appear on steampunk lists, even if they do not seem especially indebted to steampunk. Adaptations of Verne began with the dawn of cinema, with Méliès's *Trip to the Moon* (*Le voyage dans la lune*, 1902), and continued to appear at regular intervals, with adaptations of Verne such as *20,000 Leagues under the Sea* (Paton 1916), and of Wells such as *The Invisible Man* (Whale 1933). While these films appear on steampunk movie lists, they do not arouse much enthusiasm, for they feel too much of their own era to offer a steam world.

The first steampunk writers looked especially to the cinema of the 1950s, the movies of their childhood. Alongside Zeman's *Invention for Destruction*, the other movie that enjoys pride of place on steampunk lists is the Disney adaptation of *20,000 Leagues under the Sea* (Fleischer 1954). The film focuses the speculative force of Verne's novel onto one technology, Captain Nemo's submarine, the *Nautilus*, whose design combines a sense of streamlined efficiency with ornate nineteenth-century futurism. The film aims to depict the submarine's fantastic undersea adventures in as plausible or believable a manner as possible through special effects. Because special effects are used to support diegetic realism, this mode shows affinity with "narrative cinema" or "classical realism" (Gunning 1990, 59). In *20,000 Leagues under the Sea*, speculative force is largely concentrated on the *Nautilus*. Yet something strange happens. While the special effects may not have been intended to disrupt narrative form, the combination of underwater adventures, seascapes, and special effects overwhelms the narrative. Audiences tend to recall not the story but the submarine and the undersea world, or the story remembered is a clash between the *Nautilus* and the sea monster. The narrative form of cinema transforms into a world-form, in part due to the pressure of special effects. As such, while *20,000 Leagues under the Sea* shies from era-activation (only the *Nautilus* feels distinctly Vernian; the rest feels of the 1950s), it embarks on world-building.

Disney would soon reinforce and capitalize on this effect. Its first park, Disneyland, opened an attraction called the Submarine Voyage in 1959. It fell to the second park, Walt Disney World, when it opened in 1971, to transform that sort of submarine voyage into a *20,000 Leagues under the Sea* attraction, which continued until 1994. This theme-park attraction liberates "attractions" (in Gunning's sense) from their subordination to narrative form, only to subordinate them to the production of world-form. Where *Invention for Destruction* reactivates the in-camera "old tech" effects of Méliès, *20,000 Leagues under the Sea* uses contemporary technology to update the past. Where Zeman denounces militarism and entertains an alternative future for technology, *20,000*

*Leagues under the Sea* flirts with applied military technology and proclaims that the future is now—and open to visitors.

If *20,000 Leagues under the Sea* is often cited as a precursor for steampunk despite its relative lack of era-activation and speculation, it is due to its success in building a world-form through its Verne-inspired extrapolation of one speculative technology, the underwater vehicle. The technology is becoming a world, and conversely, it may be said, with a nod to Heidegger's notion of the world-picture (1997), that the actual world is becoming a technology, something to be technologically realized. The cinematic relation between submarine and ocean is instructive: the invention of a watertight enclosure for humans to explore the ocean makes the ocean feel like an enclosure, accessible to humans for exploration. The result is a world-form that can be continually updated through the application of newer technologies and expanded across media—a possibility on which the theme park seizes.

The success of *20,000 Leagues under the Sea* paved the way for a series of Hollywood adaptations of Verne, among them, *Around the World in 80 Days* (Anderson 1956), *From the Earth to the Moon* (Haskin 1958), and *The Master of the World* (Witney 1961), all titles that call attention to the construction of a world-form—circling the globe, viewing it from outer space, sky-born global domination. The contrast between these films and Zeman's during the same period is sharp. Where these Hollywood films are intent on the production of a world-picture or global order, Zeman is intent on counteracting the destructive impulse inherent in their world-picture. The fact that both sorts of film are included within steampunk cinema reminds us of the form's fundamental polarization.

The overarching continuity in the steam-world mode is due to the persistence of the extrapolative model: speculative force concentrated in one technological variable. Discontinuity arises at the level of tonality: the later worlds claimed for steampunk cinema are darker in tone, grimmer in mood, and crowded with detail. One example is Marc Caro and Jean-Pierre Jeunet's *The City of Lost Children* (*La cité des enfants perdus*, 1995). Although its historical world owes more to the nightmares of mid-twentieth-century surrealism than to the Victorian era or the Belle Époque, it is widely hailed as steampunk, surely in part due to the film's reliance on the extrapolative model. A single technological variable shapes its world—a machine that extracts dreams.

The film's directors strove for a retro-futurist vision, as eccentric technologies abound, redolent of a futurity within a past that never was. One delightfully disturbing instance is the cyborg eyewear, a technologically enhanced variation on the aviator goggles so loved by steampunk cosplayers. These stylistic flourishes follow from, and fill out, the world-form. Caro and Jeunet's style does not speculate on its film world; it does not introduce another (cinematic) technical variable to serve as counterpoint for a first technological variable. This is in keeping with their efforts to create a sense of enclosure: *The City of Lost Children* offers a nightmarish world from which there seems no escape. All paths lead back to the traumatic experience at the center of the labyrinthine city: somewhere a child's dreams are being ripped out of its head. Caro and Jeunet introduce a frisson of psychological horror, of childhood trauma, into the steampunk world mode.

While the reliance on an extrapolative model ensures an overarching continuity to the films in this lineage, the world-form undergoes significant alteration at other levels. Striking in a number of other films is the enlargement of the imaginative or fictional quality of the historical periods, that is, Victorian England (Wells) or Belle Époque France (Verne). The result is a tension within steam worlds between a sense of boundless possibilities and a feeling of enclosure, resulting in two general orientations within steam world–building.

On the one hand, in *Time after Time* (Meyer 1979), for instance, H. G. Wells pursues Jack the Ripper in a time machine. This film anticipates a larger development in steam worlds: actual historical figures mingling freely with fictional ones. Although the film sticks to the extrapolative model of one technological variable, the combination of historical figures and fictional characters creates a sensation of shattering limits, through the time machine and the mixture of actual and imagined persons. Later films, as varied as *Wild Wild West* (Sonnenfeld 1999), *The League of Extraordinary Gentlemen* (Norrington 2003), and *Steamboy* (Otomo 2004), similarly play on the interaction of fictional characters and historical figures. These three films are among the first that deliberately build on the steampunk phenomenon. While cyberpunk and cyborgs had become prevalent in cinema by the late 1980s and early 1990s, steampunk did not cross into big-budget global cinema until the late 1990s and early 2000s. None of these films spawned franchises, even though they were set up for serialization. Arguably, this was due in part to their overreliance on the action film type. The action film also creates a world, but one in which fictional characters are supposed to feel like actually existing people, confirming the reality of the historical crisis. The real world sets a limit on what is possible. In contrast, when steam worlds mix actual and fictional elements and open the historical setting to seemingly boundless interventions, the internal limit must come from elsewhere. Steampunk worlds have to sustain the speculative force on their technological variable to sustain their world.

*The League of Extraordinary Gentlemen* tries so hard to be an action film that it transforms the *Nautilus* into an improbable series of gadgets. The film also supplies each character with a "backstory trauma" to be overcome in the pursuit of justice, a mawkishly sentimental take on writer Alan Moore's sly, provocative riff on superhero comics. *Wild Wild West* also loses hold of its speculative variable, striving to hold its world together with personalities and gadgets. Of the three films, *Steamboy* comes closest to sustaining its steam world through one technological variable, the steamball (see figure 10.2). Using rarefied water that recalls the heavy water associated with nuclear power, the steamball affords such highly compressed and condensed steam that it becomes a weapon of mass (thermal) destruction. The spherical, riveted steamball provides one of the canniest depictions of the affinity of the steampunk world-form with the Heideggerian world-picture. It offers hand-held global destruction of the globe—what Rey Chow refers to as the "world target" (2006, 1)—in a bomb shaped like a globe. Despite its not insignificant virtues, *Steamboy* stumbles when it comes to sustaining the speculative force of its steamball. The gadget feels too externalized and instrumentalized to place internal limits on action. But this problem may be inherent in the project of

FIGURE 10.2: Steam technologies proliferate wildly in *Steamboy* (2004). Toho Studios.

world-building. Speculative force must, paradoxically enough, function as much internally as externally. This proposition is easier to realize with a submarine in the ocean, or even a flying machine in the sky, than with a steamball in London.

On the other hand, there are films that play up the sense of an internal limit to steam worlds, usually through psychological enclosure. Akin to *The City of Lost Children* are films such as *Dark City* (Proyas 1998) or *Vidocq* (Pitof 2001), whose dimly lit worlds circle around a violent secret that structures action like an originary trauma, but in relation to a technological variable. In *Dark City* that variable is the city itself. In *Vidocq* it is the serial killer's instrument-laden, automation-riddled lair. This orientation enters into action-oriented steampunk, too. A prime example is the 2002 adaptation of Wells's story *The Time Machine* (Wells/Verbinski). It furnishes a classic Hollywood-style traumatic backstory to explain its protagonist's motivations: when his fiancée is murdered, the time-traveler tries to alter the past to save her. Once he learns that the past is not easily altered, he flees to the future. In the distant future, he works through his grief through his love for a woman and her son, eventually liberating their people from predatory exploitation. The film offers a serviceable solution to the problematic tension within steam worlds. Steam worlds, like cyberspace, are built environments, and as such, their capaciousness requires internal material limits. If a film cannot convey a sense of those internal limits, whether it be through hackneyed evocations of some originary trauma or sustained speculation on a technological variable, the steam world gradually falls apart.

However, falling apart is not necessarily the end of things. Steampunk is a fan form, and the bits and pieces that capture fans' imaginations circulate in other ways. Fragments of a set design, elements of special effects, quirks of character, and unresolved segments of story may be put into relation in new ways. It may be for this reason that there has never been one cinematic steam world definitively acclaimed and upheld as a model by fans. Steam worlds also appeal as ruins. In any case, there is more to steampunk cinema than world-building.

# TRANSDIMENSIONAL KINEMATION

The phrase *transdimensional kinemation* is a mouthful, as unwieldy as one of steampunk's wacky flying machines yet similarly purposeful. Kinemation is intended to highlight the kinesthetic qualities of moving images in both cinema and animation. While cinema and animation are not necessarily different in kind, they have historically gone in different directions, reaching a point where difference in degree feels like difference in kind. The admixture of cinema and animation can produce unusual effects, as in *Invention for Destruction*, which draws on Czech animation. The term *transdimensional* is also indebted to Zeman, whose artful combination of 2D and 3D textures renews and expands Méliès's bag of tricks while insinuating a fourth dimension. But since transdimensional kinemation is a mouthful, let me call it the kino-mode.

Zeman set the bar high for steampunk's kino-mode. Zeman's stylistic propositions have inspired many filmmakers, among them, Tim Burton, Terry Gilliam, and Miyazaki Hayao, whose films often appear on steampunk lists, even when their films do not deal with nineteenth-century steam or speculative technologies at all. Such films are steampunk in another way, due to their use of a "retro-fantastic" style that recalls Zeman.

Gilliam's *The Adventures of Baron Munchausen* (1988) takes place during the eighteenth century, pitting the colorful tall tales of the baron against the drab and absurd Age of Reason, in which reason is linked to pointless warfare. The emphasis falls largely on zany, episodic phantasmagoria, but a parable about the dangers of modern technoscience is implied. If this Gilliam film and others, such as *Brazil* (1989), appear on lists of steampunk cinema, it is for kino-aesthetic reasons. Their speculative tone does not derive from extrapolating a technology, for their technologies feel too fantastic to be developed. In the lineage of Méliès and Zeman, Gilliam invents a cinematic style that might be characterized as "retrofantastic," whose images convey both historical weight and fantastic unrootedness. *The Adventures of Baron Munchausen* has a world, but it is more a setting than a world-form. Indeed, the film's adventures are framed by theatre, so its world-building is limited to projecting a theatrical fantasy onto the outside world. It is at the level of style instead of world that Gilliam explores something redolent of, or amenable to, steampunk—a stylistic enmeshment of never-was and might-be.

The same is true of Tim Burton. His moody yet goofy style, full of exaggerated shapes, tilted shadows, and elongated silhouettes, feels old yet modern, a past lingering in our future. Here, too, a kind of retrofantasy appears, evocative of German expressionism and the Gothic. Burton's style also shows an affinity with Zeman in troubling dimensionality, and not only between 2D and 3D textures. Across these familiar dimensional textures, Burton conjures up a fourth dimension, something like time out of joint. His evocations of pastness feel unmoored from historical coordinates, stretched uncomfortably across flat and volumetric modalities. The Burton film that most frequently appears on steampunk lists—*Alice in Wonderland* (2010), often cited with its sequel, *Alice through the Looking Glass* (2016; dir. James Bobin)—does not construct an enlarged or expandable Victorian world-form; it offers an elaborate puzzle in which stylized temporalities, fragments of impossible pasts and fantasy futures may be pieced in diverse manners.

Tim Burton's name is also associated (as producer) with other films that make the steampunk list for similarly retrofantastic reasons, like *9* (Acker 2009) and *Coraline* (Selick 2009), both feature-length animations. Neither film is steampunk in the sense of dealing with steam power, steam technologies, or the steam era. *9* takes place in an alternate 1930s world in which a robot intelligence called the Fabrication Machine has eradicated all forms of life. To counter this destruction, a scientist deploys alchemical techniques to transfer his soul into nine ragdolls called "stitchpunks" that eventually overthrow the Fabrication Machines, and microbial life begins to return to Earth. While *9* builds a world, its world is formed less through speculation about a technological variable (AI) than through the application of a retrofuturist style, wherein old tech (stitching) is infused with a punk attitude (soul). Stitchpunk is about textures, with contrasting textures evoking a sense of incongruous dimensions jockeying for position, while the stitching itself is a weird fusion of 2D and 3D textures. The contrast between AI and stitchwork evokes the tension between new technologies of digital animation and legacies of hand-crafted animation. The contrast of textures suits the story: life is situated between dimensions, as an extradimensional force holding forms together. Still, the stylistic gesture fairly overwhelms the story, as if the narrative were an excuse for experiments in transdimensional kinemation. The kino-mode is the story. Like *The City of Lost Children*, *9* comes up against the limits of steampunk, but from the other side, the kino-mode.

We might recall how *The City of Lost Children* extrapolates a technological variable with traumatic and surreal implications, a machine extracting dreams from children's brains. The resulting world-form is more psychological than historical, and the film relies more on psychic feedback loops than sociohistorical or technological imagination. The kino-mode requires different sustenance. *9* relies on character form. Over and above world and narrative, it is the ragdolls, as stitchpunks, who take on the burden of the kino-mode by embodying the stylistic confusion of 2D and 3D textures. There is a story, and it comes full circle, but the ending is a gloss on the premise built into character form—cozy textured souls versus cold slick intelligence. The kino-mode falls back on character form or design to anchor itself.

The example of *Coraline* confirms the point. *Coraline* is about a girl who journeys into a parallel world that mirrors hers in a distorted fashion, revealing uncomfortable truths, a nightmarish reflection to be confronted and resolved. This film is designated as steampunk surely due to its kino-mode, which bears some affinity to *9* in how it plays across textures and dimensions of reality. Unlike films retrofitted for 3D screening in post-production, *Coraline* used 3D techniques to enhance the strangeness of textures.

Here, too, the kino-mode of transdimensional kinemation is sustained through character form, most clearly in the button-eyes version of Coraline's family members in the other world.

The use of character form to sustain transdimensional kinemation generally holds for all of Burton's directorial efforts. The marketing and reception of his films, as well as art gallery exhibits, suggest that his characters have greater impact and staying power than his worlds. The result is a Tim Burton world instead of a steampunk world, based on characters such as Edward Scissorhands, Jack Skellington, and the Queen of Hearts. The same is true of Gilliam, albeit in a different register: the comedic ticks of his characters, their schtick, takes the place of the world-form when it comes to anchoring and extending his kino-mode. A prime example occurs in *Adventures of Baron Munchausen* when Robin Williams, as the King of the Moon, a gigantic, detached head flying on a saucer across surreal moonscapes and around tromp-l'oeil architectures, tenuously maintains this dimensional juggling with his delirious banter.

Miyazaki approaches the challenge of Zeman's speculative kino-mode from another direction. Zeman's influence is most conspicuous in *Castle in the Sky* (*Tenkū no shiro Rapyuta*, 1986), whose title sequence features a series of oddly engineered machines directly inspired by those in *Invention for Destruction*, especially flying machines featuring so many strange components that they scarcely appear sky-worthy, oversized airships whose principles of buoyancy are not apparent, and floating cities or castles in the sky. *Castle in the Sky* not only expands on Zeman's delightfully wacky designs; it also adopts Zeman's prescient post-Vernian sensibility. At the center of the film is an ancient, world-destroying technology that must be destroyed by nonmilitary means, if this postapocalyptic world is to avoid repeating the military escalation in technoscience that destroyed the prior civilization.

Of Miyazaki's animations, *Castle in the Sky* and *Nausicaa of the Valley of the Wind* (*Kaze no tani no Naushika*, 1984) feel the closest to world-building. The glimmer of a world-form arises in part because both derive from the epic fantasy world of his 1982 manga *Nausicaa of the Valley of the Wind*. Miyazaki was experimenting with Verne's worlds around the same time, for instance in his script for an anime series drawing on *20,000 Leagues under the Sea*, which became the basis for *Nadia: The Secret of Blue Water* (*Fushigi no umi no Nadia*, 1990–91). Yet *Castle in the Sky* did not turn into a Verne-inspired world-building. There is a single technological variable—a mysterious anti-gravity energy source known as the "flying stone" (*hikōseki*), but the film's premise is that the world built around this energy form now lies in ruins. The world-form is crossed out, and nature's panoramic spaces take its place, opening luminous expanses of earth and sky for action and interaction. In this respect, Miyazaki is like Burton: while his films may enter the steampunk corpus, their signature style defies a steam-tech world; the result is a Miyazaki world, one that is immediately recognizable.

Crossing out the world-form and opening a panoramic space leave Miyazaki free to explore transdimensional kinemation, to develop his signature manner of interweaving flatness and depth through movement. Working with stacks of hand-painted cels, Miyazaki produces two sensations: sliding across surfaces and dizzying yet dimensionless depth. These sensations reach their peak when action takes to the skies in scenes of banking, gliding, and circling. Such scenes reveal another manner of sustaining the kino-mode—the vehicle form. This form is implicit in the Zemanesque designs for whimsical flying machines and other eccentric technologies. It becomes explicit when simpler, low-tech vehicles, such as gliders, bicycles, or even the witch's broomstick, soar and dive. At such moments, the human being is entirely at one with the vehicle, and the result is a machine-human hybrid unlike any cyborg, for it is grace and skill, a feel for the machine, that fuse mechanism and organism. It takes only a slight push in the Studio Ghibli world to produce flying entities that feel like human-vehicle fusions—a catbus or a totoro.

After the wacky flying machines of *Castle in the Sky*, the purest expression of the steampunk vehicle form in Miyazaki's animation occurs in *Howl's Moving Castle* (*Haura no ugoku shiro*, 2004). The eponymous castle combines household and stronghold; architectural elements of each are cobbled together to form a bulbous yet angular structure supported on tiny legs on which it skitters around unevenly. The vehicle form channels a steampunk sensibility in its unsettled temporality, juxtaposing technologies old and new, but its estranged temporality is more magical than futuristic. The castle embodies a retrofantastic style. In its awkward yet agile movement, the moving castle embodies transdimensional kinemation: under conditions of movement, its architecture appears on the verge of separating into autonomous elements, as if they belonged to different dimensions of reality, magically held together.

Although steampunk typically gravitates toward retrofuturism instead of retrofantasticism, the distinction is by no means strict, and it is easy to see why *Howl's Moving Castle* has been included within steampunk cinema. But, following Zeman's kino-mode, Miyazaki's animations come to steampunk from the side of transdimensional kinemation. The evocation of a "past that never was" in films like *Howl's Moving Castle* and *Castle in the Sky* is not about building a world form to be extrapolated into a "future that might have been." Miyazaki tends more toward cautionary eco-tales. Still, it would not take much of a push to align his capricious vehicles with steampunk worlds; it would begin by doing what Miyazaki denounces: exposing and reactivating their mechanisms.

# The Adventure of Energy

As we have noted, steampunk cinema combines immiscible modes—the world mode and the kino-mode. The world mode tends toward world-building, extrapolating a single technological variable (steam power in the classic formulation) to infuse its nineteenth-century vision with speculative force. Thus, an old-tech world that never was becomes a platform for speculating on what might have been. The kino-aesthetic mode tends toward cinematic speculation, experimenting with dimensional perception under conditions of movement, reaching for dizzying experiences of an imperceptible dimension, including time. The two tendencies are polarized, representing heterogeneous orders of experience, rendering them immiscible.

Neither tendency exists in a pure form. While filmmakers such as Gilliam, Burton, and Miyazaki push toward the kino-aesthetic pole, their signature style marks an internal limit. If every gesture of their films sustained transdimensional kinemation, the result would be a purely experimental field—an interesting possibility, but not what these filmmakers aim for. They develop personalized stylistic techniques for performing the vertiginous relation between never-was and might-be, localizing and taming its wildness. For Gilliam and Burton, the deformation of character does the trick, and for Miyazaki, it is the vehicle form. For all, speculative force becomes a signature style, an auteur effect. The world mode confronts a different limit. Because the world mode relies on extrapolating a technological variable, it risks killing its world if it cannot sustain the speculative force of its technological variable. Two orientations appear. Films such as *The League of Extraordinary Gentlemen* and *Wild Wild West* draw on the conventions of action films to impart a sense of urgency to the technological variable—from countdown to showdown. Other films turn to the uncanny to lend an aura of psychological fascination to their technological variable, as in *The City of Lost Children*, but a truly speculative world cannot be merely a dark world (psychologically) or an action world (in sensory-motor terms).

Just as oil and water have some properties in common (such as liquidity and translucence), these two modes seem to have something in common, namely weird temporality—and in some instances they just feel different in degree. When the kino-aesthetic mode falls back on character form or vehicle form, it feels somehow compatible with the world mode. Conversely, the world mode seems akin to transdimensional kinemation when its worlds become darkened and agitated. Yet the two tendencies remain different in kind; they will not hold together without the application of a force. Just as oil and water must be shaken to form a colloidal suspension, so these two tendencies of steampunk are held together through the application of another force, a performative or *thematic* force.

"Thematic" is a useful term because of its multiple connotations. A thematic is theme-like, as in a musical theme or compositional theme, but it is also a problematic, in the sense of a problem that resists solution and so encourages movement in more than one direction. Raymond Ruyer uses the term "thematic" to refer to the vertical integration that arises during the process of embryogenesis (1946, 16). Like a problematic, the thematic in this sense is a sort of imperceptible principle of organization that, like memory, cannot be definitively located in space or time. In the context of genre, thematic also recalls Rick Altman's (1984) suggestion that genre is composed of syntactic and semantic elements. Like the syntactic for Altman, thematic defies simple definition. It is not narrative form, nor world form, nor film form. Still, it imparts a sense of a whole to a diverse set of semantic elements. The thematic thus shows some affinity with Altman's notion of the syntactic, but without grammatical or linguistic connotations. Instead of a set of semantic elements, steampunk cinema suggests two tendencies, sensuous orientations, meaningful but not necessarily languagelike.

The thematic of steampunk cinema hinges on *the adventure of energy*. While steam was its initial inspiration, steampunk has expanded to accommodate other forms of energy, giving rise to sub- or side-genres such as dieselpunk and clockpunk (Robb 2012, 11). A common entry on lists of steampunk films, *Sky Captain and the World of Tomorrow* (Conran 2004) is unabashedly dieselpunk. The film *Hugo* (Scorsese 2011), an adaptation of Brian Selznick's 2007 graphic book inspired by Méliès, *The Invention of Hugo Cabret*, is an example of clockpunk. Steampunk has also ventured into magical sources of energy. The number of steampunk films featuring magical energy sources is surprising. We have already noted some examples: magical talismans that power stitchpunks in *9*, the magical antigravity "flying stone" of *Castle in the Sky*, and sorcery in *Howl's Moving Castle*. Other frequently cited films with magical or quasi-magical energy sources include *Atlantis: The Lost Empire* (Trousdale/Wise 2001), *Treasure Planet* (Clements/Musker 2002), *The Golden Compass* (Weitz 2007), *The Adventurer: The Curse of the Midas Box* (Newman 2014), *The Empire of Corpses* (*Makihara Shisha no teikoku*, 2015), and *Abigail* (Boguslavsky 2019).

Different energy sources are possible in steampunk because the adventure of energy matters more than the literal source of energy. Energy does not enable adventure if it does not sustain its speculative force, so energy must be defamiliarized and reimagined. While steam power offered an expedient, dramatic source for speculation, the gradual turn toward magical sources speaks to the contemporary situation, where energy has become disenchanted. Either energy is taken for granted and ignored, or it is taken to task for the harm its extraction and consumption inflict on the planet. Energy is thus both criticized and disavowed. The steampunk adventure, however, avows energy. Avowal does not mean acceptance of the status quo or capitulation to it. The adventure of energy is complex in its implications, due to the complexity of contemporary attitudes toward energy. It is never without risk, for energy has never been a simple solution to a complex problem. For steampunk, then, energy is a problematic to be dramatized thematically. Steampunk incessantly returns to the birth of modern energy but not simply to reenchant it. This adventure is fraught with peril. Philippe Pignarre and Isabelle Stengers's observations about the "sorcery" of capitalism provide insight into the turn toward magic in steampunk: magic is a way of understanding how energy affects us, how it makes us vulnerable (2011, 42–45).

Plasticity extends to steampunk's historical setting. The usual focus for steampunk is the era of high empire—Victorian England and the French Belle Époque. Yet steampunk not only slips sideways in coeval imperial formations, such as Japan of the Meiji (1868–1912) and Taishō eras (1912–25); it also skips headlong into the interbellum or interwar period (1918–39). The actual period may be defined loosely, or periods may be combined. The Russian film *Abigail* provides a prime example of such historical plasticity. A steampunk fantasy adventure, it evokes steampunk through its clockwork mechanisms, airships, fashions, and other design features, making it feel like a nineteenth-century world extended into an alternative 1930s. But its source of power is a magical force arising in certain human beings. The story opens in a walled city where a totalitarian government is responding to an epidemic by seizing any humans who

STEAMPUNK CINEMA 157

show signs of illness. However, there is no epidemic: the government is imprisoning those who show signs of magical powers in order to eradicate them. When the heroine, Abigail, discovers her magical abilities, she learns that, even as the government strives to eradicate individual use of power, it has technologized magic to create a power source.

As the attention lavished on steampunkish fashions, airships, and clockwork mechanisms confirms, *Abigail* approaches steampunk from the side of world-building. But its technological variable is magic, which provides a compelling and expedient way to amplify the speculative force needed to sustain the world form. Like *The City of Lost Children*, *Abigail* begins with a dark city and a traumatic secret. But what initially appears to be a psychological problem of individuals and families turns out to be a political problem of governance: the manipulation of contagion to regulate populations and monopolize energy. As magic abilities replace psychological states, *Abigail* extends its world beyond the use of trauma that led to an individualizing psychological closure in *The City of Lost Children*. In its place another problematic appears: what kind of relation between "spirit" and "mechanism" is possible?

The term "spirit" here refers broadly to the nonpsychologized and nonindividualized mind, while "mechanism," equally vast in meanings, is often contrasted with organism and associated in the Cartesian worldview with a nonminded operation, something entirely artificial. Steampunk thematizes this spirit-mechanism relation, or to be precise, this relation thematizes steampunk by repeating and performing the problematic of immiscibility. The adventure of energy does not resolve the immiscible relation between world mode and kino-mode, but vigorously shakes them together, generating patterns of dispersed yet coalescing beads of spirit, while mechanisms move from the background into the foreground, intrusive and pervasive. Its adventure consists of stirring together spirits and mechanisms to find colloidal patterns worthy of prolonged exploration or observation.

The steampunk fascination with exposed mechanisms, especially clockwork mechanisms, confirms this point. A clockwork may seem to be an example of the principle of efficiency and optimization, but in steampunk it is an old technology that requires humans to wind and to recalibrate it. Its principle is one of satisfying a certain condition, performing certain tasks to keep things running in a satisfactory manner, that is "good enough." But when clockwork becomes too efficient, human bodies must intervene performatively: gracefully dancing or acrobatically dashing through the gears and levers, or even clogging the works. In the process, clockwork becomes an adventure in energy. This effect might explain why Chaplin's *Modern Times* (1936) occasionally crops up on steampunk lists.

The film *Hugo* might come as a revelation in this respect. While *Abigail* enters the adventure of energy from the world mode side, *Hugo* approaches from the kino-aesthetic side. The combination of early trick cinema and automatons provides an opportunity for director Martin Scorsese to embark on an adventure in human-spirit energy. *Hugo* centers on winding clocks, cranking movie projectors, and activating automatons. What interests him cinematically is the performative coexistence of human actors and technologies: to wind a clock is also to dwell in a clock, to crank a projector is to live in a movie, and to activate an automaton is to become its assistant. In the world of Méliès Scorsese finds a domain in which spirit and mechanism are performatively stirred together to form a variety of spiritual automatons. To convey this performance, he flirts with transdimensional kinemation, restaging the Lumières' famous *Arrival of a Train at La Ciotat* (1895) as an action sequence wherein a steam locomotive crashes through a Parisian train station. But as his side-by-side evocation of Méliès and the Lumières attests, Scorsese's style moves from transdimensional kinemation toward an homage to early cinema—an adventure in *cinematic* energy.

The other orientation of the world form, toward the action film, also undergoes a transformation in the 2000s through the adventure of energy. The films *Sherlock Holmes* (Ritchie 2009) and *Sherlock Holmes: Game of Shadows* (Ritchie 2011) might be characterized as period action films, since they involve big-budget fight sequences, chases, explosions, and visual effects, done in the predictable manner of countdown and showdown, and set in Victorian London. The films feature period technologies, ranging from steam engines and steamships to emerging forms like the gasoline-powered automobile and weapons such as machine guns and tanks, although steam power is not thematized, nor are energy sources. What contributes most to the steampunk flavor of the films are the unabashedly punkish attitudes and fashions of Holmes and Watson, who feel entirely contemporary, riffing on their Victorian context, as if playing in a past that never was.

The might-be of steampunk temporality occurs in action sequences in which Holmes's deductive reasoning plays forward. Like a chess player or martial artist who strives to envision his opponent's counterattacks several moves ahead, Holmes logically thinks through how the fight might go, seeking the best possible outcome. This process affects the temporality of cinematic action, for the fight is first depicted in a punctuated, stop-motion series of strikes and pauses, but then played through in real time. Deductive reasoning turns into speculative thinking, with Holmes's body becoming the speculative technology—a computational body that lives the abstraction of what might be or, if he miscalculates, what might have been.

Films like *The League of Extraordinary Gentlemen* and *Wild Wild West* anticipate this paradigm yet misfire because they try to build countdown-to-showdown into their technological variable. The *Sherlock Holmes* films have proved more successful because they are not extrapolating an old-tech variable in order to build a world. They introduce a different temporality into the action by placing showdown prior to countdown. This weird temporality is supported on two sides, by the punkishly contemporary stance of Holmes and Watson, and by the narrative form in which the pair prevent the future (say, World War I) from happening too soon. Holmes embodies the adventure of energy, as his *speculative-deductive* action sequences shake up steampunk's world mode and kino-mode.

The steampunk thematic is evident in Stephen Fung's *Taiji* series, *Tai Chi Zero* (*Tai ji 1: Cong ling kai shi*, 2012), *Tai ji 2: The Hero Rises* (*Tai ji 2: Ying xiong jue qi*, 2012), and a projected third film *Tai Chi 3: Dian feng zai wang* (*Tai Chi Summit*). This series mixes history and fiction to produce a nineteenth-century China that never was. It retells the story of Chen Changxing (1771–1853) adopting an outsider, Lu Chan (1799–1872), into the Chen lineage of *taijiquan* martial arts. Its version of Chen's village is stylized and idealized to the point of the magical, which intensifies the aura of nostalgia for a past that never was. The story of adopting an outsider is indicative of a larger crisis in traditions at a moment when modernization was coming to Chinese society forcibly, from without. The *Taiji* series sets up a predictable and somewhat tendentious set of oppositions: inside versus outside, tradition versus modernity, and China versus the West. Initially, steampunk elements appear to reinforce these oppositions, for steampunk mechanisms are associated entirely with militarist technologies and Western imperial powers, whose invasion of China gradually comes to center on the village.

What ensues is a battle between two ways of using energy. On one side is the spiritual intensification of bodily powers through *taijiquan*. On the other side are foreigner invaders who rely on mechanical amplification of bodily capacities through steam-powered technologies (see figure 10.3). The *Taiji* series revels in escalating this confrontation, as steam-powered weaponry becomes increasingly massive, while *taijiquan* training increases in spiritual and physical intensity. Visual effects and choreography also grow increasingly elaborate and outlandish. But then, what initially appears to be an insuperable divide, an irreparable opposition, turns into a contrast between energy and power, between *qi* (cosmological body energies) and *zhengqi* (steam power). The *Taiji* series thus shifts from combat to an adventure of energy.

This adventure emerges through a series of narrative detours. The defector, Fang Zi Ping, is treated as an outsider by the villagers due to his deficient skills in *taijiquan*, yet he is engaged to marry Chen's daughter, Yu Niang. He exemplifies the "bad son" and "false legacy," and initially sides with the Western powers in their attack on the village. Later, somewhat predictably, he sides with the village, assisting the "good son" and "true legacy." This detour passes through a steam-powered flying machine wherein *qi* energy and *zhengqi* power mesh perfectly. Another detour goes through Beijing where Lu Chan

**FIGURE 10.3**: Yu Niang defies giant Western steam-powered machinery (*Tai Chi Zero*, 2012). Huayi Brothers Media.

and Chen Yu Niang encounter a *taiji* master who fashions clocks. Such adventures of energy serve to disaggregate the monolithic unities congealed in the escalating conflict of China versus the West and tradition versus modernity. Disaggregation is achieved through a performative remixing of seemingly oppositional stances. Immiscible though they may be, they are shaken together to form a colloidal suspension, allowing new relations between tradition and modernity.

The use of graphic effects plays an integral role in this performative suspension, this adventurous "solution." For example, graphic effects are used in action sequences and transitions to help build a bridge between the fast-paced martial arts choreography and the oversized, yet intricate mechanisms featured in the action. They do not cancel out differences but rather introduce an adventure in graphism: they speculatively reactivate one of the oldest technologies—writing in its diverse forms. This adventurous "solution" is open to different interpretations. The *Taiji* series seems to envision, in the nostalgic tone of nationalism, a China that successfully confronts and assimilates Western modernity—a superior graphic regime. Yet when construed in the terms set up by steampunk, the films offer a performative suspension that creates a space that is both Chinese and Western, but also neither. The *Taiji* series thus affirms the gift of the steampunk film genre: it invites a performative reckoning with the relation between world and moving images through the adventure of energy, as the admixture of never-was and might-be unsettles not only received historical paradigms, but also the geopolitical boundaries that might be consolidated in them.

## WORKS CITED

Altman, Rick. "A Semantic/Syntactic Approach to Film Genre." *Cinema Journal* 23, no. 3 (1984): 6–18.

Blaylock, James P. "Foreword: My Steampunk Mind." In *Steampunk: An Illustrated History of Fantastical Fiction, Fanciful Film and Other Victorian Visions*, edited by Brian J. Robb, 6–7. London: Aurum, 2012.

Chow, Rey. *The Age of the World Target: Self-Referentiality in War, Theory, and Comparative Work*. Durham, NC: Duke UP, 2006.

Gunning, Tom. "The Cinema of Attractions: Early Film, Its Spectator and the Avant-Garde." In *Early Cinema: Space, Frame, Narrative*, edited by Thomas Elsaesser and Adam Barker, 56–62. London: BFI, 1990.

Heidegger, Martin. "The Age of the World Picture." In *Science and the Quest for Reality*, edited by Alfred I. Tauber, 70–88. New York: NYU Press, 1997.

Lim, Bliss Cua. *Translating Time: Cinema, the Fantastic, Temporal Critique*. Durham, NC: Duke UP, 2009.

Moine, Raphaëlle. *Cinema Genre*. London: Blackwell, 2008.

Neale, Steve. "Questions of Genre." In *Film and Theory: An Anthology*, edited by Robert Stam and Toby Miller, 157–78. Blackwell, 2000.

Pignarre, Philippe, and Isabelle Stengers. *Capitalist Sorcery: Breaking the Spell*. Trans. Andrew Goffey. London: Palgrave Macmillan, 2011.

Robb, Brian J. *Steampunk: An Illustrated History of Fantastical Fiction, Fanciful Film and Other Victorian Visions*. London: Aurum, 2012.

Ruyer, Raymond. *Éléments de psychobiologie*. Paris: Presses Universitaires de France, 1946.

Staiger, Janet. "Hybrid or Inbred: The Purity Hypothesis and Hollywood Genre History." *Film Criticism* 22, no. 1 (1997): 5–20.

Suvin, Darko. "On the Poetics of the Science Fiction Genre." *College English* 34, no. 3 (1972): 372–82.

Tibbetts, John C. "'Fulminations and Fulgurators': Jules Verne, Karel Zeman, and Steampunk Culture." In *Steaming into a Victorian Future: A Steampunk Anthology*, edited by Julie Anne Taddeo and Cynthia J. Miller, 125–43. Lanham, MD: Scarecrow Press, 2013.

# CHAPTER 11

## SUPERHERO SCIENCE FICTION

### ANGELA NDALIANIS

FROM its early twentieth-century beginnings, especially marked by the 1938 arrival on Earth of the alien who would become Superman, the superhero story has shared many conventions with science fiction (sf). Superhero stories across media have created narrative realms riddled with traditional sf themes and conventions: time travel, parallel universes, aliens, the scientific enhancement of humans, interstellar travel, robots and cyborgs, advanced artificial intelligence, and many more. As Corey Creekmur observes, superhero narratives almost invariably "show a family resemblance to sf," and "while some superheroes are explicitly magical, many are aliens (Superman, the Martian Manhunter), the products of scientific accidents (Spiderman, the Fantastic Four, the Incredible Hulk), or genetic mutations (the X-Men)" (2004, 283–84). Superhero scholars agree that the superhero story is a hybrid form and one prone to mutation, often subsuming multiple genres into its structure, while managing to retain its distinctively "superhero" character.

This chapter examines some of the reasons for this hybridity, particularly the superhero narrative's intersections with sf that have partly resulted from the pulp heroes and sf stories that influenced the form early on. While multiple generic sources may be seen as predecessors to the form (see Gavaler 2015), my primary focus here is the pulp tradition and sf magazines, which were a shared inspiration for the creators of superheroes and sf alike. After briefly outlining the origins of the superhero story and its key conventions, I examine the type's relationship with sf as its key generic contributor. The main focus will be on Marvel superheroes and, in particular, those of the Marvel Cinematic Universe (MCU) with close attention to what might be seen as its foundational entry, *Iron Man* (Favreau 2008). While DC Comics also include sf themes, stories, and characters and similarly ask us "to imagine a world where advanced science, UFOs, aliens, space exploration, time travel and high-tech gadgets are common occurrences" (Nama 2009, 133), it is the Marvel superheroes—especially those of the 1950s and beyond—that have developed a strong connection with traditional sf themes and contemporary sf cinema, such as the blockbuster films of the MCU. As I show, *Iron Man* the film (like its comic book versions) sets itself firmly within the parameters of the superhero form, but it does so by "absorbing and reworking" sf into its superhero structure (Hatfield 2013, 136).

The final section of this chapter analyzes *Iron Man* from the vantage provided by Istvan Csicsery-Ronay's *The Seven Beauties of Science Fiction* (2008). There he defines what he considers to be the genre's seven core characteristics, or "beauties"—characteristics that clearly align *Iron Man* and, by extension, the superhero story, with sf. But in examining these alignments, we shall also see how the superhero form injects a note of difference into Csicsery-Ronay's model. Given *Iron Man*'s reliance on superhero conventions, its status as "pure" sf—at least from Csicsery-Ronay's vantage—becomes complicated and suggests that we might need to extend his schema to better understand how superhero-sf works as part of the cultural and cinematic imaginary. Contemporary superhero-sf, just like the larger genre of sf, is in dialogue with issues about the morality, ethics, and trajectory of science and technology—a dialogue played out not primarily through technology but through the body of the superhero.

The superhero form is as fluid and mobile in the definition of its boundaries as is sf. Henry Jenkins has argued that for these stories genre hybridity is, in fact, the norm (2009, 17), and that, from its beginnings, the superhero form "emerged through mixing, matching, and mutating genre categories" with sf and horror the primary influences (26). The early standardization of the superhero story occurred in comic book format in the late 1930s (see Superman's 1938 appearance in *Action* #1), resulting in the formation of a narrative type that has continued to draw upon a variety of genres. Thus, Jenkins notes that "masked heroes from the pulp magazines, including the Shadow, the Phantom, the Spider, and Zorro, modelled the capes and masks iconography and the secret identity [that have become] thematic" (27) for superhero narratives.

Yet while the superhero narrative is notable for its hybridity, some of its characteristics appear so frequently that it also seems quite stable in terms of character types and stories. Peter Coogan in *Superhero: The Secret Origin of a Genre* describes this figure as

> a heroic character with a selfless, pro-social mission; with superpowers—advanced abilities, extraordinary technology, or highly developed physical, mental, or mystical skills; who has a superhero identity embodied in a codename and iconic costume, which typically express his biography, character, powers, or origin (transformation from ordinary person to superhero); and who . . . can be distinguished from characters of related genres . . . by a preponderance of generic conventions. Often superheroes have dual identities, the ordinary one of which is usually a closely guarded secret. (2006, 30)

Coogan further notes that the superhero's mission is selfless and equates with the "mores of society and must not be intended to benefit or further his own agenda" (2006, 31); these figures possess powers that are far greater than the hero of other adventure or action genres; and their identity is often but not always secret, forcing them to adopt a persona that is coded through costume (including mask), a heroic codename, and an origin story that leads to the formation of the superhero identity. Furthermore, the superhero's "creation" usually results from a trauma—murdered parent/s, an accident with radioactivity, bodies used for scientific experimentation, even a deal with the devil (which extends the story's possible supernatural/horror link).

The *Iron Man* narrative incorporates all of these characteristics. The original comic book version was published in *Tales of Suspense* #39 in 1963 with the character receiving his own title a few years later in *Iron Man* #1 (1968). Tony Stark is first introduced as a scientific genius who has invented a "tiny transistor" that boosts the power of magnets—an invention developed for the military. In the film version, the character's story and its context have been updated to mirror more contemporary cultural issues and technological inventions. Thus the film begins in Afghanistan where Stark—under the protection of a military convoy—unveils Stark Industries' new invention, the Jericho missile, which uses his newly invented arc reactor. After an explosive display of Jericho's destructive capacity, there is much backslapping and gloating—by Stark himself—about the weapon's success. However, the celebration turns sour when terrorists attack his convoy, seriously wound Stark, and take him captive. Waking, he finds that a fellow captive, Professor Yinsen, has saved his life by fashioning a battery-powered magnet to stop the shrapnel in Stark's chest from entering his heart. Pretending to build a Jericho missile for the terrorists, Stark instead creates a miniature arc reactor to power his chest magnet and adds to this a suit of armor that becomes the Mark I Iron Man suit. As Iron Man, Stark escapes and later holds a press conference to announce that his company will no longer manufacture weapons for the military or other organizations. Like other Marvel heroes of the 1960s and 1970s, Iron Man's superhero origins are thus intertwined with an sf concept—the technological enhancement of the human body. While Stark is both the object of scientific innovation and the scientist who innovates, his "mighty electronic body" suggests that "the armor is, in effect, an extension of his body" (Patton 2015, 9), as he becomes a superhero who is essentially a cyborg—the stuff of sf.

In short, we have: the birth of the superhero who, following a traumatic event, has greater powers than an average hero; the establishment of a new identity as Iron Man (refined further with the technical enhancements Stark later adds to the armored suit); Stark's realization that he has failed society in creating the Jericho warhead, which he corrects by ceasing to manufacture weapons and taking an active role in stopping their misuse; a traumatic event that makes him adopt the role of Iron Man; and, finally, a secret identity, which he unveils to the media at the film's end. In this final move, Stark proves to be one of the early Marvel superheroes—like the Fantastic Four—who are not that secretive about their superhero identity.

In examining what is often seen as the ground zero for the superhero, Richard Reynolds begins with the "original" appearance of Superman in comic book form in *Action Comics* #1 in 1938 (soon followed by Batman in 1939, Wonder Woman in 1940, and an array of others in the late 1930s and 1940s). This "first" superhero, Reynolds explains, was "conceived by the teenage Jerry Siegel as early as 1934" (102), with his character and backstory drawing heavily on multiple generic influences, and with his arrival creating "a wholly new genre out of a very diverse set of materials" (103). However, the one element that stands out is sf. Superman (Kal-El) is from the dying but scientifically advanced planet Krypton and, as a baby, he is placed into a spaceship by his scientist father and, to save his life, is propelled into space to eventually crash-land on Earth where he is raised by the Kents as their son, Clark Kent. From this first appearance, Reynolds notes, a " 'scientific' explanation of Clark's superhuman abilities" is offered, "comparing his strength with the proportionate strength of ants and grasshoppers" (2003, 102–3).

Hatfield, Heer, and Worcester pose an appropriate question about this hybrid origin, asking if the superhero is an "offshoot of science fiction or a meta-genre that imaginatively fuses together material from a variety of literary traditions?" (2013, xviii). But there is no easy answer to this question, since there are many instances where sf and the superhero influences converge, particularly in pulp magazines of the 1920s and 1930s. The origins of sf (or a proto–science fiction) have been traced back to Mary Shelley's *Frankenstein, or, the Modern Prometheus* (1818) and the later "scientific romances" and "exotic adventures" created by figures such as H. G. Wells, Jules Verne, and Edgar Rice Burroughs, although Wells is most often labeled "the father" of sf. However, during the early twentieth century the pulp magazines heavily impacted the formation and solidification of sf as a genre. As Brian Attebery explains, "Magazines such as *Astounding Science Fiction* were chiefly responsible for creating a sense of sf as a distinctive genre" (2003, 32), with Hugo Gernsback's *Amazing Stories* (1926) as the first magazine "not only to limit its fictional contents to stories of scientific extrapolation and outer-space adventure but also to attempt to define the genre which the editor initially called 'scientifiction,' but began to refer to as 'science fiction' by 1929" (Attebery 2003, 33; see also Luckhurst 2005, 14–15). Because of this publishing influence, Gernsback too has been called "the father" of the genre. Pulp magazines like *Science Wonder Stories, Air Wonder Stories,* and *Science Wonder Quarterly,* all of which appeared in 1929, were soon followed by *Astounding Stories* (1930), *Thrilling Wonder Stories* (1936), *Marvel Science Stories* (1938), *Startling Stories* (1939), and *Astonishing Stories* (1940), all of them disseminating the new fiction that would come to be labeled sf, while also publishing older tales—by Wells, Verne, Edgar Allan Poe, and others—that would be rechristened as sf.

Many of those stories appearing in the sf and pulp magazines featured larger-than-life heroes like Buck Rogers and Flash Gordon. Flash Gordon first appeared in comic strip form in 1934 but was adapted for the pulp *Flash Gordon Strange Adventure Magazine* in 1936, with the Universal serial based on his adventures appearing in the same year (with sequels in 1938 and 1940). Buck Rogers followed a similar trajectory, appearing first in the pulp *Amazing Stories* in 1928, as a comic strip the following year, and as a serial in 1939. Edgar Rice Burroughs's time traveler John Carter of Mars also migrated to the pulp magazine *The All-Story* as early as 1912. The market was dominated by these sf adventurers, along with the masked men of fantasy and detective pulps, including such figures as Zorro, Captain Future, Doc Savage, the Shadow, and the Spider. These larger-than-life heroes quickly became part of the cultural imaginary.

Many of the early writers, artists, and editors of superhero comics were also writers, artists, editors, and fans of the emerging sf genre. Jack Kirby, Jerry Siegel, and Stan Lee, for example, were all sf fans who were associated with the broader pulp tradition.

Jerry Siegel (co-creator of Superman), began by writing sf stories that he published in his own magazine, *Science Fiction: The Advance Guard of Future Civilization* (Gresh and Weinberg 2002, xiv–xv). The first articulation of Superman, "The Reign of the Superman," was published as a short story with illustrations by Joe Schuster in this magazine in 1933. This version of the superman, however, was a telepath named Bill Dunn who was given an experimental drug by a mad scientist. Far from being a selfless superhero, he becomes a supervillain who finally loses his powers when the drug wears off. As another example of this cross-breeding of superhero and sf, Gerard Jones points to the example of *The Girl from Mars*, a novel by Jack Williamson that was published by Gernsback in 1929. The plot "explained the superbeing in a science-fictional way . . . and it featured a strange visitor from another planet with powers far beyond those of normal men" (Jones 2013, 18).

The economics of both pulp and comics production was such that writers worked within multiple genre traditions. As Henry Jenkins (2009) explains, in early comics a writer like Jack Kirby

> might produce work across the full range of pulp genres in the course of his or her career and thus would be able to draw on multiple genre models in their superhero work. The intensity of comics production—new stories about the same characters every month and, in some cases, every week—encouraged writers to search far and wide for new plots or compelling new elements, while the openness of comics . . . made it cheap and simple to expand the genre repertoire. (27–28)

Marvel, which began as Timely Comics in 1939, was created by Martin Goodman, who began his career as a pulp magazine publisher. Timely's first issue in 1939, *Marvel Comics* #1, included the Human Torch, an android superhero inspired by sf. In 1941, another sf-linked figure was born, when the frail Steve Rogers was given an experimental serum that transformed him into a super soldier who became known as Captain America (*Captain America Comics* #1). In the 1950s, tapping into the era's fascination with both "science fiction fantasies and creature features," Goodman instructed his editor Stan Lee to create "a series of fantasy anthologies such as Journey into Mystery, Tales to Astonish, and Tales of Suspense—all of which would linger well into the self-proclaimed Marvel Age of Comics, long after the bug-eyed monsters and presumptuous scientists had given way to Marvel's new generation of superheroes" (Darowski 2015, 6).

According to Peter Coogan (2006), superpowers and identity, in conjunction with the mission, are the main differentiators "between Superman [and fellow superheroes] and his pulp and science fiction predecessors" (31–32). While there is a clear distinction between the pulp mystery men and most sf heroes (e.g., Doc Savage, Buck Rogers, Flash Gordon) and the superheroes who followed them, in the case of some superheroes—especially those created by Marvel—the character results from an act that itself could be considered sf. Thus Marvel brought together sf and the superhero figure in ways that more thoroughly embedded, transformed, and aligned sf with the generic requirements of the superhero story. Not only were the new superhero stories set against the backdrop of sf stories and their generic themes, but the super bodies of these figures were typically products of a specifically scientific circumstance: cosmic rays (the Fantastic Four), an exploding gamma bomb (the Hulk), radioactive blood (She-Hulk), experimental super-soldier serum (Captain America), radioactive spider bite (Spider Man), the chemical substance "Pym Particles" (Ant-Man), genetic mutation aided by scientific experimentation (the X-Men), blood fusion with alien blood / power of the cosmic stone known as the Tesseract (Captain Marvel), and so on. And as we have noted, in the case of Tony Stark, his traumatized body is transformed and reborn through his own scientific genius into Iron Man, so like most Marvel heroes of the 1960s and 1970s, his superhero origins are intertwined with an sf concept, with Stark both the object of scientific innovation, and the scientist who innovates.

As noted earlier, Csicsery-Ronay's schema of sf identifies a set of major concepts that mark the genre. By considering *Iron Man* and its place in this schema, we can see not only how the film sits within sf, but also how that vantage might be adjusted to allow for the hybridity found in a film like *Iron Man*. Csicsery-Ronay has suggested that sf texts share "a desire to imagine a collective future for the human species and the world" (2008, 1). Despite the difficulty in pinning down the "meaning" of the genre, he argues that sf typically engages in a dialogue between "conceiving of the plausibility of historically unforeseeable innovations in human experience (novums) and their broader ethical and social-cultural implications and resonances" (3). Thus Csicsery-Ronay sees sf placing possible "technoscientific conceptions" within a narrative world that examines the social, ethical, and moral implications of their realization. He states that "the resulting fictions may be credible projections of present trends or fantastic images of imagined impossibilities. Usually, they are amalgams of both" (3). Moreover, he emphasizes that "because sf is concerned mainly with the role of science and technology in defining human cultural value, there can be as many kinds of sf as there are theories of technoscientific culture" (4). This multiplicity includes diverse media—novels, pulp stories, films, television shows, videogames—and diverse approaches, subgenres, or themes such as space operas, alien invasion, cyberpunk, parallel universe, alternate realities, time travel, and so on.

Within this diversity, we might consider the superhero form—as seen in *Iron Man* and other MCU films—as complicating Csicsery-Ronay's model by linking various sf themes to the key trope at the center of the superhero tale: the superhero's quite different body. Drawing on the work of Darko Suvin, Csicsery-Ronay argues that sf should be considered mainly as a "mode of thought" that succumbs to what he calls "the seven beauties" of sf—categories that intellectually address the "restless technological transformations" that are "active in an age" (2008, 4). Those "seven beauties" are fictive neology, fictive novums, future history, imaginary science, the science-fictional sublime, the science-fictional grotesque, and the Technologiade. These seven beauties can all be identified in *Iron Man*, underscoring its strong connections to sf; however, the superhero's conventions also force the sf concepts into what we might think of as a superhero structure, or superhero-sf. Below I outline the characteristics of each beauty and consider their significance in relation to superheroes, *Iron Man*, and the MCU.

# Fictive Neologies and Fictive Novums

According to Csicsery-Ronay, in sf—as with all fantasy genres—we "encounter new words and other signs that indicate worlds changed from their own" (2008, 13). The difference compared to other fantasy types, however, is that, through various such fictive neologies, sf constructs the logic of its "science-fictional worlds" that are "symbolic . . . of radical newness"; as "fictive novi," or signs of the new, these elements "connote newness and innovation vis-à-vis the historical present of the reader's culture" that are embedded in a new discourse around them (13). Thus *Iron Man*, as the first film in the MCU series, must set up the conditions for this discourse, which is then further extended and expanded in the shared universe of MCU films.

Some of these neologies are familiar but projected into the future, while others are new and become emblematic of the future. For example, AI (artificial intelligence) is a term we are familiar with today, but in *Iron Man* it has advanced dramatically, and we see it represented in the neology that is J.A.R.V.I.S. (Just A Rather Very Intelligent System), Stark's AI assistant. J.A.R.V.I.S. controls most of the technologies in Stark's home and lab—the robot arm, the computers, and even the Iron Man suit. In *Avengers: Age of Ultron* (2015), Ultron—the supervillain robot who is also a fictive neology—destroys J.A.R.V.I.S. and uploads his intelligence onto the internet so he can access information. He also has an android body made out of indestructible synthetic vibranium into which he plans to transfer his own consciousness, but Stark and Bruce Banner manage to download J.A.R.V.I.S. instead into the android body, thus creating the new character Vision. In the context of this character appropriation, we might recall Csicsery-Ronay's explanation that fictive neologies are "double-coded" since they represent "cultural collisions between the usage of words familiar in the present . . . and the imaginary, altered linguistic future asserted by the neology" (2008, 19); so while the words may be familiar to us, they can be "appropriated by imaginary new social conditions to mean something new" (19). In *Iron Man*, variations and the synthesis of artificial intelligence with androids, cyborgs, and robots—all familiar to us today—can generate powerful sf conceptions of future artificial life.

Other examples found in the MCU include the arc reactor created by Tony Stark to keep him alive; the Tesseract, which houses the power stones; the substance "vibranium," from the fictional African nation Wakanda, that is known for its capacity to store and release kinetic energy, and for its indestructible nature (see Black Panther's suit and Captain America's shield); Hydra, the terrorist organization that relies on a never-ending supply of mad scientists to create new weapon technologies; S.H.I.E.L.D. (Supreme Headquarters, International Espionage and Law-Enforcement Division), the "good guys" who have their own never-ending supply of scientists working for the greater good; and Stark Industries, a hi-tech corporation that manufactures advanced weapons and military technologies.

To these fictive neologies found in the MCU, we should also add the superheroes themselves, many of whom are products of advanced science or sf concepts, such as

Iron Man, the Hulk (exposure to gamma radiation), Rocket (a cybernetically modified racoon), Vision (an android), Ant-Man (shrinking technology), Captain America (supersoldier serum), Groot (a flora colossus from Planet X), and Thor (an alien prince). Thus the named superhero also becomes a fictive neology. The superhero's inclusion within this sf structure points not only to the conventions and themes shared across both "genres," but also the superhero form's point of departure. As we will see with the other beauties, the superhero body becomes central to combining and delivering sf and superhero themes, resulting in the amalgam superhero-sf. The scientific or technological tropes familiar to sf—cyborgs, gamma radiation, shrinking technology—can thus be examined through the superhero fictive neology.

The second category, the fictive novum or new thing, is closely aligned with the fictive neology. While the neology introduces a new word that represents a different object or thing in the imaginary, sf world, the fictive novum embodies the changes that can be brought about. Csicsery-Ronay reminds us that sf can "provide imaginary models of radical transformations of human history initiated by fictive novums," with the novum denoting "a historically unprecedented and unpredicted 'new thing' that intervenes in the routine course of social life and changes the trajectory of history" (2008, 5–6). The novums of sf are "radically new inventions, discoveries, or social relations," but bound by conventional scientific logic (47). He further explains that the novum "is a narratological mega trope, a figural device that so 'dominates' ([Darko] Suvin's term) its fiction, that every significant aspect of the narrative's meaning can be derived from it: the estranged conditions caused by a radically new thing, the thematic unity of the work, and even changes in readers' attitudes toward their own world after reading" (49). For example, within the MCU the Tesseract functions as a fictive novum. As a discovered object of alien making, its very existence threatens the survival of the universe and, as the MCU film sequels progress, the narratives revolve around the Avengers capturing the supervillain Thanos who plans to use the Tesseract for mass destruction. When he uses his Infinity gauntlet to activate the Tesseract's Infinity Stones, half the population in the universe is wiped out of existence, producing a radical change in "the trajectory of history." In this respect, the Tesseract functions within the parameters of Csicsery-Ronay's schema, but as with the fictive neology, the body of the superhero expands upon that schema.

Iron Man too is a fictive novum, the "narratological mega trope" that represents a specifically superhero-sf articulation of a fictive novum. His trauma—nearly losing his life but also realizing his failings in creating weapons of mass destruction (a report calls him the Merchant of Death)—leads him to cease weapon production and adapt Stark Industries' most innovative technologies. He decides to create a mini reactor and body armor suit that encases him and that he connects to his AI assistant J.A.R.V.I.S. Scientific advancement and the invention of new technologies thus create a new, mechanized man who has super strength, can fly, has weapons embedded in his suit, and can connect his thoughts and actions to computers and his AI. In fact, in his more recent incarnations, using nanotechnology, Stark implants the suit into himself and controls it with his thoughts. In this way Iron Man projects into the future the possibilities and fears about our current advances in human mechanization, with his body exploring the ethical and social questions about the transhuman body and encapsulating the film's sf themes.

# FUTURE HISTORY AND IMAGINARY SCIENCE

As noted earlier, Csicsery-Ronay believes that sf places possible "technoscientific conceptions" within its story world in order to critically engage with their social, ethical, and moral implications (2008, 3). In *Iron Man* those social, ethical, and moral issues cluster around the body of Iron Man as it embodies the technologically advanced weaponry produced by Stark Industries, thereby embodying a superhero-sf logic. Stark's moral dilemma, along with his realization that his company's weapons are being used by terrorists and have destroyed countless innocent lives, is a turning point in the film that is marked by Stark becoming Iron Man, sacrificing the personal for the benefit of the social. In doing so, he not only fulfills one of the characteristics expected of the superhero, but he exemplifies a third beauty, future history. As Csicsery-Ronay notes, sf "constructs micromyths of the historical process" (75), often by positing imaginary science, a fourth category. Iron Man as novum is what initiates the film's questions regarding future history and imaginary science. In fact, future history is here represented by the imaginary science that is the Iron Man—"a history transforming novum" (76). Central to this relationship is the concept of "techno-evolution" (93), seen in Stark's re-creating himself as Iron Man, thereby tampering with evolution by advancing his own human potential. The ethical question that arises is whether he will allow further evolution (and duplication) of the armor-suit to benefit the military, but he becomes the film's moral core by refusing that option. However, Obadiah Stane becomes Tony Stark's dark double. He represents a future history that can potentially threaten humanity through the science that has been imagined: cyborg technology in the form of Iron Monger—Stane's revised version of Stark's technology. While Stark's motivation is selfless, Stane's is ego and financially driven.

While Scott Jeffrey views such transhuman mergings as an example of "how humanity might be improved though technology," he also sees Iron Man as forecasting the Military-Industrial Body where the body-as-machine serves the needs of military and political masters (2016, 18). The future history of this imaginary science is thus signaled as having a potentially negative impact on humanity—a concern that the film builds into its structure through the superhero/supervillain confrontation, something specific to the superhero tale. Following his brush with death, Stark announces that he will no longer allow his company to produce hi-tech weaponry. Rather than presenting a posthuman questioning of the mechanized—which Obadiah Stane does through his misuse and abuse of robotics and AI—Iron Man represents its idealized double: the perfect human as superhero. The symbolic dilemma of what future technology can hold for humanity is thus played out in specifically superhero terms: the battle between superhero and supervillain that resolves the ethical dilemma.

## The SF Sublime and the SF Grotesque

The science-fictional sublime and science-fictional grotesque—the fifth and sixth sf beauties—are flip sides of the same coin, as the sublime invites astonishment and amazement, while the grotesque engages the audience with horror. Both evoke states of wonder and heightened emotion, but their affective response to the object inducing wonder differs. Csicsery-Ronay argues that sf, of all genres, is most likely "to evoke the experience of the sublime" (2008, 7), particularly what has been described as the American technological sublime: the human ability to master and control nature through techno-scientific creation. This effect is embodied in Stark's creating a new human through his mastery of science and technology, with specific sublime moments occurring with the unveilings of the Mark I, II, and III Iron Man suits. Mark I is revealed in the terrorist camp at the film's start. As Stark emerges from the cave in his chunky metal suit and defeats the terrorists, the audience marvels not only at the technological wonder he has created, but also at the birth of this "new man." We witness the unveiling of Mark II in Stark's lab, when he tests a sleeker and more agile, chrome-colored version of the suit (figure 11.1). The blunders and errors with propulsion and smooth motion add further delight, especially when Stark finally gets it all right. The Mark III reveals what the audience has been waiting for—the familiar Iron Man suit with its signature red and gold coloring. It is in the Mark III that Iron Man performs his first civic duty as superhero by flying to Afghanistan, destroying the Terror Ring, and saving the villagers; it is also in the Mark III that he defeats Obadiah Stane. Tony Stark asking J.A.R.V.I.S. to add a final touch, a hot-rod-red color to the suit, adds another sublime moment for the fan, which is different from the sublime wonder experienced when in the presence of Iron Man as an embodiment of science. For the Iron Man fan, the affective response

**FIGURE 11.1:** Tony Stark (Robert Downey Jr.) and the technological sublime: human mastery through techno-scientific creation in *Avengers: Age of Ultron* (2015). Walt Disney Studios.

arises out of this final touch—the color—that makes this model the one audiences most readily recognize.

But more than just the general sublime effect outlined by Csicsery-Ronay, the film generates a specifically superhero-sf version of the sublime. The sf cinema, as I have elsewhere noted, "has always revelled in displaying the technological capabilities of the film medium, while thematically deliberating on the future effects of such technology" (Ndalianis 2004, 181). Consider the arrival of the alien spaceship in *Close Encounters of the Third Kind* (1978), the representation of the massive black hole in *Interstellar* (2014), the mesmerizing dance of the camera through space in *Gravity* (2013), the futuristic view of the city of Los Angeles in *Blade Runner* (1982). All of these visual effects bring to life overpowering imaginings of future science and technology. But *Iron Man* reveals an sf sublime that is doubled: these scenes of Stark's transformation speak specifically to the wonders of the mechanized human of the future (made possible through visual effects technologies), while they also ask the audience to revel in the creation of the wondrous superhero—seeing Iron Man fully come into his superpowers and, at the film's end, revealing his identity to the world.

The same pattern can be seen in how the sf grotesque also aligns with the conventions of the superhero form. Csicsery-Ronay notes that the sublime and the grotesque are related ways of feeling and expression (2008, 147), and in the superhero-sf merger this relationship is played out across the bodies of both the superhero and supervillain. Where Iron Man embodies the sublime, Obadiah Stane / Iron Monger represents the grotesque. Stane's Iron Monger is signaled as an aberration not only through his efforts to kill Tony Stark / Iron Man, but in the use of dark lighting and shadows that suggest the horror genre whenever he is present. Thus Csicsery-Ronay observes that the "sf film has evolved into an apparatus for rendering affects through special-effects technology," just as "capturing, reproducing, and foregrounding the violence of sublime and grotesque shocks has become one of the main purposes of f/x technology and sf film in general" (147). He further notes that the sf sublime "emphasizes the dramatic arc of the technosublime: recoil at the unutterable power and extension of technology, and recuperation through ethical judgments about its effects in the future" (160). In this respect, the awe-inspiring appearances of Iron Man and the terror-invoking ones of Iron Monger similarly play across the spectator's body, with the affective response to each helping the film deliver its ethical judgments about future cybernetic technology. In effect, *Iron Man* forms a dialogic relationship between the sublime and grotesque cyborg bodies, with Iron Monger posing a warning about the possibility that Iron Man technology could lead to chaos.

## THE TECHNOLOGIADE

Csicsery-Ronay's final category is the Technologiade, which refers to "miniature myth-structures" that illustrate "the transformation of human societies as a result of innovations attending technoscientific projects" (2008, 8). It thus represents "the epic struggle surrounding the transformation of the cosmos into a technological regime" (219). Drawing upon established myth and archetype models, Csicsery-Ronay maps an sf variant onto the traditional romance, as "human heroes prove their powerful virtue through a series of trials, many of which take place in anomalous spaces where normal laws do not apply" (216). The Technologiade follows this pattern by using two distinctive sf narrative forms: the space opera (scientifically driven adventures set in space) and the techno-Robinsonade (a modification of the classical Robinsonade adventure tale, named after Daniel Defoe's hero in *Robinson Crusoe*). The former can most clearly be seen in MCU's *Guardians of the Galaxy* (2014), *Captain Marvel* (2019), and *Avengers: Endgame* (2019) films. The latter especially characterizes *Iron Man*, and on these features the following analysis focuses.

The Robinsonade is an adventure story with recurring narrative functions represented by various narrative agents—the Handy Man, the Fertile Corpse, the Willing Slave, the Shadow Mage, the Tool/Text, and the Wife at Home—all of which have sf variants. The characteristics of each can be readily observed in *Iron Man*. The sf Handy Man (or woman) is, as Csicsery-Ronay suggests, the "ideological core of sf" (2008, 246), and is usually represented by "scientists, engineers, space pilots, cyberneticists, tycoons, hackers, astronauts, anthropologists, and so forth, whose technical prowess is matched fully by their moral qualities" (246). Like Odysseus, the Handy Man is an adventurer whose story is prone to serialization, which echoes in the comic book and film serialization of Iron Man's adventures. The character possesses skills in handling and manipulating tools to "fashion new ones . . . [and has the ability] to extend his power over the environment through technological control" (227). Stark is clearly a Handy Man—the man of science who creates technological wonders that extend beyond nature's creative capacity (even as he sometimes threatens to slip into the role of mad scientist). One of the early images that underscores this role is seeing him in action looking like a blacksmith (figure 11.2), covered in dirt and sweat, as he swings his hammer down on the metallic helmet he is forging in the terrorist camp.

The sf Fertile Corpse is similarly a key part of *Iron Man*. While Csicsery-Ronay argues that the "Fertile Corpse is the scene of the Handy Man's performance" (2008, 227), the figure rarely appears in human form but is often associated with female reproduction. In Frank Herbert's *Dune* (1965) the Fertile Corpse is the Arrakis desert that is mined for spice, but it can also be embodied in a variety of other things: "a code, a mainframe, the cyberspace matrix, or virtuality itself" (Csicsery-Ronay 2008, 228). In *Iron Man* computer technology, its coding, software, and hardware help form Stark's Fertile Corpse. He manipulates and shapes it, and it is the means through which he gives birth to Iron Man. In many respects, the Tool Text and Fertile Corpse serve similar functions in the film, as the Tool Text equates to the resources that allow the Handy Man to attain his goal. Stark's Tool Text is computer technology and science, both of which succumb to his genius and allow for his technological inventions.

The other agent that assists the Handy Man is the Willing Slave, which in sf is often a robot or cyborg. This figure accepts the Handy Man's guardianship and protection,

FIGURE 11.2: Tony Stark (Robert Downey Jr.) constructs his new superhero identity in *Iron Man* (2008). Walt Disney Studios.

and learns "to use the Handy Man's techniques and science (cognition), in order to consolidate a sense of autonomous identity and to aid the Master's further extension of his power" (229). In *Iron Man*, this role is most clearly performed by J.A.R.V.I.S., Stark's AI assistant. With his aid, Stark is able to refine the Mark I armor suit and make it into an agile, advanced piece of technology that then connects to J.A.R.V.I.S., further developing the relationship. Eventually, in trying to protect J.A.R.V.I.S. from destruction in *Avengers: Age of Ultron*, the Willing Slave is granted freedom when his consciousness is transferred to an android body and becomes Vision.

The sf Shadow Mage is the antagonist who obstructs the Handy Man, while competing with him for the Fertile Corpse and the Willing Slave. Clearly, *Iron Man*'s Shadow Mage is Obadiah Stane, who wants Stark's knowledge and inventions in order to create his own body suit, which he can then sell to the highest bidder. This figure is shadowy, dark, and evil, and he stands in stark contrast to the morally grounded Handy Man. That contrast shows most clearly in Stane's mad scientist figuration and his desire to use Stark's technology for self-interest and mass destruction. The Shadow Mage is also often depicted as having magical powers. Pitting Stane's misuse of technology against Stark's rational and respectful adoption of the powers of technoscience, the film presents Stane as dark and demonic—particularly when he becomes Iron Monger and his representation succumbs to the conventions of horror. In deforming the Iron Man technology, he becomes an almost supernatural force, transcending the human to become something evil and otherworldly.

Finally, we should consider the sf Wife at Home figure, "the flesh-and-blood woman who is married or betrothed to the Handy Man hero" and is "left behind when the Handy Man embarks on his adventures" (Csicsery-Ronay 2008, 258). In *Iron Man*, this figure is

Pepper Potts, Stark's assistant and love interest, who offers stability to Stark's universe. She provides the human and humane balance for Stark's obsessive, techno-scientific superhero. Csicsery-Ronay makes the point that the Wife at Home agent is often muted in sf and even represented as a desire for home or a yearning for planet Earth. Throughout the *Iron Man* films, Pepper Potts is a constant presence who embodies stability, humanity, and home for Tony Stark, and who affirms *Iron Man*'s fit within the mythic structure of the Technologiade. But this last beauty is also in dialogue with the others—a dialogue that, in the case of *Iron Man*, creates its own specifically superhero-sf path.

# CONCLUSION: THE SEVEN BEAUTIES OF SUPERHERO-SF

*Iron Man* is typical of many recent blockbuster superhero films, works wherein sf themes come to the fore thanks to the film industry's investment in effects technologies that have made visible the amazing, imagined technoscience of the future and especially the equally amazing bodies of superheroes who test the parameters of the human. In *Iron Man*'s case, the superhero story's connection to sf is a historical one that traces back to the origins of the form and whose dimensions can be seen in its reflection of many of sf's seven beauties. Similar reflections can be observed in many other superhero stories that also foreground the transformed body and make it a site where sf's social, ethical, and moral concerns are played out, and we might especially consider *Black Panther*'s (Coogler 2018) treatment of the Black body and *Black Widow*'s (2021) presentation of the female. Moreover, as *Iron Man* and these other films demonstrate, those beauties, when extended, also provide us with a more detailed understanding of the generic hybrid that is superhero-sf.

Superhero-sf engages in a dialogue with the science and technology of today, which it projects into an imaginary future in order to contemplate, critique, and wonder at their impact. The gap between today and the future is closing with, as Csicsery-Ronay suggests, sf now informing many facets of our life—actual robots, advanced communication technologies, AI used by social media to recognize faces and target users, "deep fakes" in which machine learning and AI manipulate audiovisual content to replace someone's likeness with another's. Csicsery-Ronay labels this situation "science fictionality" (2008, 2). Similarly, Brooks Landon claims that sf has expanded the conventional boundaries of the genre, generating what he calls "science fiction thinking" (2002, xiii). With superhero-sf this sort of thinking is embedded in its structure, in both the technological advances it imagines and the methods used to imagine the future. We might consider a real-life example, the scientist and CEO of Oblong Industries, John Underkoffler, a consultant on *Iron Man* who also created the gesture technology and multi-touch augmented reality interfaces for Steven Spielberg's *Minority Report* (2002). In the latter film, pre-crime investigator John Anderton (Tom Cruise) famously uses his hands like a music conductor, their motions shifting screens into and out of his central vision. Underkoffler has created his own company—Oblong Industries—to advance today's version of this technology, using it in *Iron Man* to design a number of Tony Stark's user interface technologies—his personalized keyboard, the gestures that command his self-assembling suit, the AR screens that he works with, and his interface with technology while in the Iron Man suit (Murphy 2015). Here we see both science fictionality and science fiction thinking in action and, while *Iron Man* can fit into the schema offered by Csicsery-Ronay, adjusting its parameters to encompass superhero-sf can help us see how sf might, as Underkoffler's real-life development illustrates, alter its own structure through its dialogue with the superhero tale.

## Works Cited

Attebery, Brian. "The Magazine Era: 1926–1960." In *The Cambridge Companion to Science Fiction*, edited by Edward James and Farah Mendlesohn, 32–47. Cambridge: Cambridge UP, 2003.

Coogan, Peter. *Superhero: The Secret Origin of a Genre*. Austin, TX: Monkeybrain, 2006.

Creekmur, Corey K. "Review: Superheroes and Science Fiction: Who Watches Comic Books?" *Science Fiction Studies* 31, no. 2 (2004): 283–90.

Csicsery-Ronay, Istvan, Jr. *The Seven Beauties of Science Fiction*. Middletown, CT: Wesleyan UP, 2008.

Darowski, Joseph J. "Introduction." In *The Ages of Iron Man: Essays on the Armored Avenger in Changing Times*, edited by Joseph J. Darowski, 1–4. Jefferson, NC: McFarland, 2015.

Gavaler, Chris. *On the Origin of Superheroes: From the Big Bang to Action Comics No. 1*. Iowa City: U of Iowa P, 2015.

Gresh, Lois H., and Robert Weinberg. *The Science of Superheroes*. Hoboken, NJ: John Wiley, 2002.

Hatfield, Charles. "Jack Kirby and the Marvel Aesthetic." In *The Superhero Reader*, edited by Charles Hatfield, Jeet Heer, and Kent Worcester, 136-54. Jackson: UP of Mississippi, 2013.

Hatfield, Charles, Jeet Heer, and Kent Worcester. "Introduction." In *The Superhero Reader*, edited by Charles Hatfield, Jeet Heer, and Kent Worcester, xi–xxii. Jackson: UP of Mississippi, 2013.

Herbert, Frank. *Dune*. Boston: Chilton Books, 1965.

Jeffery, Scott. *The Posthuman Body in Superhero Comics: Human, Superhuman, Transhuman, Post/Human*. New York: Palgrave Macmillan, 2016.

Jenkins, Henry. "'Just Men in Tights': Rewriting Silver Age Comics in an Era of Multiplicity." In *The Contemporary Comic Book Superhero*, edited by Angela Ndalianis, 16–43. New York: Routledge, 2009.

Jones, Gerard. "Men of Tomorrow." In *The Superhero Reader*, edited by Charles Hatfield, Jeet Heer, and Kent Worcester, 16–22. Jackson: UP of Mississippi, 2013.

Landon, Brooks. *Science Fiction after 1900: From the Steam Man to the Stars*. New York: Routledge, 2002.

Luckhurst, Roger. *Science Fiction*. Cambridge: Polity Press, 2005.

Murphy, Mike. "The Scientist Who Designed the Fake Interfaces in 'Minority Report' and 'Iron Man' Is Now Building Real Ones." *Quartz*, May 31, 2015. https://qz.com/415418/the-scientist-behind-the-fake-ui-in-minority-report-and-iron-man-has-built-a-real-one/.

Nama, Adilifu. "Brave Black Worlds: Black Superheroes as Science Fiction Ciphers." *African Identities* 7, no. 2 (2009): 133–44.

Ndalianis, Angela. *Neo-Baroque Aesthetics and Contemporary Entertainment*. Cambridge, MA: MIT Press, 2004.

Patton, Brian. "'The Iron-Clad American': Iron Man in the 1960s." In *The Ages of Iron Man: Essays on the Armored Avenger in Changing Times*, edited by Joseph J. Darowski, 5–16. Jefferson, NC: McFarland, 2015.

Reynolds, Richard. *Super Heroes: A Modern Mythology*. Jackson: UP of Mississippi, 2003.

# III

## NEW SLANTS ON SCIENCE FICTION FILMS

CHAPTER 12

····································································································

# THE ANTHROPOCENE
# AND ECOSOPHY

····································································································

## GERRY CANAVAN

THE birth of the contemporary environmental movement is frequently dated to 1962, the year Rachel Carson published *Silent Spring*, a clarion call against the use of pesticides that linked political concerns about nuclear radiation and fallout-compromised global fertility to the even larger threats that modern culture posed to the futures of both human society and the natural world. The ensuing decades saw a sea change in the way contemporary nations considered the environment, resulting in such reforms as the 1970 establishment of the Environmental Protection Agency (EPA) and the Clean Air Act, the 1972 Clean Water and Endangered Species Acts (in the United States), and the 1973 Environmental Action Programme (in the European Economic Community, the precursor to the EU). These and later, similar reforms all fall under the general rubric of liberal environmentalism, in the sense that they sought to strike a balance between the frictionless functioning of the economy and environmental protection that put (some) brake on unlimited economic growth to protect (some) portions of the environment from despoliation. However, the precise constitution of this balance would become a point of intense political contestation between various factions in the worldwide political economy, with the well-heeled partisans of "growth" frequently having the upper hand, and the environmentalists often being reduced to seeking harm reduction and damage control. This mode of liberal environmentalism can be registered quite clearly today in a film like the Al Gore documentary *An Inconvenient Truth* (2006), which announced the imminent threat of climate catastrophe and potential civilizational collapse only to, at its end, instruct its viewers to turn off the lights when they leave a room, write their Congresspeople, and pray.

But more radical strains of environmentalist thought refuse, in various ways, the bargain between human economy and natural protection that those liberal environmentalists have posed. The school of thought generally called "deep ecology," originated by Norwegian philosopher Arne Næss, in a foundational paper, defines the "deepness" of deep ecology against the "Shallow Ecology" of the liberal environmentalists, who *only* "fight against pollution and resource depletion" and are centrally concerned with "the health and affluence of people in the developed countries" (1973, 95). For shallow ecologists, nature ultimately remains a resource to be tapped for human use, simply in ways that are better or worse for the long-term sustainability of human society. In contrast, deep ecologists refuse the notion of the liberal balance between ecology and environment in favor of a different model of balance that understands humankind to be part of the environment, not radically separate from it, and not necessarily more intrinsically valuable or worthy of protection than other forms of life. Næss therefore called for an *ecosophy,* that is, an ecophilosophical system of "ecological harmony or equilibrium" that can provide a single "unified framework" for ecological action, rather than viewing the domains of pollution, overpopulation, resource depletion, mass extinction, and so on as separate domains, each requiring unique and precise interventions (99). Such a system of first principles could steer human interactions with the environment away from pure instrumentality toward genuine coexistence and co-thriving.

What distinguishes an ecosophy, then? According to Næss, "it is the global character, not preciseness in detail"—a system that "articulates and integrates the efforts of an ideal ecological team, comprising not only scientists from an extreme variety of disciplines, but also students of politics and active policy-makers" (1973, 100). This scope of effort prevents ecosophy from becoming bogged down in complex local disputes over any particular resource or practice, and it becomes, instead, the overarching system of valuation from which we might then determine the proper course of action on any specific local question of environmental policy. Næss, with George Sessions, would later identify eight basic principles of such a deep ecology:

1. The well-being and flourishing of human and nonhuman Life on Earth have value in themselves (synonyms: intrinsic value, inherent value). These values are independent of the usefulness of the nonhuman world for human purposes.
2. Richness and diversity of life forms contribute to the realization of these values and are also values in themselves.
3. Humans have no right to reduce this richness and diversity except to satisfy *vital* needs.
4. The flourishing of human life and cultures is compatible with a substantial decrease of the human population. The flourishing of nonhuman life requires such a decrease.
5. Present human interference with the nonhuman world is excessive, and the situation is rapidly worsening.
6. Policies must therefore be changed. These policies affect basic economic, technological, and ideological structures. The resulting state of affairs will be deeply different from the present.
7. The ideological change is mainly that of appreciating life quality (dwelling in situations of inherent value) rather than adhering to an increasingly higher standard of living. There will be a profound awareness of the difference between big and great.

8. Those who subscribe to the foregoing points have an obligation directly or indirectly to try to implement the necessary changes. (1984)

Deep ecology thus names a state of affairs in which political liberalism and the capitalist marketplace it uses to produce and distribute all goods, services, and wealth are intrinsically incompatible with the idea of an ecologically rational, sustainable futurity (or what sf author Kim Stanley Robinson and others have called a *permaculture*) ("Permaculture"). Deep ecology announces the fragility of our contemporary institutions, against their claims to ahistorical immortality, while promising both a *temporal* and an *ideational* rupture with the present. This break in time might be utopian in character—a better, more just, healthier future, ecosocialism—or, alternatively, the horror and collapse of mass immiseration and civilization (as the old socialist slogan suggests, the choice to return to barbarism). But the key to understanding this moment of intertwined crisis-opportunity is disassociating ourselves from a fixation on contemporary wants, contemporary needs, and contemporary cultural practices, in favor of a wider understanding of the possibilities of time—a revolution in perception that, as we shall see, the special effects regime of contemporary sf film is well positioned to produce.

However, ecosophy is by no means a philosophical monolith. Félix Guattari takes up the term in his volume *The Three Ecologies* and begins from a similar place with respect to the need for harmony between the human and nature, calling for an ecosophy that will unite environmental ecology with social ecology and mental ecology (41). But he ends by positing that this result may ultimately require humankind to take direct, active, and permanent control of the planetary environment through geoengineering:

> There is a principle specific to environmental ecology: it states that anything is possible—the worst disasters or the most flexible evolution. Natural equilibriums will be increasingly reliant upon human intervention, and a time will come when vast programmes will need to be set up in order to regulate the relationship between oxygen, ozone and carbon dioxide in the Earth's atmosphere. ([1989] 2000, 66)

Elsewhere on the spectrum of radical ecology, leftists like Slavoj Žižek and Kim Stanley Robinson have also recently suggested that a properly ecological mindset might at this world-historical conjuncture require constant, deliberate, and permanent manipulation of the planetary environment by humankind, rather than the more hands-off attitude suggested by the Næss-style deep ecology. What unites the radical ecologists is not so much any particular political program but rather a single ecosophic foundational principle, expressed perhaps in its purest form by Enrique Dussel in his *Twenty Theses on Politics*: "The critical ecological principle of politics could be expressed as follows: We must behave in all ways such that life on planet Earth might be a perpetual life!" (2008, 87).

The most singularly important ecosophic framework for thinking through environmental consciousness in the present moment may be the twenty-first-century concept of "the Anthropocene," which has been formative in the academy as a tool for understanding non-/post-/inhuman ecological perspectives both inside and outside the constraints of liberal environmentalism. Here the "depth" of the Anthropocene is produced through an effect of temporal dislocation often called "deep time"; when we think of the Anthropocene, we recalibrate our sense of temporal scale from the decades of human life, the centuries of empires, and the millennia of recorded history, to a much wider scale, that of the human species and of life on Planet Earth writ large. Popularized in the early 2000s by Nobel Prize–winning atmospheric chemist Paul J. Crutzen, the concept of the Anthropocene was proposed as a new geologic epoch that denotes the extreme, mostly negative effects of humankind on the planetary environment. If, the argument goes, humankind first evolved in the Pleistocene (a period ranging from approximately 2.5 million to twelve thousand years ago), and entered into its full agricultural and technological mastery in the Holocene ("the warm period of the past 10–12 millennia" beginning with the end of the last Ice Age and the subsequent glacial retreat), then in the Anthropocene the human species itself becomes a superhistorical force, leaving traces of itself in the geologic record of the planet (2002, 23). From an ecosophic perspective these traces may be better characterized as scars: extending beyond climate change to ocean acidification, desertification and deforestation, and mass extinctions. Furthermore, that perspective suggests that some far-future nonhuman intelligence, reconstructing the history of the Earth from the sort of scientific evidence we now use to periodize the deep past, will recognize this era as a radically disruptive moment of rapid ecological change, leading, as most such moments do, to significant habitat destruction and extreme biomass loss in both plant and animal life.

The extremity of such a proposal may not be immediately appreciable to those outside the academic fields that typically think of time on a geological scale. The "deep time" of geologic formation takes place over millennia, if not over hundreds of thousands or even millions of years, on a planetary scale in which the spatial and temporal measure of an individual human life is barely recognizable. Because geochronologic boundaries between eras in stratigraphy are organized around superhistorical changes to the climate, nearly all of which took place in the many millions of years before humankind had even evolved, much less the scant six or so thousand years that constitute recorded history, the idea that humankind itself now constitutes a geologic actor is in some sense contrary to the habits of thought required to think of time geologically—to think "deep time" at all. Indeed, debate around the possibility of the Anthropocene rapidly zeroes in on our era specifically as a hinge point, a recentering of not only human timescales but *our* personal lifespans in particular that would be viewed as deeply suspect by geologists prior to the late twentieth century. Crutzen's proposed boundary, for its part, is nearly coterminous with the historical existence of the United States as a political entity. As he suggests, the Anthropocene "started in the latter part of the eighteenth century, when analyses of air trapped in polar ice showed the beginning of growing global concentrations of carbon dioxide and methane. This date also happens to coincide with James Watt's design of the steam engine in 1784" (2002, 23). Others working with the concept push the date somewhat later, as late as the first detonation of the atomic bomb in 1945, or still beyond—suggesting the possibility that many people alive today have lived through, or have even personally orchestrated, the transition between geological epochs.[1]

Although this concept is not yet adopted by either the International Commission on Stratigraphy or the International Union of Geological Sciences, the Anthropocene Working Group of the ICS aims to formalize it, officially dating the start of the Anthropocene to the mid-twentieth-century. But the notion of the Anthropocene was originally intended to serve as much as an environmentalist or political provocation, as a provable scientific proposition. The point was to use the rhetoric of stratigraphy to shock scientists into recognizing the extent of humankind's effect on the global climate and to goad them into taking action. As Crutzen ends his article in *Nature*,

> Unless there is a global catastrophe—a meteorite impact, a world war or a pandemic—mankind will remain a major environmental force for many millennia. A daunting task lies ahead for scientists and engineers to guide society towards environmentally sustainable management during the era of the Anthropocene. This will require appropriate human behaviour at all scales, and may well involve internationally accepted, large-scale geo-engineering projects, for instance to "optimize" climate. At this stage, however, we are still largely treading on *terra incognita*. (2002, 23)

The disorienting shock of the Anthropocene, the way it forces us to think on two incompatible time scales at once, is intended to push us in the direction of ecosophy (here defined in terms rather unfriendly to deep ecology as "environmentally sustainable management").

Jason W. Moore's influential critique of the concept of the Anthropocene and his coining of the alternative term "Capitalocene" in response is a humanistic attempt to further this provocation, bringing the idea of the Anthropocene more fully into the ecosophic discourses with which this chapter began. Moore notes that the Anthropocene blithely suggests "humanity as an undifferentiated whole" (2017, 2) is the culprit behind our present ecological crisis, rather than the specific forces of capitalist expansion and technological acceleration that have brought us to the brink of catastrophe, despite decades of warnings about the consequences. Moore suggests that "Capitalocene" is a terminology better informed by ecosophy, which he refers to as "Green Thought," and which he says calls on us to push past the "human/nature binary" to "consider human organizations—like capitalism—part of nature" (2).

The aim of this chapter is to use sf films to illustrate these sorts of ecosophic provocations and shocks as they have been similarly deployed. By using both cinematic effects and special effects to push the boundaries of human cognition past its usual limitation, the sf cinema has found specific strategies to make deep time seem comprehensible within human time, as well as to dramatize (however literally or allegorically) the destructive consequences of technological modernity on both the wider world and its own necessary conditions for continuation. If we consider, for instance, Stanley Kubrick's *2001: A Space Odyssey* (1968), we can see it recasting the future of spaceflight and presumed near-term human colonization of the solar system into a larger part of a story of intelligence that spans hominid species, the human- machine boundary, civilizational history, terrestrial life, and ultimately even the limits of human cognition, forcing the viewer to disinvest in the situated assumptions of the contemporary political horizon in favor of a much wider range of possible concepts and values. Kubrick's project is precisely to force us to follow him on this journey, both through unexpected juxtapositions (like the famous match cut of a bone/club into a space station that inaugurates the main action of the film) and nonmimetic imagery (like the equally famous and much-parodied psychedelic "Star Child" sequence that ends the film).

Film techniques like zoom motion and time lapse date back to cinema's origins, and their use has frequently been trumpeted as a means to push forward the boundaries of human perception. Stephen Kern uses the metaphor not of deepening but of "thickening" to describe how "any moment could be pried open and expanded at will . . . by directors who spliced time as they cut their film" (2003, 88). Indeed, as the original mechanism of film is simply the rapid presentation of still images so quickly that humans perceive it as movement, we might well say that all film is time-lapse cinematography, only usually happening at the ordinary pace of human experience rather than this sort of hyperaccelerated velocity. The contemporary viewer is intimately familiar with the visual logic of the "fast forward." As David Lavery notes, the radical potential for defamiliarization and estrangement tokened by cinematic manipulation of time quickly became routinized; juxtaposing Colette's ecstatic reception of slow motion and time lapse in 1920 with the forgettable, almost invisible use of the same techniques in a United Airlines logo, Lavery observes how "time-lapse, co-opted for use by modern advertising, had become mundane, commonplace" (2006, 2).

Time lapse, Lavery further notes, has had the most success in scientific education, recording everything from the life cycle of insects to the controlled demolition of a building to the growth and death of bacteria colonies to "glaciers, blood corpuscles, blossoming flowers (hundreds and hundreds of flowers in bloom), cell division, sea creatures, cloudscapes, celestial mechanics, construction projects, rotting fruit, the sun rising and setting, puddings baking, storm fronts, traffic patterns" (2006, 2). As this list suggests, time lapse is an especially rich site for investigating nonhuman phenomena moving at nonhuman spatial and temporal scales. Because humans move so quickly, the sped-time eye of time-lapse cinema can barely take note of them, recognizing their moments in a herky-jerky, stop-motion logic, if indeed the speed of the film is slow enough to capture their presence at all.[2] Time lapse is also highly attractive to ecological cinema for precisely this reason; the erasure of human beings leaves behind only their negative impact on the environment. In this way time-lapse cinematography can help upend traditional debates about the mimetic nature of the cinematic apparatus, as it gives us access to the world in a way that is simultaneously real but totally removed from human experience. Such a disassociative effect can help the mind understand the world in nonanthropocentric (and more properly "deep ecological" terms). As Rudolf Arnheim offered, "Watching a climbing plant anxiously groping, uncertainly seeking a hold, as its tendrils twine around a trellis, or a fading cactus bloom bowing its head and collapsing almost with a sigh" via time lapse can offer the "uncanny discovery of a new living world in a sphere in which one had of course always admitted life existed but had never been able to see it in action" (qtd. in Lavery 2006, 5). And the long zoom does for space what time lapse does for time, making accessible the connections between disparate locations by zooming the camera out to a scale where their connection is immediate and undeniable. Perhaps the most famous examples for the history of ecological photography are the stills and film of the Earth as seen from space, which—for better and for worse—erase all question of identity particulars in favor of a pale blue dot on which, as many have suggested, "we are all in this together."

As many observers have noted, this sort of totalizing spatial and temporal thinking can be pernicious as well as galvanizing, erasing differences that are important to preserve as we seek sustainability and ecological justice—for instance, by eliding the difference between the nations whose power elite have perpetrated the climate crisis versus those nations who are climate change's unwilling and undeserving victims, or by encouraging a sort of maximum passivity and helplessness in the face of overawing natural forces and the inexorable march of time. But the deep ecologists would remind us that this sort of thinking has an important educational function as well, by shocking viewers out of their complacency and making them receptive to new understandings of the interconnectedness of all things—a crucial insight necessary for breaking the confines of liberal ecological bargaining.

Special effects from sf cinema that mimic zoom motion and time lapse can similarly disrupt our perceptions, even as they push past the "meta-mimetic" representation of "the real" (Lavery 2006, 5) into the presentation of speculative or fully imaginary representation. In *The Seven Beauties of Science Fiction,* Istvan Csicsery-Ronay Jr. identifies the sublime as the fifth and in some ways most important "beauty" of sf, linking it to the vaunted "sense of wonder" that has long characterized aesthetic appreciation of the genre; "the sublime," he says, "is a response to a shock of imaginative expansion, a complex recoil and recuperation of self-consciousness coping with phenomena suddenly perceived to be too great to be comprehended" (2008, 146–47)—precisely the effect the time lapse and the zoom-out produce. In his influential essay "A Key to Science Fiction: The Sublime," Cornell Robu (1988) similarly develops a generalized theory of science fictional aesthetics along these same lines, arguing that sf is in essence about achieving the dis- and reorienting "pleasure in pain" that is the Kantian or Burkean sublime: "The anchoring into science, the assimilation and appropriation of the image of the universe as it results from the incessantly developing twentieth-century science ensures for sf unlimited resources to *figure infinity,* resources more remote for the other forms of literary expression" (21–22). These sorts of infinite tableaus—whether produced by unaltered photography or produced by computer simulations—are not somehow at odds with mimesis but rather grant us access to spheres of knowledge not ordinarily accessible; as Robu notes, it is actually science itself that "enforces the sublime" (23).

A sequence from Simon Wells's largely forgotten (and mostly forgettable) 2002 remake of *The Time Machine* perhaps provides the most vivid demonstration in recent decades of ecosophic and Anthropocentric thinking. Fresh from the discovery that he is unable to use his time machine to rescue his murdered fiancé Emma (Sienna

Guillory) from death, the despondent time traveler Alexander Hartdegen (Guy Pearce) decides to travel instead to the future to see if anyone has been able to use time travel to change the past. In the novel the experience of time dislocation is characterized by seasickness, nausea, and the "horrible anticipation . . . of an imminent smash" (Wells [1895] 2002, 20)—typical effects associated with an encounter with the sublime—but the experience in the film is much more pleasurable. As Hartdegen engages his machine and time accelerates around his time bubble, the film begins to move at a sped-up rate far faster than the usual pace of human cognition (figure 12.1). We see a spider craft a web at superspeed, then a storm quickly pass over Hartdegen's workshop's windowed roof; days and nights smear together in a single blur. Outside, the ivy grows and flowers in seconds; an instant later, it is dead again, and the building is covered by snow (which just as quickly recedes with the ivy then returning). The pace of temporal acceleration continues to increase, and soon the workshop is cleared out entirely (the house is presumably sold to new owners after Hartdegen's mysterious disappearance) and becomes storage for cars, which rapidly progress through the style development of early automobiles. Across the street the same thing happens with fashion, as decade-by-decade trends in skirt length and style pass in the blink of an eye.

At this point Hartdegen accidentally drops a cherished locket containing a picture of Emma out of the time bubble; the photograph rapidly withers and disintegrates. Now in a crane shot, the point of view of the acceleration moves from what Hartdegen can personally see to all of New York City. We see his workshop has been dismantled, while nearby skyscrapers are built that dwarf the scale of the original buildings. As we zoom out farther and farther, we see a twin-engine plane flying by, and a few seconds later, commercial jetliners. We continue to zoom away from Hartdegen's point of view into low Earth orbit, finding satellites and, ultimately, a modified version of the space shuttle arriving at an outpost on the Moon. When we return to ground level, Hartdegen is intrigued by a moving advertisement appended to the building across from his

FIGURE 12.1: In a zoom-out, the world changes around the time traveler Hartdegen in *The Time Machine* (2002). Dreamworks Pictures.

machine and slows the device back to the pace of normal time, emerging in May 2030, nearly a century and a half after his own time. He discovers that the notice is advertising "Lunar Leisure Living," suggesting that outpost is, in fact, a thriving city colonized not simply by scientists or soldiers but by ordinary citizens; indeed, the advertisement's appeals to fishing, golf, and the reduced gravity of the Moon seems targeted specifically at retirees.

While this temporal dislocation is certainly immense, and the future shock that afflicts Hartdegen is undoubtedly quite serious, its scale does not yet deliver the full shock of the Anthropocene. We remain not only in a recognizably modern urban environment, but fully within the space of endless expansion and boundless technological progress that characterizes the futurological logic of capitalism. Only in the next major fast-forward sequence does the viewer experience a truly vertiginous encounter with deep time. After a brief visit to 2030, Hartdegen attempts to go further into the future, but his machine is quickly rocked by an explosion and he exits the time bubble only seven years later, in 2037. The thriving futuristic city he had left only seven years before is ruined and abandoned, marked by fires and smoke; he is immediately confronted by armed police officers in riot gear instructing him to return to his designated evacuation center or face arrest under martial law. In the distance we hear the screech of an alarm, sounding like an air-raid siren or tornado warning; looking up, he sees that the Moon is breaking up. The police—wondering how anyone could be unaware of this seismic event—explain that the demolitions for the construction of the lunar settlements have disrupted the Moon's orbit and have led to its slow destruction, apparently causing massive gravitational disruptions on Earth as well. Following several explosions, including the eruption of a lava fissure in the street, Hartdegen is able to escape the police and return to his craft, where he is knocked out by falling debris, and another explosion slams his unconscious body into the machine's levers, sending it into maximum acceleration.

Only then do we see something like a fully realized view from the Anthropocene. The destruction of the surface, caused by human activity, apparently produces massive climactic destruction, ultimately killing all life on the surface and turning New York into a desert. The buildings disappear and soon leave no trace behind. While he sleeps, the land transforms around him; new rivers, inlets, and islands form and disappear, and soon a massive canyon appears, as the river erodes the surface. Entirely novel geological formations materialize and, soon after, vegetation returns, as the area transforms from something like a steppe to a grassland to new forest. We even pass through something that appears to be a renewed Ice Age as a massive glacier rolls in, completely covering the time machine and everything around it in a massive wall of ice. Afterward a renewed period of vegetation again fills the world with green. By the time Hartdegen regains consciousness, the time machine has reached AD August 27, 802701, and he emerges into a radically transformed world. Only then does Hartdegen find himself truly unmoored, simultaneously inhabiting the deep geologic time of the Anthropocene that is radically independent of human causation and human timescales, while still bearing its traces and scars. Crucially, Hartdegen must become unconscious here, and not simply because the plot of the source material demands he must go much further into the future than the character would ever choose to go on his own. As with the zoom-out to space in the previous sequence, the full scope of what is being represented must erase the human at its center if it has any hope to produce a full picture of the world-transformation being produced. But as viewers we are able to catch a glimpse of the immensity of a deep relationship with the history of life, if highly mediated through the visual language of sf cinema, and if only for the briefest of moments before returning again to the human scale.

While this adaptation omits the most Anthropocenic moment of Wells's original novel—the time traveler's trips to even further futures of maximum entropy and a dying, emptied Earth—it does retain a glimmer of this gesture in the way Hartdegen defeats the Morlocks. Inverting the polarity of the time bubble, he causes the time machine to implode, shooting out some sort of temporal energy that reduces the Morlocks to skeletons and erodes their fortress to dust in a matter of seconds. A coda shows Hartdegen living with the Eloi in a jungle environment that was once New York City. "There's nothing here," one of his Eloi companions says. "Well, it was different then. My laboratory was all around here; the kitchen was up there where that tree is." (figure 12.2). And as he speaks, the film uses a new special effect to show temporal dislocation; in

FIGURE 12.2: Hartdegen experiences a new Earth and a new relationship to the world in *The Time Machine* (2002). Dreamworks Pictures.

the background, we see a ghostly version of Hartdegen's maid and friend enter his laboratory, hundreds of millennia earlier. This composite shot continues until nearly the end of the film; for one eerie, uncomfortable moment, we are able to exist in both time periods at once, inhabiting the radical extremity between them. We even see in this final shot that this extremity need not be antihuman—humanity still flourishes in both temporal locations—so long as we, like Hartdegen, are able to properly attune ourselves to the full flux and churn of time.

While this chapter primarily focuses on only a single cinematic example, this sort of temporal and spatial dislocation is nearly ubiquitous as an environmentalist virtual rhetoric in contemporary sf film, from the zoom-in from deep space at the beginning of *WALL-E* (Stanton 2008), in which the spaces we first interpret as cities turn out to be gigantic mounds of trash; to the time lapse of Ramin Bahrani's short *Plastic Bag* (2009), which uses time lapse photography to give us a sense of what the million-year decomposition of a discarded plastic bag might actually *feel* like to the jaundiced, quasi-immortal bag (delightfully voiced in the somber, melancholic timbre of Werner Herzog); to the haunting zoom-out and speed-up at the end of Wanuri Kahiu's short film *Pumzi* (2009), which depicts a living future for the planet seemingly at the cost of suicide for the human race; to the wide orbital shots, vertiginous superspeed, ubiquitous macro- and microscale satellite surveillance, wormholes, time portals, hyperdrives, and the many other modes of spatiotemporal play that have become de rigueur across our space opera, espionage, and superhero narratives. The colossal ruins of Star Destroyers and Death Stars that litter the J. J. Abrams—led *Star Wars* sequel trilogy (2015–19) are certainly a registration of the Anthropocene, as are the shattered cityscapes of American zombie fantasy; the climate-changed Americas of *Blade Runner 2049* (Villeneuve 2017), *Interstellar* (Nolan 2014), and *Snowpiercer* (Bong 2013); the indescribably alien weird ecology of *Annihilation* (Garland 2018); and the life-annihilating "snap" that marks *Avengers: Infinity War* (Russos 2018) and *Avengers: Endgame* (Russos 2019)—among other examples. Our sf cinema has increasingly helped us build a collective visual and narrative vocabulary for understanding the fragility of human institutions against the depth of ecological time.

To the extent that these cinematic innovations break through our ordinary habits of cognition, helping us to understand time and space outside the limited standpoint of ordinary human life, they can be thought of as doing ecosophic work—often in spite of the intended politics of their creators. Film, and perhaps especially sf film, has trained us to think with radical new flexibility about the way we perceive time and space, and by extension the way we understand our relationship with the planet and its other inhabitants. The rupture of the ordinary terms of the twenty-four-frames-per-second cinematic exchange, despite the overall "disappointing career of time-lapse photography" (Lavery 2006, 2) and the ease with which we have all become habituated to these once-shocking visual effects, still retains some power to disorient us in ways that can be generative, and may even still help to shock us into a renewed recognition of nature's vital, thriving existence, radically independent of our use, exploitation, and brutal despoliation of it.

## NOTES

1. Others working with the concept of the Anthropocene push the concept backward in time. Moore (2017), discussed later, views the Anthropocene as being more or less coterminous with the last five centuries of modern capitalism, beginning around 1492. Others have suggested a wider scope, aligning the Anthropocene with pre-first-contact indigenous and/ or prehistoric agricultural practices that potentially push the moment of the emergence of the Anthropocene quite close to the full length of the Holocene.
2. Writing of William Raban, Mike Leggett, and Chris Welsby's time-lapse *Park Film* (1972–73), Catherine Elwes (2015) notes how the "kaleidoscopic portrait of a place" populated by ghostly "twitching, stick-like people" who appear and disappear with the weather has an "environmental resonance" in the "assimilation of the 'stuttering' figures into the landscape," and quotes Emily Richardson's ecstatic description of this effect as "the breath of nature" (117–18; Richardson's quote is from her catalog entry in a London exhibition on landscape photography and the moving image).

## WORKS CITED

Crutzen, Paul J. "Geology of Mankind." *Nature* 23 (January 2002). https://doi.org/10.1038/415023a

Csicsery-Ronay, Istvan, Jr. *The Seven Beauties of Science Fiction*. Middletown, CT: Wesleyan UP, 2008.

Dussel, Enrique. *Twenty Theses on Politics*. Trans. George Ciccariello-Maher. Durham, NC: Duke UP, 2008.

Elwes, Catherine. *Installation and the Moving Image*. New York: Columbia UP, 2015.

Guattari, Félix. *The Three Ecologies*. Trans. Ian Pindar and Paul Sutton. London: Athalone Press, [1989] 2000.

Kern, Stephen. *The Cultures of Time and Space, 1880–1918*. Cambridge, MA: Harvard UP, 2003.

Lavery, David. "'No More Unexplored Countries': The Early Promise and Disappointing Career of Time-Lapse Photography." *Film Studies* 9 (Winter 2006): 1–8.

Moore, Jason W. "The Capitalocene, Part I: On the Nature and Origins of Our Ecological Crisis." *Journal of Peasant Studies* 44 (2017): 1–37.

Næss, Arne. "The Shallow and the Deep, Long-Range Ecology Movement. A Summary." *Inquiry* 16 (1973): 95–100.

Næss, Arne, and George Sessions. *Basic Principles of Deep Ecology* (1984). theanarchistlibrary. org. 2011. https://theanarchistlibrary.org/library/arne-naess-and-george-sessions-basic-principles-of-deep-ecology.

"Permaculture." *KinStanleyRobinson.info*. https://www.kimstanleyrobinson.info/content/permaculture.

Robu, Cornell. "A Key to Science Fiction: The Sublime." *Foundation* 42 (Spring 1988): 21–37.

Wells, H. G. *The Time Machine*. New York: Signet, [1895] 2002.

# CHAPTER 13

........................................................................................

# BIOPOLITICS AND BIOETHICS

........................................................................................

## SHERRYL VINT

BIOPOLITICS refers to theoretical work that considers the modes and aims of governance when the state takes an interest in managing the bodies of its citizenry at the biological level, that is, governing people as both political subjects and living beings. Most centrally associated with the work of Michel Foucault, "biopolitics" builds on and expands his earlier expression "biopower," which describes how power creates specific subjectivities through the management of the body, often through establishing a norm and policing bodies that deviate from it. At its heart, biopolitics is about which lives are valued and what criteria we use to make distinctions between valued and disposable life.

In his late lectures, Foucault increasingly theorized links between biopolitical governance and capitalism, especially in its neoliberal form, whose emergence he describes in *The Birth of Biopolitics* (2004). Later theorists have both built upon and reoriented Foucault's work, with Giorgio Agamben (1998; 2004) and Roberto Esposito (2008) proving to be especially prominent voices in distinct branches of biopolitical theory. Agamben's work focuses on the distinction between those fully recognized as human and those deemed mere "bare life," that is, outside civic protections and rights, while Esposito focuses on the politics of immunity through which mechanisms designed to protect and foster life can turn threatening. Other theorists, most significantly Achille Mbembe (2003), Sylvia Wynter (2003), Jasbir Puar (2017), and Alexander Weheliye (2014), have critiqued these European biopolitical thinkers for their failure to account sufficiently for colonialism as a constitutive governance, and especially for its violence. Theorists approaching this topic from a decolonial framework often prefer the term "thanatopolitical" over "biopolitical," the different prefix indicating "death" and thus stressing that the more relevant frame of reference for understanding this mode of governance is to examine those lives that are devalued, shortened, or damaged by the discriminatory, predominately racist structures of such governance.

Biopolitics has thus emerged less as a field than as a methodology for asking questions about power, subjectivity, embodiment, and the management of life at the biological level. Embodiment in this context connotes not only the simple fact of having a body, but the larger political framework by which certain kinds of bodies and bodily practices are normalized or criminalized via the dominant culture, creating strong links between biopolitical theory and research in such areas as queer and disability studies. Foucault's interest in biopolitics emerged first in relation to medicine, and the field of biomedicine typically takes Foucauldian concepts as its starting point. Paul Rabinow (2009) and Nikolas Rose (2006) have been influential in fusing Foucauldian concepts with social theory to study biotechnology and medicine, especially changing regimes for managing life through genetics, synthetic biology, pharmacotherapy (that is, medical treatment through drugs), and similar interventions. One result of this fusing is a shift away from sovereignty expressed through the power to "make die or let live" (Foucault 2003, 240)—as in executions—and toward strategies to "make live and let die" (241), that is, toward policies and institutions that foster life in particular configurations and toward distinct ends, while withdrawing resources and support elsewhere. While biomedicine focuses on various regimes of "making live" in distinct ways, decolonial theorists engaging with biopolitical issues draw attention instead to how systemic racism and an ongoing legacy of colonialism "lets die" the people and places occupied and marginalized by the West.

Biopolitics often focuses on people at the level of the "population," what Foucault calls the "species body, the body imbued with the mechanics of life and serving as the basis of the biological processes: propagation, births, mortality, the level of health, life expectancy and longevity, with all the conditions that cause these to vary" (1978, 139). That is, biopolitics is interested in thinking about people and bodies in an aggregate sense and according to specific categories, rather than in the individualized mode of policing subjectivity that Foucault explored in his earlier work under the term "biopower." It is precisely this question of "the conditions that cause these [biological conditions] to vary" that drives most biopolitical critique, as we see in the observation that systemic racism routinely channels vitality from colonized peoples toward those who align better with a default "human," one historically presumed to be white, male, heterosexual, and property-owning. Agamben's work, especially that which distinguishes between *zöe* (the biological, living being) and *bios* (the "proper" life recognized as the citizen/subject), focuses on this issue of political recognition. For Agamben, the concentration camp and its protocols for making people vulnerable to violence by removing their human status is paradigmatic of the modern biopolitical state. Contra Agamben, scholars such as Wynter, Puar, and Weheliye point to colonialism and its racialization of people into human and subhuman categories as the original thanatopolitical face of biopolitical governance.[1]

Given its focus on governing both legal and biological subjects, biopolitics makes distinctions among people that touch on belonging not only to a citizenry but also to humanity. From this point of view, Hannah Arendt's (2003) work on totalitarianism and the insufficiency of liberal concepts like "human rights" to protect stateless people can be understood as part of this intellectual history. Following this lead, Sylvia Wynter rejects an overly narrow concept of Western bourgeois man that "overrepresents itself as if it were the human itself" (2003, 260), while offering another genealogy that outlines the history by which enslaved Africans, colonized peoples, and the indigenous have been "dysselected" (267) as insufficiently human. Wynter argues that we need a more inclusive definition of the human, and from this point of view, biopolitics intersects with critical posthumanist thinking in a project to consider the foundations of human ethics and identity beyond the damaging binaries that have grounded Western philosophy.

The human/nonhuman boundary thus underpins biopolitical thinking in a substantial way, a factor that has taken on new significance now that technology can modify genomes, create novel DNA sequences in synthetic biology, or manufacture chimeras (i.e., beings made of cells from more than one "individual"), including fusing human and nonhuman genes. The bioethical challenges such entities impose reiterate the political history through which the human species has been racialized into "fully" and "less-than" human groups. Drawing attention to the fact that "letting die" can be just as violent as "making die," Mbembe has thus coined the term "necropolitics" to describe how colonial methods of governance took the form of an "expulsion of humanity altogether" (2003, 169) for designated subjects. The aim was not to foster the lives of slaves, nor simply to let them die; rather, they were valuable "kept alive but in a *state of injury*" (170), as a source of free labor-power without civic rights. Jasbir Puar (2017) similarly focuses on what the wounded body accomplishes for biopolitical governance in her concept of the "right to maim," arguing that it constitutes a strategy of "debilitation" (xi), draining energy and political affect from subjugated populations. Since medicalization has long been recognized as a technique to solve social or political problems by managing specific populations and their bodies, Puar draws attention to the reality that this includes actively damaging—not simply neglecting—the health of nonvalued segments of the population.

Economics has also been shown to be a crucial element of biopolitical governance, as Mbembe adroitly demonstrates in his focus on slavery. Liberal governance arises to manage this relationship between economic imperatives and political strategies, and in its neoliberal form prioritizes conforming with market logics. Ongoing research in biotechnology and legislative regimes that commodify life—not only nonhuman species, but also immortal cell lines derived from human tissues, human gametes in the fertility industry, and sequences of human DNA in diagnostic medicine—recognize this management as a kind of governance for the bioeconomy. Moving beyond either fostering or fatally neglecting life, this biopolitics structures life to maximize its economic value, which may be detached from individual organisms. Bioeconomic governance shapes life both materially (in genomics) and conceptually (rewriting what we culturally value about life). Thus, scholars such as Kaushik Sunder Rajan and Kalindi Vora theorize how markets in bodily tissues and capacities (such as gestation) reinscribe the racialized exploitations of colonial expropriation in their geographical parameters. The result is that the natural world comes to be seen more and more as a site of economic productivity, extending the potential scope of biopolitical governance from the living bodies of human subjects to living biomass overall.

As this brief overview should begin to suggest, there are myriad points of intersection between biopolitical theory and motifs surfacing in the contemporary sf film. Indeed, theorists have often turned to metaphors associated with sf aesthetics in trying to concretize some of these abstractions. Melinda Cooper describes bioeconomics as a project that strives to "overcome the ecological and economic limits to growth associated with the end of industrial production, through a speculative reinvention of the future" (2011, 11). Her use of the term "speculative" readily suggests financial and economic instruments such as derivatives, yet connects to fictional speculation as well.[2] Similarly, Kaushik Sunder Rajan speaks of the biotech industry as based on "promissory futuristic discourse" (2006, 116) because of the importance of narratives for securing venture funding to conduct research. In its bioeconomic mode especially, then, biopolitics has strong parallels with sf, as the industry itself uses sf terminology and even techniques for projecting the future.

Biopolitics is equally valuable for thinking about sf film, providing a social context through which to grasp the reasons that certain icons and motifs recur across multiple texts. The ongoing work to differentiate valued from nonvalued life that is at the core of biopolitical governance echoes in the many films about human/nonhuman relations, from robots to aliens to augmented humans. The term "alien," for example, has a political history in reference to immigration, as much as a genre history in reference to extraterrestrials, and biopolitical theory can help us see that this parallel is more than a coincidence: species belonging and citizenship are historically entwined. Similarly, stories of robots emerged from a history of exploited labor and strategies of dehumanization via which such cheap labor was made available. Critiques made by biopolitical theorists regarding liberalism's refusal to recognize some people as fully human resonate in myriad sf films that position their audiences to sympathize with marginalized and nonhuman protagonists, who can be understood to inaugurate a different, posthuman logic of personhood. In our era of the Anthropocene, concerns about anthropogenic climate change and its potential threat to any future life on earth, another frequent sf film topic, emerge from biopolitical attempts to calibrate futurity and extinction.

As a specific illustration of how biopolitical theory can provide tools for understanding sf, we might consider Claire Denis's oblique and poetic film *High Life* (2018) from this perspective. The film is about a mission into space, with the narrative largely told via flashbacks, a few about times on earth but mostly about earlier periods of the voyage. Details about the mission's objectives emerge slowly, and often only in retrospect can we comprehend characters' motivations. As the film begins, only two characters remain on the spacecraft, Monte (Robert Pattinson) and his infant daughter Willow (Joni Brauer/Johann Bartlitz). The film's title is not even projected until seventeen and a half minutes into its run time, after we have watched an extended sequence establishing Monte's routine with Willow, and then observed an exhausted-looking Monte put his deceased former crewmates into their spacesuits and eject them from the airlock. As they fall slowly down the screen, in front of a backdrop of stars that connotes the cosmic scope of the universe as compared to the fragility of humanity, the words "HIGH LIFE" appear in sharp white contrast. The film passes the thirty-minute mark before we see any extended scenes of these characters alive, although frequent cuts juxtapose Monte's care of Willow with brief images of earlier encounters with the crew. Eventually these central facts are established: the people on the ship were prisoners, offered the chance to participate in an experimental mission in lieu of serving their term; they traveled several years to approach a distant black hole; and they had two missions: reproductive experiments to create a baby able to withstand high levels of ambient radiation, and harnessing the energy of the black hole using the Penrose Process, thereby supplying massive energy to a dying earth economy/ecology.[3]

There are many correspondences between the film's themes and biopolitical framings. The mission to find an energy source hints at issues of sustainability and the looming exhaustion of fossil fuels, speaking to tensions between neoliberal priorities and any possibility for human flourishing. What place is there for most of humanity in a future so saturated by market values that the depletion of the environment that sustains life is justified in the name of economic growth? Images of space travel have long been bound up with questions of the future of the human species, including ideas about the next stage of humanity as explored in Stanley Kubrick's *2001: A Space Odyssey* (1968)—a film whose aesthetics echo in Denis's work, including the bodies of dead astronauts floating in space. Where once *2001* and its starchild pointed to the possibility of a next stage of evolutionary transcendence comparable to the leap from primate to human, now space voyages often symbolize an escape hatch from a dying planet, a chance to start over for the selected few—evident in proselytizing about missions to Mars by corporations such as SpaceX. It is precisely this kind of conflation of sf imaginaries with venture-capital projections that *High Life* critiques, focusing as it does on those whose lives have been deemed expendable.

The film particularly references issues of ecology and sustainability in its production design and use of color. The ship is divided into two areas of sharply contrasted aesthetics, separated by a plastic curtain that at times also serves as a barrier between the audience and the action. In one setting, a lush, green garden is lit by warm, yellow light. In the opening sequence before the title credit, several shots focus in close, loving detail on the delicate curves of leaves or the rich colors of growing fruit (figure 13.1). At one point the film cuts from Monte carrying harvested vegetables up the ladder from the garden to a bowl of pureed baby food that he feeds to Willow, as if to underscore the necessary relationship between human and other kinds of life. Even in space humans are entirely dependent on the living world of other species as our external support system, as much as on the technological systems of the ship.

In contrast to the verdant expanse of this garden enclosure, the rest of the ship seems harsh and sterile. It is lit by a cool blue light, the corridors are angled and institutional, without any sense of personalized space, and the machinery looks worn and cobbled together through improvised strategies, especially in scenes of Monte alone with the infant Willow. The plastic separating the garden from the other spaces calls to mind *Silent Running* (Trumbull 1972) and its fragile terrariums filled with earth's remaining botanic life—habitats ordered destroyed in the name of efficiency in that film's central conflict. Here, however, the garden eventually overgrows the barrier that separates it from the rest of the ship, which is most evident in the final minutes of the film as Monte and Willow leave the ship to enter the black hole. By having the garden exceed its containment, *High Life* seems to prioritize the ecological world over technological environments as the means to sustain human futures.

FIGURE 13.1: Tcherny (Andre Benjamin) marvels at how the garden proliferates in outer space (*High Life*, 2018). A24 Films.

Yet in the context of space, humanity is also deeply dependent on this artificial environment. The opening sequence of Monte and Willow underlines this dependence by foregrounding how life in space puts a strain on embodied being. The soundtrack is dominated by sounds of Willow babbling and later crying. She is alone inside a jerry-rigged crib, while Monte is outside the ship performing a repair. She watches him on one monitor, while images of an indigenous ceremony from the documentary *In the Land of the Head Hunters* (Edward S. Curtis 1914)—which documents traditional life among the Kwakwaka'wakw people of Canada's west coast—play on another.[4] Monte attempts to comfort Willow through a voice link, but at one point her distress about his physical absence causes her to shriek loudly, which makes Monte drop his tool. The camera looks on as it drifts irretrievably away into the darkness of space, a stark reminder of their isolation and lack of any resources beyond themselves and this finite ship. This setting is clearly not designed to accommodate an infant, and Monte must continually invent ways of reconfiguring available equipment to create a safe place for Willow. Images of technology thus repurposed creates an atmosphere in which technology itself seems somewhat alien to human existence, an imposition rather than an extension as some traditions would theorize it. The frequent emphasis on bodily contact between Monte and Willow, and the prevalent sound of her crying when they are not in contact, reinforce this impression. The camera often lingers on intimate details of Willow's body, such as her tiny, curled hand, and warmer light often highlights the parts of the screen where Willow and Monte are in proximity, signifying both the vulnerability and preciousness of human life.

The contrast between a sensuous human life and one mediated by technology is also suggested by the film's most provocative sequence: an extended masturbation scene in "the box," a sealed container housing a machine that includes an apparatus that rhythmically thrusts a steel phallus. The main research scientist Dibs (Juliette Binoche) brings herself to orgasm with this machine, and in the final shots of the sequence the mechanical chair seems to have morphed into an animal body. After her session, the box leaks out white fluid, but the image suggests mechanical plumbing rather than organic ejaculation. Other scenes of Dibs calibrating a series of tanks that filter the ship's water through "black," "grey," and "white" stages draw attention to this artificial circulation of fluids as part of their life support system, in barren contrast to more vital flows of fluids within human bodies, just as the medicalized spaces of the ship's interior contrast with its garden.

Here, the biopolitical management of life becomes an overt focus of mundane activities. In space everything militates against life's extension, as the cold and airless vacuum of space surrounds this tiny capsule of aging systems that filter waste to create water. In the opening sequence, after Monte feeds Willow and puts her to sleep, he goes to a computer monitor where he is required to make a report to an automated system: once he assures it that all the equipment is functioning and both he and the child are healthy, it renews life support systems for another twenty-four hours and we see sprinklers mist the garden. The relationship between cultivating desired life (making live) and disallowing what is deemed unhealthy or nonnormative (letting die) is aptly captured by this system and its rationing out of futurity in twenty-four-hour increments. So long as the lives on board the ship correspond to parameters programmed into the mission objectives, resources will enable them to continue living; should they deviate from these directives, their deaths will be assured, not through overt violence but by the withdrawal of resources.

In later scenes—later in screening, earlier in diegetic chronology—we see that this system was also designed to be controlled by the captain alone through a device implanted into his index finger. When the captain dies, Dibs transfers this device to herself. The reports were also meant to be a failsafe against any change in the hierarchical arrangements that structure society on board, and when Dibs makes her first report, she notes not only that the equipment remains functional, but also that "all crew members are complying with experiments." Later, just before she dies, she makes a point of passing this implant on to Monte. We see in this protocol an echo of the analysis of biopolitical governance as a means to create inequality within human communities, privileging some types of people over others, often through racialization as discussed earlier. In this case, the fact that some of the crew members are coerced into serving as experimental subjects suggests also the human/nonhuman boundary (*bios/zöe*) which Agamben (1998) contends is the foundation of biopolitical governance.

Such representation of technology offers a means of understanding this film as a riposte to *2001*: there technology is linked with humanity's evolutionary transcendence, as in its triumphant cut between the bone first improvised as a tool and a functioning orbital station.[5] *High Life* offers a different perspective through the logic of montage. We see Monte and Willow lying together in bed, one of many scenes that convey the depths of his gentleness and patience with her. Next, we cut to very brief shots (less than five seconds each) of a path in a country road, and then a girl lying on the ground with a wound on her head. Through similarly short flashes to the past, we understand that this is Monte's memory of the crime that resulted in his presence on board the ship—as an adolescent, he was out with this girl and his beloved dog (at one point in the film, he tells another crewmember he was "raised" by his dog). The nonlinear sequence withholds precisely what happens, but earlier flashes to this time include the very friendly dog playing with both children, the drowned body of the dog in a stream nearby, a hand picking up a rock, and then finally this shot of the girl lying on the ground.

The film cuts from her body to a train speeding across a desolate European landscape, and then to the train's interior where a woman interviews a professor from India (Victor Banerjee) who speaks about "the final age of man." He is traveling to a conference that will address "radical experiments" in which "death row inmates" are to be used as experimental subjects. "Is this really how occidental governments hope to deal with criminals?" he asks, thereby framing his opposition in terms that echo biopolitically framed critiques of the violence of colonial governance. Part of what distresses him is that these people are not being told the truth: not only will they never return to earth, but the distance they must travel means contact is not even possible: "Their messages and reports now take years to get back to us. We will be bone dust while they are still hurtling through space." His mention of bone dust perhaps subtly references the famed bone from *2001*. As he speaks, the camera cuts to look up at the sky—the high life of space travel—and then cuts to a close-up of Monte's and Willow's feet on the ship, where he is teaching her to walk. Like most of the film, this sequence of shots of Willow's tentative steps is intercut with images from earlier times—one of Dibs sitting in a clinical room, another of Monte sleeping by the computer that authorizes their continued existence. Willow squats down due to her unstable footing, and Monte encourages her "up, up" as he motions her to try to stand. Here the film's title references both the "high life" of existence in space and the "high life" of erect, evolved hominids, that is, humans differentiated from other species. Contra *2001*'s technological leap, Denis emphasizes bodily, animal life even in a highly technological context: the tactile comfort of contact between Willow and Monte, the physical vulnerability of infants who must each individually learn this erect, "human" posture.

Setting the interview with the professor on a train connects it with a scene that follows of Boyse (Mia Goth)—Willow's mother—on earth before she was imprisoned. She and several others ride on a freight train, dirty and disheveled, calling to mind crises of homeless and displaced migrants familiar from news screens. Near the end of this sequence, we hear Monte in voiceover explaining their recruitment: "We were scum, trash, refuse who didn't fit into the system. Until someone had the bright idea of recycling us. Some of us got life sentences; some of us were on death row. The agency made us an offer to serve science." Partway through this voiceover, the imagery shifts from the train to scenes of a younger Monte (Mikolaj Gruss) committed to the justice system, then the adult, incarcerated Monte (Pattinson), soon to be transferred into this program. The recruitment logic succinctly captures neoliberal elements within biopolitical governance: the requirement that life prove its economic value as justification for its continued existence. As people, Boyse and her compatriots have no future—no jobs, no family, no hope. Rather than continuing as a drain on the state via the costs of their incarceration, they can be transformed into productive value by becoming experimental subjects. They have already been given life sentences, and so the lethal risks of the mission do not raise ethical concerns—or at least not from a market valuation point of view. Here we have another possible meaning of the title "high life"—life that is valued more highly if servicing economic ends.[6]

A central part of the film's engagement with biopolitical governance, then, is its critique of how life is stratified into valued and nonvalued instantiations. The numeral "7" prominent on the ship and on some items of clothing further reinforces that the society that sent these people into space had no difficulty sacrificing some lives for the sake of others deemed more valuable: theirs is one of multiple such experiments. The dehumanization that fuels this sorting is emphasized all the more near the end of the film when Monte and a now-adolescent Willow (Jessie Ross) encounter an identical ship labeled "9," this one filled with dogs, another set of experimental victims—moreover, the numbering suggests that humans may have been sent up first. Monte finds puppies on this craft who, like Willow, demonstrate the fecundity of life that exceeds attempts to constrain it through biopolitical governance: even in the dangerous environment of space, life finds a way.

The film thus interrogates the issue of equivalences among kinds of living beings: Monte is condemned for killing the girl who killed his dog, a punishment that makes sense only within an anthropocentric frame. Moreover, Monte's gentleness with Willow, his tender care of her despite his own isolation, establishes him as a humane and caring person, not someone who easily and callously takes other lives. Is he wrong, then, to prioritize the life of the dog he cares for over that of someone able to engage in the casual disregard of other life that the girl presumably displayed? Yet the possibility that Monte also might be dangerous haunts the film. In an early sequence he coos to Willow, "could have drowned you like a kitten, would be so easy," and it is not clear if this is his temptation or his memory of things his own father might have said to him. Such nonchronological sequencing raises other doubts: we see Monte dispose of his crewmates' bodies long before we learn how each died. Most of the deaths are violent and several due to interpersonal violence, but Monte is not responsible for any on the ship. Murders during the voyage are graphically violent, whereas the death of the girl on earth is shown only via still images, most separated by other, longer sequences. The film seems to ask, if Monte was wrong to kill the girl who took his dog's life, why is it acceptable for the state to put his life in jeopardy as a form of criminal justice?

If perhaps it seems an easy choice to privilege caring for a baby over caring for a dog, *High Life* complicates this calculation with its depiction of the mission's second purpose: reproduction. Dibs harvests sperm from the men in exchange for drugs and coercively impregnates the women via artificial insemination. Her techniques are implicated in an attempted rape—Dibs ties Boyse to her bed to prevent her from douching, leaving Boyse vulnerable to an attack that ends with several injuries and the murder of the rapist. What's more, the first death (in the diegetic linear timeline) is the Black inmate Elektra

(Gloria Obianyo), who dies during childbirth (the infant dies soon after). Looking at her body in the morgue, Tcherny (André Benjamin),[7] the only other character of African descent, remarks, "Even up here, Black ones are the first to go," linking the film's themes to critiques of racialization as an aspect of biopolitical governance. Elektra's death in a fertility experiment also maps to the reality of global markets in reproductive services that typically involve women of color providing surrogacy and other services—which can compromise their own fertility—to augment reproductive possibilities for predominantly white clients.[8] Does the promise of future human life via the reproductive experiment warrant the sacrifice of the imprisoned women's lives? Dibs, who calls herself "devoted to reproduction," is a far more sinister and damage-inflicting character than Monte, although she is also involved in nurturing new life via this experiment. The tension between what may seem just according to abstract notions of how life is to be valued, as compared to what seems just based on dispositions and choices, suggests the failure of biopolitical categories to account for living realities.

The final conundrum is that Dibs is a prisoner as well, condemned for suffocating her own children. Her aggressive role in manipulating the others to participate in reproductive experiences suggests that she is also motivated by a logic of life-for-life equivalence—new babies making up for her murdered children. The damage she does shows how easily biopolitical strategies of governance can turn thanatopolitical in relation to devalued aspects of life: the babies and the mothers are all expendable, with Willow the only baby who lives. In an exchange with Boyse, Dibs calls her a "filthy little crackhead" when Boyse is reluctant to participate. When they finally near the black hole, Boyse murders a crewmember to hijack the small vessel intended for the Penrose Process and then dies by suicide by subjecting herself to gravitational forces that make her body implode. Willow is Monte and Boyse's daughter, but her birth is the result of a bizarre rape: Dibs drugs the crew and she rapes a barely conscious Monte, collects his sperm, and uses a tool to inseminate a comatose Boyse (figure 13.2). The film depicts conception using an abstract visualization that is both an image of cell division in Boyse's uterus and an image of the cosmos—this footage is modeled on the Orion Nebula, an interstellar formation visible to the naked eye from earth. Willow's specific future is thus intermingled with the idea of human futurity on a cosmic scale.

While my reading of *High Life* has focused on how the film illuminates the damage done by biopolitical governance in its mode of "let die," the film ends on a more hopeful note, as biopolitical theory also helps clarify. Dibs rapes Monte because he chooses to withhold his "fluids," an action that can be seen as refusing the normative logic that demands we reproduce more of the same—more babies, more humanity despite ecological collapse, more energy to continue global capitalism. Monte knows the prisoners are disposable in the state's view, and while he never directly challenges this judgment, he repudiates perpetuating the system. At the same time, he does not let the violation of the rape determine his relationship with Willow. Twenty minutes from the end, Dibs dies, leaving the infant Willow in a shielded incubator, at which point we see the first physical encounter between Monte and his daughter (in diegetic not cinematic time). When he opens the incubator in response to her screams, they are the only two people left alive

FIGURE 13.2: Dr. Dibs (Juliette Binoche) experiments on life in *High Life* (2018). A24 Films.

on the ship. The interior is suffused with a cool light that reflects on Monte's face as he pauses for an extended period, looking down at the baby and contemplating, before he finally takes her hand. As we have already seen, a deep and loving relationship follows. Here, the film cuts to an older Monte and adolescent Willow, both working to keep the ship functional. Although Willow continues to watch the images from earth in an attempt to understand human culture, it is clear they are alien to her, lacking any context. After the encounter with Ship 9 and its dogs, they see another black hole nearby and decide to enter it using the smaller ship. They must leave everything else behind, including all the data gathered on the trip, suggesting a complete break with earth, its past, and the mission's objectives.

Willow can be read as a new kind of human creature, one who does not root her identity in hierarchy and difference from other life, a point made clear by their banter (he calls her "rat faced"; she responds by praising rats' intelligence). As they suit up, Monte tells Willow she is "like no one else, and I love that." He has raised her with a tenderness denied to him as a child, and she has never experienced the devaluation enacted within biopolitical frameworks. In the darkness of space, a golden light illuminates their faces as they get closer and closer to the event horizon, marked by a band of glowing particles. In the final images, they stand side-by-side, no longer separated from the viewer by their suit helmets, as he takes her hand and asks, "Shall we?" The camera cuts to the reverse-angle shot of the golden light of the event horizon, which grows to fill the screen as the film ends. Rather than death, this moment embodies the possibility of a new beginning, a humanity defined by love and connection rather than what Roberto Esposito characterizes as the "autoimmune illness" that typifies a biopolitics in which "the protection of life" turns into "its potential negation" (2008, 116). Rather than an image of the transcendent next stage of human evolution, *High Life* offers a celebration of embodied connection and life's fecundity, a metaphor for human futurity that, in the clear language of biopolitical film, asks us to rethink what we value about "life itself."

## Notes

1. We should note that Foucault does theorize that a new "modern" racism can arise with biopolitical governance, a splitting of the population into "healthy" and "sick" based on biological identity. He further observes a dialectical relationship between the two, such that fostering the health of the overall community is accomplished by the disappearance of unhealthy elements within it. To understand this logic, one need only think of anti–welfare state rhetoric which contends that the overall health of the citizens can only be secured by removing the burden that welfare recipients are perceived to be. Foucault did not, however, focus on skin color and systemic racism as more conventionally understood, thus motivating this more pointed critique.
2. See my "Promissory Futures" (Vint 2019) for a more detailed discussion of this point.
3. The Penrose Process is an actual theory in physics, and Denis consulted with physicist Aurélien Barrau on this and the visualization of the black hole in her film. The idea is to decrease the angular momentum of the black hole, a loss of momentum that is emitted as energy. The physics has to do with distinct ways that entropy operates near the event horizon of a black hole, as described by the physics of quantum entanglement. Unlike the rest of the universe, black holes maintain constant entropy.
4. The choice of this film as intertext is intriguing. Often anthropological representations of indigenous cultures are framed in terms of their presumed coming extinction, as an attempt to capture aspects of such lifeways before they disappear. This film thus might also be understood as reinforcing imagery that suggests the potential end of all human life that this mission suggests.
5. My reading of this film sees many of its images as ironic, but *2001* is typically interpreted as supportive of a teleological narrative about human advance.
6. Several online commentaries also note that the term "high life" is used in parts of Africa to refer to the life of white people and thus intersects with the subtext of racialization that is central to a biopolitical sorting of valued and nonvalued life. Denis was raised in French colonial Africa and thus would be familiar with this usage.
7. Benjamin is also known as Andre 3000, part of the Afrofuturist music group Outkast, a link between the film and afrofuturist sf.
8. These global markets and their racialized character have been analyzed by Kalindi Vora (2015) and by Melinda Cooper and Catherine Waldby (2014).

## Works Cited

Agamben, Giorgio. *Homo Sacer: Sovereign Power and Bare Life*. Trans. Daniel Heler-Roazen. Stanford, CA: Stanford UP, 1998.

Agamben, Giorgio. *The Open: Man and Animal*. Trans. Kevin Attell. Stanford, CA: Stanford UP, 2004.

Arendt, Hannah. "The Perplexities of the Rights of Man." In *Biopolitics: A Reader*, edited by Timothy Campbell and Adam Sitze, 82–97. Durham, NC: Duke UP, 2003.

Cooper, Melinda. *Life as Surplus: Biotechnology and Capitalism in the Neoliberal Era*. U of Washington P, 2011.

Cooper, Melinda, and Catherine Waldby. *Clinical Labor: Tissue Donors and Research Subjects in the Global Bioeconomy*. Durham, NC: Duke UP, 2014.

Esposito, Roberto. *Bios: Biopolitics and Philosophy*. Trans. Timothy Campbell. Minneapolis: U of Minnesota P, 2008.

Foucault, Michel. *The Birth of Biopolitics. Lectures at the Collège de France, 1978–1979*. Trans. Graham Burchell. New York: Picador, 2004.

Foucault, Michel. *The History of Sexuality: An Introduction*. Volume 1. Trans. Robert Hurley. New York: Vintage, 1978.

Foucault, Michel. *Society Must Be Defended: Lectures at the Collège de France, 1975–1976*. Trans. David Macey. New York: Picador, 2003.

Mbembe, Achille. "Necropolitics." In *Biopolitics: A Reader*, edited by Timothy Campbell and Adam Sitze, 161–93. Durham, NC: Duke UP, 2003.

Puar, Jasbir. *The Right to Maim: Debility, Capacity, Disability*. Durham, NC: Duke UP, 2017.

Rabinow, Paul. *Anthropos Today: Reflections on Modern Equipment*. Princeton, NJ: Princeton UP, 2009.

Rose, Niklas. *The Politics of Life Itself: Biomedicine, Power and Subjectivity in the Twenty-First Century*. Princeton, NJ: Princeton UP, 2006.

Sunder Rajan, Kaushik. *Biocapital: The Constitution of Postgenomic Life*. Durham, NC: Duke UP, 2006.

Vint, Sherryl. "Promissory Futures: Reality and Imagination in Finance and Fiction." Ed. David Higgins and Hugh O'Connell. *Centennial Review* 19, no. 1 (2019): 11–36.

Vora, Kalindi. *Life Support: Biocapital and the New History of Outsourced Labor*. U of Minnesota P, 2015. Kindle ed.

Weheliye, Alexander. *Habeas Viscus: Racializing Assemblages, Biopolitics and Black Feminist Theories of the Human*. Durham, NC: Duke UP, 2014.

Wynter, Sylvia. "Unsettling the Coloniality of Being/Power/Truth/Freedom: Towards the Human, After Man, Its Overrepresentation—An Argument." *New Centennial Review* 3, no. 3 (2003): 257–337.

# CHAPTER 14

# CULT BEHAVIORS

## JEFFREY ANDREW WEINSTOCK

THIS chapter uses the idea of the "cult science fiction (sf) film" as a means for analyzing a type of film that has tended to trouble sf categorization: what I refer to as the quirky dystopian sf film. These films, including Terry Gilliam's *Brazil* (1985), Alex Cox's *Repo Man* (1984), Yorgos Lanthimos's *The Lobster* (2015), and Boots Riley's *Sorry to Bother You* (2018), are particularly useful for thinking about the relationship between sf and cult movies because, through various characteristics—including absurd plots, surreal elements, low-affect delivery of lines, black humor, and cynical themes—they amplify the cultic features that are consistently inherent in sf; that is, they make an already strange genre seem even weirder. By undercutting or exaggerating generic features, these quirky dystopian films thus exemplify how cult films "behave" and offer illustrations of the process of cult sf film formation.

Before turning to this subgenre, we should first acknowledge that the parameters of a cult film in general are far from firmly fixed. Although the term developed organically in the twentieth century to refer to nonmainstream films with extremely loyal fan bases, the rubric in the twenty-first century often functions largely as a marketing device, used to designate any film perceived as antagonistic (intentionally or not) toward "mainstream" values or established conventions of Hollywood filmmaking.[1] This is a wide category and includes films such as Ed Wood's *Plan 9 From Outer Space* (1959) and Tommy Wiseau's *The Room* (2003), both of which are unintentionally "bad" for technical reasons, such as unconvincing effects and sets, choppy editing, and amateurish shots, as well as wooden dialogue and stilted deliveries. But it also takes in films such as David Lynch's *Eraserhead* (1977) or Spike Jonze's *Being John Malkovich* (1999) that are intentionally "weird" because of their absurd or surreal imagery, themes, and situations; movies with nonlinear or absent plots that make them difficult to follow, like Richard Kelly's *Donnie Darko* (2001); or films that freely mix up genres, as in the case of the Coen Brothers' *Barton Fink* (1991). The category also includes films that are deemed "tasteless" or "offensive" because of camp, gore, explicit sexuality, or a "permissive" attitude toward "nonnormative" sex/sexuality, or other violations of cultural decorum, such as in John Carpenter's *Pink Flamingos* (1972). Related to tastelessness, the cult film category also includes works that are construed as "juvenile" because of silliness for its own sake, such as Terry Gilliam and Terry Jones's *Monty Python and the Holy Grail* (1975). We might also file under this rubric various movies whose messages run contrary to prevailing ideologies. They may seem naïve or alarmist from a contemporary perspective and thus ridiculous (Lawrence Meade's *Reefer Madness* [1936]), they may champion generally discredited conspiracy theories (Peter Joseph's *Zeitgeist: The Movie* [2007]), or they may be perceived as discouragingly cynical, disturbing, or "dark," as in the case of Lars von Trier's *Dancer in the Dark* (2000). However, this is not a complete list of characteristics that incline a film toward cultdom; moreover, these aspects often overlap.[2]

The "cult film" thus often seems a kind of catch-all category—a miscellaneous folder or island of misfit films. An IMDb list of cult films captures well in its title this heterogeneity of the type: "2,100+ Weird, Bizarre, Strange, Extreme, Odd, Atrocious, Random, Outrageous, Trippy, WTF, Extraordinary, Silly, Crazy, Intense, Unusual, Wild, Surreal, Disturbing, Fantastic, Audacious, Incredible, Amazing, Freaky, Awesome, Cult-Classic, Must-See Movies!" (redmiatazoomzoom 2013). The question of whether films categorized or characterized as cult films in the twenty-first century actually command a cult following is in fact curiously difficult to answer. Some movies famous for being cult works such as Jim Sharman's *The Rocky Horror Picture Show* (1975), *Monty Python and the Holy Grail*, or *Plan 9 from Outer Space* obviously do, as one can point to such obvious cultural markers as campus screenings, midnight showings, communal forms of celebration, and allusions or references to these films in other places. However, how might one then conclude, as the title of Shane Scott-Travis's article for *Taste of Cinema* (2019) puts it, that Isaac Ezban's *Parallel* (2018) or Bodo Kox's *The Man in the Magic Box* (2018) might be among "great recent cult movies you may have missed." It seems likely that what Scott-Travis means is that these films, reflecting qualities outlined earlier that make them bad, weird, challenging, tasteless, austere, gloomy, and so on, are—as Colin McCormick's *Screenrant* article (2019) proposes—"*destined* to become cult classics" because, "while niche, [they] are bound to make their mark with the right audience" (italics added). We might thus think of the twenty-first-century cult film as a movie that, for one reason or another, has selective ("niche") appeal but is assumed to command (or will someday command) intense affection from an appreciative minority. Ironically, the contemporary cult film, sf or otherwise, may not (yet) actually have a cult following.

Of course, many films can be described as having "selective appeal," and not all of them end up on lists of cult films (or lists of cult films in the making), so it is reasonable to ask what distinguishes existing or nascent cult films from the immense body of films created since the end of the nineteenth century. To a certain extent, there will always be an element of chance involved in the attainment of cult status—*Rocky Horror* arguably might not have achieved canonical cult status in the way that it did (or at all) had it not been repackaged as a "midnight movie" and shown at the Waverly Theater in Greenwich Village, or if, as the story goes, audience member Louis Farese Jr. hadn't shouted vulgarly at the character Janet (Susan Sarandon) to get an umbrella (Weinstock 2007, 33).

There is, however, a general quality that arguably characterizes all films that achieve cult status: an element of *excess*. The cult film may be so bad, it is good; the cult film may be one that exceeds the conventions of a particular genre by being so gory, violent, campy, confusing, or explicit that even fans of a particular genre are surprised and potentially alienated; cult films often overflow the constraints of genre altogether and thus become hard to categorize; or the cult film may evoke a powerful—and generally disturbing—affective response through specific imagery and scenarios, the general plot line and conclusion, or overall attitude and implications. The appeal of the cult film thus usually derives from a pattern involving *amplification* resulting in *inversion*: the film exceeds what is considered appropriate, acceptable, or conventional for a given genre, and this amplification of genre qualities alienates the majority while being embraced by a minority who find it central to their enjoyment of the film. What the cult film fan loves represents an inversion of values: the weirdness, badness, cynicism, absurdity, and so on that alienate other viewers are what the cult film fan embraces about the film. The excessive element of the film elicits an excessive response from particular viewers.

Fantasy and sf films, as J. P. Telotte appreciates, are arguably born with one foot on the path toward such a cult status, both because they cater to a niche (although increasingly broad) audience from the start and because the picture of the world they present inevitably deviates from the world with which we are familiar. Thus Telotte notes that the "relationship between the cult film and the genre of science fiction . . . has always seemed closer, somehow more *natural* . . . than in the case of most other film genres" (2015, 1). This situation is arguably because the "novum" at the heart of the sf work—the critical difference from the world as we know it around which the plot revolves—ensures that the world of the work will be something other than the one we know. This difference may involve all the conceits of sf from faster-than-light travel and alien races to time travel, artificial intelligences, and so on. The consequence of this difference is what Darko Suvin famously refers to as "cognitive estrangement" (1979, 4). As Sherryl Vint summarizes, science fiction "is defined by the skewed perspective it encourages us to take on both reality as experience and reality as it is represented in realist fiction"—and, by extension, film and other media; "sf," she continues, "forces us to confront ideas and conventions that have been *made to appear* natural and inevitable, by giving us a world founded on other premises" (2014, 39). Put simply, sf is implicitly weird or even cultlike because the picture it presents of the world includes something beyond our ken. It shows us a world that, in some way, does not reflect the world we know; that is what makes it *speculative* fiction.

And yet, for an sf film to become a cult film, it still needs to be in some way excessive, even for a genre that by definition exceeds the real (at least beyond the real at the time of the text's creation). In their deviation from the reality we know, sf films are, as we have noted, inherently weird. The cult sf film is one that, in some way, is excessive in relation to understandings of the genre and thus stands out by being weirder, stranger, quirkier, or more different in quality than most other sf films.

As with cult films in general, there is no single formula for the cult sf film, and cult sf films clearly evidence the same kinds of characteristics that mark cult films overall. One, therefore, might speak of the campy cult sf film (Roger Vadim's *Barbarella* [1968] or Mike Hodges's *Flash Gordon* [1980]), the "so bad it's good" cult sf film (*Plan Nine from Outer Space* or Phil Tucker's *Robot Monster* [1953]), the extremely weird or trippy cult sf film (Slava Tsukerman's *Liquid Sky* [1982] or John Boorman's *Zardoz* [1974]), the cerebral cult sf film (Andrei Tarkovsky's *Solaris* [1972] or Kubrick's *2001*), the gross-out sf film (David Cronenberg's *The Fly* [1968] or Stuart Gordon's *Re-Animator* [1985]), the graphically violent cult sf film (Stanley Kubrick's *A Clockwork Orange* [1971] or Kinji Fukasaku's *Battle Royale* [2000]), and so on. It is notable as well how many cult sf films are ones that fuse genres—for example, Ridley Scott's *Blade Runner* (1982) is a mash-up of sf and film noir; John Carpenter's *The Thing* (1982) and Ridley Scott's *Alien* (1979) are sf horror films; Michael Crichton's *Westworld* (1973) fuses sf with the Western. In some cases, what differentiates the cult sf film is an overall "attitude," such as the flippancy of John Carpenter's *They Live* (1988) or, as discussed later, the punk refusal of Alex Cox's *Repo Man* (1984).

Despite the fact that cult sf films are excessive in the same ways as are films from other genres, this excessiveness is obviously filtered through the familiar conceits of the genre, which, of course, is what most differentiates the cult sf film from those in other genres. Thus, cult sf films involve themes such as dystopian societies and the place of the individual, the encounter with the other (artificial intelligence or alien), time travel, the perils of science, transformation of the self, and so on. In this way, cult sf films, in keeping with sf in general, are well poised to serve as "critical commentary." As Vint explains, the "gap" between the reader's world and the fictional one prompts us "to see given reality in a new way" (2014, 54). This is the primary "cultural work" performed by the conceits of sf as their participation in imagining a new world allows us to reflect critically on the one we know (Vint 2014, 5)—we look back on our own world as from an outside perspective. In the cult sf film—and particularly in ones featuring dystopian societies—this critical commentary is often caustic indeed.

## QUIRKY DYSTOPIAN CULT FILMS

With this description in mind, I would now like to address a very specific sf film subcategory that has been one of the most fertile sites for cult contributions—the "quirky dystopia" film. Dystopian societies, of course, have often been featured in sf literature and film and are usually presented as cautionary tales. The audience for dystopian sf such as Margaret Atwood's *The Handmaid's Tale* (1985), Michael Moore's *V for Vendetta* (1982–85, 1988–89), George Orwell's *1984* (1949), and their various adaptations is supposed to appreciate the society represented as oppressive and to reject its premises. While it is always risky to extrapolate authorial intent, we are on fairly safe ground conjecturing that, as readers or viewers, we are not supposed to approve of the oppression of women in Atwood's Gilead, for example, or the invasive surveillance culture of Oceania in Orwell's novel. Similarly, readers and viewers of postapocalyptic dystopian narratives are not supposed to desire worlds in which everything boils down to a constant quest for survival against discouraging odds.[3] As a result, such narratives are deeply ideological, naturalizing a particular set of assumptions about the role of government, and the relationship between the individual and the state. The state in *The Handmaid's Tale, V for Vendetta,* and *1984* is too oppressive, rigidly circumscribing opportunities for individual growth, expression, and happiness in draconian ways; in contrast, the absence of any central government in postapocalyptic dystopian wastelands such as Jack Smight's *Damnation Alley* (1977), the Mad Max films, or the Hughes Brothers' *The Book of Eli* (2010) catalyzes lawlessness and the rise of warlords and thugs. Consumers of dystopian narratives are thus, on the one hand, positioned to reject "undue" constraints on personal autonomy by unelected government officials while, on the other hand, appreciating that some measure of government is necessary to secure personal liberty—because the nature of humanity in postapocalyptic narratives seems invariably Hobbesian.

Dystopian narratives, whether involving invasive government or the absence of government, are generally pessimistic in tone; in some cases, a kind of "chosen one" character can incite a revolution that remakes the society—the rebels in the *Hunger Games* series prevail in the end—or society starts to rebuild itself at the end of a postapocalyptic story. More commonly though, any victory is severely circumscribed by the insurmountable obstacles of the setting: one man cannot overthrow the state in *1984,* nor does Max (Tom Hardy) and Furiosa's (Charlize Theron) victory over Immortan Joe (Hugh Keays-Byrne) in George Miller's *Mad Max: Fury Road* (2015) somehow restore water to a desiccated landscape. The bleakness of such narratives—invested in the infuriating impotence of the protagonists to make real change or contend with overwhelming structural forces—is arguably part of the reason such films are "niche." While a lot of action, an optimistic conclusion, some sexy starring actors, and a big publicity budget can offset the alienating frustrations of a dystopian setting, more cynical or grittier versions seem destined for the cult film category.

What I am calling the quirky dystopian film is a type of black comedy that uses absurdity to offset the despair of the dystopian setting and often to amplify the film's critique of contemporary life by exaggerating features perceived as negative. Into this category, I would place films such as Rachel Talalay's *Tank Girl* (1995), John-Pierre Jeunet's *Delicatessen* (1991), Mike Judge's *Idiocracy* (2006), John Carpenter's *They Live,* and L. Q. Jones's *A Boy and His Dog* (1975), as well as the four films I discuss here: Gilliam's *Brazil,* Cox's *Repo Man,* Lanthimos's *The Lobster,* and Riley's *Sorry to Bother You.* The first two films are mid-1980s works that undeniably fall into the category of cult sf films; the second two, *The Lobster* and *Sorry to Bother You,* evidence all the qualities of cult films in the making. All four are defined by a particular kind of excessive weirdness, as their absurdity creates a jarring tonal dissonance between the depressing setting and message, and their ridiculous comic elements. As such, they exemplify the features that commonly mark the cult sf film.

Gilliam's *Brazil* is a useful starting point for any discussion of quirky dystopian work because it encapsulates so much of the preceding discussion. Reflecting one of the titles considered for the film, "1984½," *Brazil* is obviously indebted to Orwell's novel for its depiction of an excessively bureaucratic, totalitarian government—and, indeed, what arguably stands out about the film most are its impressive sets featuring brutalist government architecture and the infuriating Kafkaesque labyrinth of bureaucratic rules and regulations to which citizens are subject (Matthews 1998, 45). The world of *Brazil* is one in which bureaucracy exists for its own sake—there are forms for everything, receipts for the forms, and receipts for the receipts. And it is this frustrating world of surveillance and forms that government employee Sam Lowry (Jonathan Pryce) must negotiate as he attempts to win the literal woman of his dreams. The complication is that this woman, Jill Layton (Kim Greist), having witnessed the mistaken arrest of her neighbor, is now considered a terrorist for attempting to report the government's error.

What differentiates *Brazil* from a more conventional dystopian film are its ridiculous elements, which introduce a tension with the film's overall grimness. The true object of the government's attempt at apprehension was not Archibald *B*uttle, a cobbler, but Archibald *T*uttle (Robert DeNiro), a renegade air conditioner repairman who does not play by the rules and exits from surreptitious repairs via zipline. Sam's mother, Ida (Katherine Helmond), and her crony Alma (Barbara Hicks) are addicted to plastic surgery, and the viewer is treated to a scene of Ida's face literally being stretched back like plastic. Once transferred to the Information Retrieval Department (so that he can dig up information on Jill), Sam fights for half a desk, which he shares with another employee on the other side of his office wall; meanwhile, his boss roves the hallways pursued by an entourage of form-waving employees as he barks orders. After various twists and turns, the film ends with an extraordinary, and increasingly surreal, chase sequence. Sam, having been arrested by the government for his attempts to assist Jill, is about to be tortured by his former friend Jack Lint (Michael Palin) on a platform in what seems to be the interior of a nuclear reactor cooling tower when he is rescued by Tuttle and members of the resistance who rappel down from above and then blow up the Ministry of Information building; Tuttle is swallowed up by a whirlwind of papers, and Sam, pursued by police, stumbles into the funeral of his mother's friend Ida, who has died from excessive plastic surgery—her body seemingly having dissolved into a mass of slush and bone (see figure 14.1). Sam tumbles into her casket and lands in a street from his daydreams, where he is attacked by zombielike monsters on one side and cornered by the police on the other. Just in the nick of time, he is rescued by Jill, and the two leave the city together. But this happy ending is abruptly revealed to be only fantasy; in reality, Sam remains strapped to the torturer's chair, lobotomized and insane, humming "Aquarela do Brasil," the song from which the film takes its title.

This ending is somber, the sequence leading up to it increasingly surreal, and the overall tenor absurd. *Brazil*—which, in some respects, is a greatly expanded *Monty Python's Flying Circus* sketch—uses absurdity to highlight the dangers of unchecked government power and, as Kafka understood, the ways in which bureaucracy functions as a façade to conceal malevolent intent and frustrate citizen resistance. With its manic chase sequences, steampunkish technology, imposing sets, and fantasy sequences, the film removes us from the world we know; however, always lurking in the background is the real world of bureaucratic forms and surveillance. *Brazil* is thus simultaneously disheartening and ridiculous in its presentation of its doomed hero, an everyman seeking love and contentment in a world that seeks to reduce individuals to cogs in a

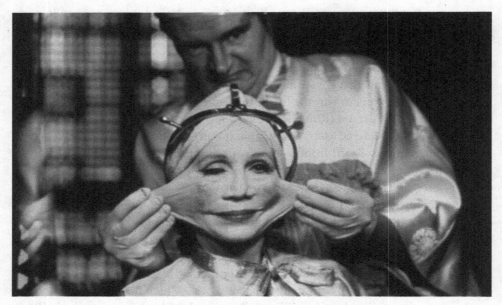

FIGURE 14.1: Preparations for plastic surgery in *Brazil* (1985). Universal Pictures.

machine. At 143 minutes in length (the director's cut), the film is indulgent, leading its viewers, like its protagonist, down seemingly endless corridors and through labyrinthine streets. In his dreams, Sam can fight the monsters, rescue the damsel, and soar through the clouds; yet in reality, the world is gloomy and the monsters (who are not zombies or giants but men in suits) win—and this arguably is not a message with broad appeal.[4] As a consequence of all its excesses, the film serves as a prime example of the cult sf film, but especially one that exemplifies the quirky dystopia.

Released a year prior to *Brazil*, Alex Cox's *Repo Man* is also a quirky dystopian sf black comedy that interweaves absurdity and surreal elements for comic effect, but it also does so to highlight ridiculous aspects of everyday existence. In some respects, the films are curiously similar: in place of renegade air conditioner repairmen, *Repo Men* gives us repo men, car repossessors operating on the margins of or indeed outside the constraints of the law. Similarly, Jill's punk aesthetic in *Brazil* is a reflection of *Repo Man*'s punk aesthetic, the film's dominant mode. The government in *Repo Man*, typical of dystopian and conspiracy-theory films, is shown to be concealing its operations and surveilling citizens, and it is through a labyrinthine urban landscape that the protagonists of each film move in search of something that will provide liberation. Each film even has a character who has been lobotomized![5]

Yet tonally, *Brazil* and *Repo Man* are quite different—both literally and figuratively. As noted, *Brazil*'s title is taken from the song "Aquarela do Brasil" by Ary Barroso. In keeping with the fantasy scenes in which it is featured most prominently, this samba is soothing and uplifting. Granted, it is used for ironic effect, especially at the end, where it contrasts jarringly with Sam's actual situation, suggesting that comfort and liberation are available only in one's dreams. However, the centrality of this song in particular, and music in general, to the overall meaning and experience of the film, as noted by W. Russel Gray, "surprises and delights" (1999, 148).

In contrast, *Repo Man* signals its much more aggressive posture toward the confining effects of contemporary culture through its punk rock soundtrack, which meshes with its overall punk aesthetic. *Repo Man* is, put concisely, a punk sf film, and the songs used in the film, several of which work in diegetic fashion as protagonist Otto (Emilio Estevez) attends a party or a nightclub, are not only a primary means through which the film expresses its refusal of modern values, but also very much implicated in its cult film status.[6] Black Flag's "TV Party," for example, is a tongue-in-cheek satire on the vapidity of 1980s television. The Circle Jerks' "When the Shit Hits the Fan" expresses the joylessness of economic disparity in the midst of Reaganomics. Suicidal Tendencies' song "Institutionalized" is an angry screed about a frustrated young man misunderstood by his parents and subject to other people's evaluations of his mental state. Less angry but more absurd is Burning Sensations' version of "Pablo Picasso," which explains that, despite his short stature, the famous painter was a hit with the ladies and was "never called an asshole—not like you." Music, both diegetic and nondiegetic, is thus central to setting the mood and conveying the film's antagonistic position vis-à-vis its 1980s setting. The film then furthers this critique by having a listless Otto confront the "ironic deflation of American mythology" at every turn (Bould 2020, 63). His zombified parents have been duped into donating their money to a television preacher, grocery store shelves are stocked with wholly generic products, and the streets are trash-strewn and graffiti-covered (see figure 14.2).

FIGURE 14.2: Otto (Emilio Estevez) amid the generic goods at the Pik 'n Pay (*Repo Man*, 1984). Universal Pictures.

But despite its angry punk tone, *Repo Man*, like *Brazil*, softens its aggressive edge through the interweaving of absurdity, which is intrinsic to the plot, as well as through the characters' dialogue. The plot focuses on a contest among repo men and government officials to retrieve a stolen 1964 Chevy Malibu driven by renegade scientist J. Frank Parnell (Fox Harris). Unbeknownst to the repo men, the trunk of the car apparently contains the radioactive corpses of aliens that vaporize anyone who comes into direct contact with them. The ridiculousness of the plot is accentuated by supporting characters who deliver bizarrely comic lines. For example, when Otto's former friend Debbi (Jennifer Balgobin) suggests to her boyfriend Duke (Dick Rude) that they "go do some crimes," his enthusiastic response is, "Yeah. Let's go get sushi and not pay." Later, when Duke is shot, he dramatically laments to Otto that "the lights are growing dim, Otto. I know a life of crime has led me to this sorry fate, and yet, I blame society. Society made me what I am"—to which the stoic Otto replies, "That's bullshit. You're a white suburban punk just like me." Much of the film's absurdity, however, is invested in the repo compound's mechanic and resident sibyl, Miller (Tracey Walter), who expounds a kind of weird Zen philosophy through lines such as the following:

> A lot o' people don't realize what's really going on. They view life as a bunch o' unconnected incidents 'n things. They don't realize that there's this, like, lattice o' coincidence that lays on top o' everything. Give you an example; show you what I mean: suppose you're thinkin' about a plate o' shrimp. Suddenly someone'll say, like, plate, or shrimp, or plate o' shrimp out of the blue, no explanation. No point in lookin' for one, either. It's all part of a cosmic unconsciousness.

Whether it is Agent Rogersz (Susan Barnes) telling Otto that people just explode from time to time due to natural causes, or the other repo men explaining to Otto with religious solemnity the repo man code, the film avoids taking itself too seriously—which finally is perhaps its most punk rock quality: it refuses even to commit to refusing, choosing a kind of absurdist nihilism over the pursuit of revolution or perhaps even celebrating the disappearance of conventional meaning.

In the end, *Repo Man* gives way to silliness entirely—and, in this, it is the antithesis of *Brazil*, offering a literal inversion of *Brazil*'s conclusion. In *Brazil*, Sam has always dreamed of flight, soaring above the clouds on golden wings. In the final sequence of *Brazil*, Sam and Jill escape the city together in her truck—and then reality comes crashing down as we discover none of it was real. Sam remains strapped in his chair, lobotomized and insane. At the end of *Repo Man*, Miller invites Otto into the Malibu, which, glowing green, takes flight, zooming across the city and then upward to the stars. It is a plainly preposterous and wholly campy conclusion to a film that (much like John Carpenter's *They Live* of three years later) has been simultaneously giving the finger to Reagan's America while winking at the viewer throughout—and it is this interweaving of comic absurdity with sf tropes and social criticism that arguably gives the film its cultic appeal.

# FUTURE CULT FILMS

*Brazil* and *Repo Man* find their contemporary parallels in *The Lobster* (2015) and *Sorry to Bother You* (2018), both of which are quirky dystopian sf films that introduce a jarring contrast between a grim dystopian society and comic absurdity, and both of which are arguably "cult films in the making." The premise of Lanthimos's *The Lobster* is as simple as it is strikingly bizarre: all adults must be in monogamous relationships; single individuals are changed into animals. This is the challenge confronting the film's protagonist David (Colin Farrell), whose wife has left him and who, as a consequence, has been required to check into a hotel for single people. He has forty-five days to find a partner with whom he shares some attribute or skill or he will be transformed into an animal. After David fails at masquerading callousness to partner with a notoriously cruel woman, he escapes into the woods and joins a group of "Loners" whose rules are the antithesis of the hotel's: romance among the Loners is forbidden under punishment of mutilation or death (see figure 14.3). David, however, falls in love with another Loner, identified in the credits only as "Shortsighted Woman" (Rachel Weisz). When their plans to escape together are discovered, she is blinded by the leader of the Loners (Léa Seydoux), and at the end of the film, David is seen preparing to blind himself so that they

FIGURE 14.3: David (Colin Farrell) resists the threat of animal transformation in *The Lobster* (2015). Sony Pictures.

will again share a fundamental quality. The film ends with the knife poised in front of his eye, but also with his obvious hesitation; whether he completes the act is left unknown.

*The Lobster*, which won the Jury Prize at the 2015 Cannes Film Festival, is the kind of film often described as a "critics' darling"—which tends to mean that it has primarily "niche" appeal. Characterized as a "dystopian sci-fi love story" (Hestand 2016), the film interweaves absurdity throughout, from the novum of the film—enforced partnering under threat of transformation into an animal—to smaller details such as hotel guests hunting Loners in the woods with tranquilizer darts, prohibiting masturbation for hotel guests while hotel maids stimulate guests (with lap dances for the men) as part of their duties, and Loners holding silent raves in the woods, dancing on their own to electronic music through headphones. The surreal humor of the film is encapsulated in the film's title, *The Lobster*, which reflects David's choice of animal should he fail to find a partner. When asked about his decision by the hotel manager (Olivia Colman), he explains, "Because lobsters live for over one hundred years, are blue-blooded like aristocrats, and stay fertile all their lives. I also like the sea very much. I waterski and swim quite well since I was a teenager." "I must congratulate you," replies the hotel manager. "The first thing most people think of is a dog, which is why the world is full of dogs. Very few people choose an unusual animal, which is why they're endangered. A lobster is an excellent choice." Never considering that he might end up on someone's plate, David's choice to be transformed into a crustacean and the hotel manager's approval reflect the absurd logic that prevails throughout the film—in a world in which transformation is a given for those who fail to partner, personal choice is expressed through picking the animal. It's a bit like being allowed to choose your method of execution for something not recognized as a crime.

Two other elements that stand out in terms of *The Lobster*'s cultic appeal are the hyperbolic awkwardness of the characters and the inconclusive ending. In the world of *The Lobster*, nothing is fluid or easy, and this constant underlying tension is conveyed through the stiff movements of the characters and their clipped, affectless delivery of lines. Unlike Otto in *Repo Man*, the characters in *The Lobster* seem desperate to connect, but they remain trapped within themselves, unable to speak or move easily. This awkwardness is perhaps most on display during David's unwieldy signing to the Shortsighted Woman in the woods. Since romance is forbidden among the Loners, David and the woman invent a complicated system of bodily contortions to communicate their desires. The motions are so obvious and clumsily performed that it would be difficult for any observer not to take notice of this form of communication. As for the ending, the film concludes with an unresolved moment of incredible tension—the protagonist poised to blind himself but hesitating: precisely the kind of inconclusive conclusion mainstream releases eschew when test audiences complain.

The animal transformation "novum" of *The Lobster* is, of course, preposterous from the start; however, it is also easily seen as a satire on contemporary cultural expectations concerning dating and marriage that assume monogamous partnering as the "natural" human desire. Through this ludicrous premise, the film highlights the extreme pressure exerted on individuals to find their "perfect match" and the ways not partnering can feel like failure—as though one were somehow less than human. The awkwardness evinced by the characters reflects the difficulty of negotiating cultural mandates that insist on dating and partnering as the only road to fulfillment. *The Lobster*, like *Brazil* and *Repo Man*, thus uses its absurdist premise as a form of cultural critique, highlighting the rigidity of cultural expectations and the ways they can alienate those whose desires flow in different channels.

While the "high concept" premise of *The Lobster* can be stated succinctly and the object of its critique precisely identified, the plot of *Sorry to Bother You* is messier and its social import broader. *Sorry to Bother You* is a film in two parts, focused on the experience of protagonist Cassius "Cash" Green (Lakeith Stanfield). In the first part of the film, Cash takes a job as a telemarketer for the RegalView company. Observing his lack of success, coworker Langston (Danny Glover) advises him to use his "white voice"—and not "Will Smith white" but a comical nasal parody of uptight whiteness. Once Cash lets his inner white voice out, the "green" starts rolling in. Management takes note of his success and he is soon promoted to "power caller," an elite team of telemarketers with their own elevator to the top floor. However, what the RegalView power callers are selling are not encyclopedias, but slave labor provided by a company called WorryFree.

While elements of fantasy are present in the first half of the film (notably, Cash is shown as though he is literally in the same room with those to whom he is attempting to sell things), it is in the second part of the film that the science fictional component is introduced. WorryFree CEO Steve Lift (an obvious jab at Apple's Steve Jobs), taking note of Cash's success, invites him to party at his mansion, where he explains that his company has created a human/equine hybrid to meet labor needs and they want a transformed Cash to be a kind of double agent among the "Equisapiens," serving as a leader for the horse people but really reporting to Lift. Cash refuses and, instead, leads a mission to liberate the Equisapiens from Lift's mansion, but at the end of the film he is transformed into a human/horse hybrid.

Like *Repo Man*, *Sorry to Bother You* takes capitalism as a focus of criticism—in this case highlighting the immorality of a system that places profit above all else. RegalView will sell anything, including slave labor and weapons, as long as there is a profit to be made, and WorryFree is engineering a race of genetic hybrids as slaves. Like *The Lobster*, *Sorry to Bother You* also uses the idea of the transformation of the human into an animal as a metaphor for thinking about social categorization. The top floor of RegalView is reserved for those who are—or who can pass as—white. In the basement of Lift's mansion are dark bodies engineered to provide slave labor. Playing on racist ideology associating Black bodies with animality, *Sorry to Bother You* in the end offers a messy critique of the racist underpinnings of specifically American capitalism. The film does this, however, through comic exaggeration and absurdity: Cash's "white voice" is performed by David Cross and overdubbed onto Cash; the special elevator taking "power callers" to the top floor of RegalView has a ridiculously long security access code; the most popular television program in the world of *Sorry to Bother You* is something called *I Got the Shit Kicked Out of Me*, in which "contestants" are beaten and humiliated; and, of course, there are Equisapiens. As in *The Lobster*, the absurd and comic elements of *Sorry to Bother You* convey social criticism via exaggeration. For Black bodies still coded via sedimented racist ideology as closer to animality than white bodies, attempting to pass as white remains a strategy and in some cases a necessity for social and professional advancement.

All four films addressed here fuse science fictional conceits with comic and absurd elements within a dystopian setting to convey more or less explicit social critiques. Because these films defy conventional expectations for genre and Hollywood film-making, they present challenges to viewers seeking comfort in the familiar—that is, they raise questions about how to consume and enjoy the film. Their "quirkiness," finally, becomes a marker of and for cultists who might find in the quirkiness a reflection of themselves and their desire for individuality. What unites Sam in *Brazil*, Otto in *Repo Man*, David in *The Lobster*, and Cash in *Sorry to Bother You* is that all four of them—in keeping with dystopian narratives in general—are fighting against a system that seeks to force them into a rigid mold. Their resistance is encapsulated in sf narratives that themselves resist conventions of genre; and those viewers who embrace the resistance and find enjoyment in tales told differently form the basis for such films' cult followings.

## Notes

1. Importantly, this does not mean that marketing a film as a cult film somehow conjures into being a zealous fan base. Rather, it means that the film is similar in kind to established cult films. It is to designate a film as cult based on qualities rather than the actual existence of a cult.
2. The understanding of cult film I advance here, contra Umberto Eco (always a perilous position to take), excludes classic or contemporary mainstream films that have dedicated fan bases. For better or worse, I consider these mainstream films with cult followings rather than cult films. See Eco (1985) on the cult film.
3. However, one must add that there is often an element of desire associated with the postapocalyptic landscape that liberates individuals from the stultifying conventions of modern life and allows them to pursue an authentic, if precarious, destiny.
4. As Jack Matthews (1998) chronicles, the ending of the film tested poorly and Universal Pictures insisted on a substantial reedit to give it a happy ending—a move that Gilliam vigorously contested, resulting after a lengthy battle in a modified 132-minute version with Gilliam's preferred ending.
5. J. Frank Parnell (Fox Harris), the scientist driving the prized Malibu, tells Otto (Emilio Estevez) before he dies that he has been lobotomized.
6. Mark Bould in "Cult Science Fiction Cinema" observes that he first encountered *Repo Man* by way of a bootleg cassette of the soundtrack (2020, 63).

## Works Cited

Bould, Mark. "Cult Science Fiction Cinema." In *The Routledge Companion to Cult Cinema*, edited by Ernest Mathijs and Jamie Sexton, 59–68. New York: Routledge, 2020.

Cohen, Anne. "Why *The Lighthouse* Should Be the New *Rocky Horror Picture Show*." *Refinery 29*, October 18, 2019. https://www.refinery29.com/en-us/the-lighthouse-halloween-horror-movie.

Eco, Umberto. "*Casablanca*: Cult Movies and Intertextual Collage." *SubStance* 14, no. 2 (1985): 3–12.

Gray, W. Russel. "Taking *Nineteen Eighty Four* Back to the Future, or There's an Awful Lot of Orwell in *Brazil* (Not to Mention Python)." *Utopian Studies* 2, no. 1–2 (1999): 147–56.

Hestand, Zac. "*The Lobster*, a Dystopian Sci-Fi Love Story." *Film Criticism* 40, no. 3 (2016). DOI: http://dx.doi.org/10.3998/fc.13761232.0040.325.

Matthews, Jack. *The Battle of Brazil: Terry Gilliam v. Universal Pictures in the Fight to the Final Cut*. Lanham, MD: Applause, 1998.

McCormick, Colin. "10 Movies of 2019 That Are Destined to Become Cult Classics." *Screen Rant*, June 27, 2019. https://screenrant.com/2019-underrated-movies-destined-cult-classics/.

redmiatazoomzoom, "2,100+ Weird, Bizarre, Strange, Extreme, Odd, Atrocious, Random, Outrageous, Trippy, WTF, Extraordinary, Silly, Crazy, Intense, Unusual, Wild, Surreal, Disturbing, Fantastic, Audacious, Incredible, Amazing, Freaky, Awesome, Cult-Classic, Must-See Movies!" *IMDb.com*, July 11, 2013. https://www.imdb.com/list/ls053942167/.

Scott-Travis, Shane. "10 Great Recent Cult Films You May Have Missed." *Taste of Cinema*, August 26, 2019. https://www.tasteofcinema.com/2019/10-great-recent-cult-films-you-may-have-missed/2/.

Suvin, Darko. *Metamorphoses of Science Fiction: On the Poetics and History of a Literary Genre*. New Haven, CT: Yale UP, 1979.

Telotte, J. P. "Introduction: Science Fiction Double Feature." In *Science Fiction Double Feature: The Science Fiction Film as Cult Text*, edited by J. P. Telotte and Gerald Duchovnay, 1–20. Liverpool: Liverpool UP, 2015.

Vint, Sherryl. *Science Fiction: A Guide for the Perplexed*. London: Bloomsbury, 2014.

Weinstock, Jeffrey Andrew. *The Rocky Horror Picture Show*. London: Wallflower, 2007.

# CHAPTER 15

........................................................................

# DIGITAL SCIENCE FICTIONS

........................................................................

## CHUCK TRYON

IF we were to look for the origin point for the birth of digital cinema, it might be the Wachowskis' science fiction (sf) trilogy, *The Matrix*. The first film of the series was released in 1999, at the very moment that millennial fears and fantasies were beginning to merge with the newly emerging worlds of the World Wide Web. Breathless accounts of virtual worlds gave life to fantasies that digital technologies would transform our lives, our selves, our world, and even our films. By imagining a virtual world—the matrix—where humans were plugged in to an online reality that mirrored our own world, only with cooler clothes and sunglasses, the Wachowskis used a distinctly science fictional language to speculate about the impact of technology on our future lives, including the possibility that these online spaces could, for good or ill, make identity itself into something malleable. *The Matrix* also invoked a cornucopia of new special effects, using computer-generated imagery (CGI) and other digital technologies to depict the film's future world. While its notorious "bullet-time" effects had appeared previously, *The Matrix*'s iconic use of that visual effect helped to provide viewers with the fantasy that, in the future, we might all slow down time in much the same way that users, wielding the remote controls of digital devices, could speed up, slow down, or even stop the movie they were watching (see figure 15.1). While these effects astonished theatergoers, they also became a key marketing tool for the DVD, which contained multiple making-of documentaries, depicting the special effects work that made the look of the film possible. In turn, these documentaries helped market the various Matrix films as technological cutting-edge works.

Besides these effects, *The Matrix* also helped to launch the emerging practice of what Henry Jenkins has called "transmedia storytelling," wherein a single story world sprawls across multiple media (2006). In addition to the film trilogy, the world of *The Matrix* included an animated film, a graphic novel, a companion website, *Matrix* video games, and more. As Jenkins explains, this expansive world also made it difficult for casual viewers to grasp certain story elements that appeared in the film sequels. Key storylines, including the unexpected appearance of Niobe (Jada Pinkett Smith) in *The Matrix Reloaded* (Wachowskis 2003), are mentioned in the video game but never fully

FIGURE 15.1: In the digital world of *The Matrix* (1999)—bullet-time effect. Warner Bros.

explained within the movies. Thus the transmedia character of the Matrix films inspired various fan interpretations, with websites dedicated to disentangling all of their narrative threads and intertextual references. Particularly attentive fans traced out allusions to philosophers, including the work of Jean Baudrillard, to world religions ranging from Christianity to Buddhism, and, of course, to other movies. As Jenkins readily admits, these practices of transmedia storytelling, including the explicit engagement with fan cultures, are largely driven by economic motives, as fans feel practically compelled to engage with their favorite story worlds across multiple platforms and even multiple media (Jenkins 2006, 102–6).

More crucially, *The Matrix* situated itself as a theoretical text, one that was working both to make sense of the emerging world of cyberspace and to amplify some of the digital regime's key myths. As Nick Couldry reminds us, digital refers to media that are "convertible into information bits of basically the same type" (2012, 15). Because of this malleability, the digital seemed to hold out the promise that we could transcend the physical world and our material, often deteriorating bodies, in much the same way that humans who jacked into the matrix were vibrant, energetic, and youthful, unlike their real, dormant bodies outside the artificial Matrix world. They could, as the cyberhacker Neo (Keanu Reeves) learned, bend spoons with their minds or, in an instant, master kung fu just by downloading it from a computer. In this sense, *The Matrix* was one of the most powerful articulations of the myth of the digital. According to Vincent Mosco, this idea of the digital as a "myth" holds that digital technologies can enable us to transcend history, geography, politics, even industrial circumstances (2004, 13). Digital technologies seem to promise that the barriers—geographical or technological—that prevent information from freely circulating have been eliminated. These myths seemed to be confirmed when images from a political revolution halfway around the globe, from ostensibly authoritarian governments in Egypt or Iran, could almost immediately show up in our Twitter feed, or when an unknown YouTuber could suddenly become an international celebrity.

These myths, as Mosco explains, are neither true nor false. Instead, they are, in his words, either living or dead, and as long as a myth continues to live, it holds enormous power: over the ways we see our lives and the ways we think about digital technologies, including a digital cinema (2004, 29). Digital cinema, seen in this context, seems to hold out the potential of new forms of storytelling through the use of digital special effects. The malleability of digital media also promotes the myth that films can circulate more widely than ever before, as well as the notion that digital tools can democratize access. Through the digital, it seems, fans can participate in the production process, as anyone with access to a digital camera and editing tools like iMovie can potentially become a filmmaker. This chapter engages with these myths, particularly as they surface at the intersections of sf and digital cinema, and it explores the ways in which these two categories have begun to mutually impact each other at each stage of the production-distribution-exhibition-reception chain. Each stage in this chain has been shaped by the myth of the digital, seeming to allow us to transcend—in what we might term sf fashion—constraints of time, space, and resources. As a primary example, we consider the low-budget but digitally dependent sf series *Iron Sky* as a key reference point.

The 2012 film *Iron Sky* and its 2019 follow-up, *Iron Sky: The Coming Race*, were both directed by Finnish filmmaker Timo Vuorensola, who had cut his teeth—and built a following—directing the prominent *Star Trek* fan film, *Star Wreck: In the Pirkinning* (2005). Like many contemporary sf films, both of the *Iron Sky* films made extensive use of digital special effects—the CGI producer Samuli Torssonen estimates the original film involved over eight hundred effects shots—in order to depict the primary setting, a space station built on the dark side of the moon (Energia 2011), a process that might have been significantly more difficult in the era of analog media. Additionally, the filmmakers used innovative distribution and exhibition strategies that exploited tools such as streaming video and digital downloads to make *Iron Sky* and its sequel accessible to an international audience, while allowing the filmmakers to engage in a crowdfunding campaign that helped raise money across geographic borders (Tryon 2015). While the sf film has always foregrounded the representational possibilities of film as a medium, the concept of the digital, as the *Iron Sky* films might suggest, has extended the track of those possibilities across the entire range of production and consumption, providing us with a new tool for interrogating how contemporary sf functions, and especially how, in so many ways, it shapes its narratives for a contemporary audience.

While the *Matrix* films were marketed as revolutionizing sf film, a quieter transformation was already taking place alongside them. During *The Matrix*'s exhibition, a small number of theaters were converting to digital projection systems, replacing the mechanical film projectors that had been used for over a century. In June 1999, select theaters in Los Angeles, New Jersey, and New York used digital projection to show the much-anticipated *Star Wars Episode 1: The Phantom Menace* (Lucas 1999) and *An Ideal Husband*, Oliver Parker's adaptation of the Oscar Wilde play. News reports alternated between technological hype and alarm, striking both utopian and dystopian notes. The *New York Times'* Andrew Revkin described a test screening of *The Phantom Menace*, portraying digital projection as a technological solution to the limitations of film, specifically the "unavoidable accumulation of dust flecks, scratches, and pops that come with using a century-old medium" (Revkin 1999). And the lack of bulky film prints, he suggested, would solve the "problem" of quickly releasing low-budget films to a wide, potentially appreciative audience. At the same time, Godfrey Cheshire (1999) worried that the flexibility of digital projection would negatively impact the space of movie theaters, turning them into interactive spaces that would erode the boundaries between cinema and television, changing the ways in which cinematic and televisual images circulate. In both cases, predictions about the film were linked to larger utopian or dystopian myths about the impact of technological change on our everyday experiences.

In this light, *The Matrix* phenomenon and the introduction of digital projection in theaters can be seen as giving rise to both digital cinema and what we might term digital thinking about cinema. Historically, the idea of cinema overlapped neatly with the medium of film. Cinema described the institutions, protocols, and values associated with a physical format—a reel of film going through a projector that displays an image on a screen. Lisa Gitelman (2006) has defined *media* as "socially realized structures of communication, where structures include both technological forms and their social protocols" (7). These social protocols include the norms, habits, and conditions that shape how we use a specific medium. For moviegoing, some of these habits and norms have historically included the expectation that audiences remain silent while watching a movie, that they arrive at theaters on time, and that they otherwise refrain from distracting other viewers. Other norms might include not spoiling key plot points, whether in casual conversation or in published movie reviews. However, as Francesco Casetti has argued, encounters with film are no longer guided by a "well-structured set of norms, constraints, and intentions" (2015, 207). Rather, screens proliferate in our daily lives, as most of us carry smartphones in our pockets with personalized video screens. We also watch films on iPads while reclining in bed, on home theater screens, and on big screens in theaters, with the implication that we now inhabit a world of spectacle. Casetti, in particular, seems to reinforce the myth that the digital enables cinema to transcend geography and history; thus a "relocated cinema" (2015, 10) now becomes available anywhere and at any time. And as the examples of *The Matrix* and *Iron Sky* illustrate, fans can become participants, shaping the reception and in some cases the production of films, whether by taking footage from one film and reediting it or by contributing to the making of a movie through the techniques of crowdsourcing and crowdfunding. These "norms" also exist at the level of production. On the one hand, digital tools provide filmmakers with greater flexibility when it comes to creating special effects. On the other, these new tools have changed the labor practices of movie production. John Thornton Caldwell, for example, has extensively documented the rise of the "digital sweatshop," a dystopian situation in which digital artists work long hours in difficult labor conditions characterized by "vocational volatility and impermanence" (2008, 193). As a result, the digital corresponds to a transformation of cinema at all levels of the production-distribution-exhibition-reception process.

These broader questions about the definition of cinema as a medium are frequently explored within the realm of sf film and television, which historically have been concerned with technologies of vision. As J. P. Telotte has offered, "When we watch a science fiction film, we see as well a narrative about the movies themselves" (2001, 25). In fact, the role of sf in theorizing cinematic ways of seeing dates back to the origins of film. Georges Méliès' whimsical *Trip to the Moon* (1902) used crude special effects, including substitution splices, multiple exposures, simulated tracking shots, and other techniques, to imagine a group of Frenchmen being shot to the moon by cannon, and its techniques became a convenient way of thinking about film—as a device for producing a different reality. Over a century later, sf still plays a crucial role in allowing us to rethink how we look at the world and how cinema can help to imagine different possible futures, especially through those digital technologies that are compelling audiences to rethink the cinema's role in envisioning our future and theorizing our present.

# PRODUCTION

Digital effects have obviously allowed filmmakers to find new ways to tell stories cinematically. The malleability of digital images seems to reinforce the myth that digital cinema can let us transcend our frail human bodies, pushing to new lengths sf's tradition of exploring the limits of identity. This entails the use of motion capture to create alien characters, such as the humanlike N'avi depicted in James Cameron's *Avatar* (2009), the use of motion capture to age the actors in a fantasy film such as *The Curious Case of Benjamin Button* (Fincher 2008), or even the ability to bring dead actors back to life, as J. J. Abrams did after Carrie Fisher, who played Princess Leia Organa, died suddenly while filming *Star Wars: The Rise of Skywalker* (2019). It can also involve the ability to create artificial spaces, such as the space stations depicted using green screen and digital compositing in the *Iron Sky* films. On the one hand, digital effects are linked to the production of the consumer-oriented spectacles that especially mark sf cinema. As Phillip Rosen has argued, digital innovation is part of "consumer culture, whose advertising rhetoric is in perpetual search of product differentiation and hence a rhetoric of the new" (2001, 304). In the case of sf, innovations in digital effects have been used to entice audiences into theaters to experience new forms of spectacular imagery, as with the hauntingly beautiful Stargate sequence in Stanley Kubrick's *2001* (1968), the bullet-time effects noted in *The Matrix,* or the digital 3D images of *Avatar* that immerse viewers in the lush world of Pandora. In fact, Cameron promoted *Avatar* with the promise that he was creating the most immersive cinematic experience in history, that viewers would feel as if they were inhabiting Pandora, the planet of the Na'vi. On the other hand, digital technologies have made film production—including special effects—more accessible than ever to low-budget filmmakers. These tools have, in some ways, democratized cinematic production, making it possible for independent sf filmmakers to create memorable and compelling images and stories, using consumer-grade cameras and editing on platforms such as Apple's iMovie, as with Shane Carruth's time-travel film *Primer* (2004), which was made on a budget of just seven thousand dollars yet won the Sundance Grand Jury Prize. Such films can take advantage of consumer-grade tools, such as Adobe Creative Suite, to digitally alter images or create convincing special effects.

Digital tools also allow filmmakers to revisit and, in some cases, radically reedit older movies, often in ways that have elicited backlash from fans. Nicholas Rombes argues that digital media are characterized by what he calls "incompleteness" and explains that films in the digital era "exist as multiple versions (in director's cut and so forth) and across numerous platforms" (2009, 43). Of course, the ability to reedit films to create new versions has always been available—think of network television–era edits of Hollywood films or versions of classic movies done to satisfy local censors—but digital tools make it easier to think of film texts as "incomplete" and available for constant reediting. For the twentieth-anniversary edition of *E.T. the Extra Terrestrial* in 2002, Steven Spielberg digitally altered the film's key chase scene, replacing the guns federal agents carried with walkie-talkies. Spielberg acknowledged that he made the edits in response to parents' groups who objected to some of the language in the film and to the presence of guns in general. That decision to tinker with a classic sf film was, not surprisingly, met with objections from fans, eventuating in the film's thirtieth-anniversary edition on DVD and Blu-Ray, released in 2012, having the original 1982 footage restored (Leitch 2011).

Similarly, the tense conflict between Han Solo (Harrison Ford) and Greedo (Paul Blake) in the Mos Eisley cantina, one of the most iconic scenes in George Lucas's *Star Wars: A New Hope* (1977), has been reedited multiple times as the film has moved across platforms. In the original theatrical release, Solo shot first. However, in the 1997 Special Edition release, the scene was reedited so that Greedo shoots first. Finally, when *Star Wars: A New Hope* was added to Disney's streaming platform, Disney+, Lucas again edited the scene, adjusting it so that Greedo and Han Solo appear to shoot in the same frame (Alexander 2019). Adding complexity, the sprawling *Star Wars* universe—with its countless sequels, prequels, and side stories—has allowed others to impose interpretive frameworks on these scenes. In *Solo: A Star Wars Story* (Howard 2018), writers Lawrence and Jonathan Kasdan deliberately built up Han Solo's backstory so that viewers would come to the conclusion that Han shot first. These changes are far from trivial, with the digital shaping and reshaping our sense of character. Solo's initial reputation as a morally ambiguous gun-for-hire is altered considerably if he shoots in self-defense rather than preemptively firing on his rival, and for many fans, the idea that Solo acted in self-defense simply dilutes his moral redemption at the end of the original *Star Wars* film.

As these examples suggest, there appear to be two distinct motivations for this form of reediting. First, filmmakers have gained the opportunity to rethink or reinterpret the meanings of their films. Spielberg, for example, implicitly suggests that digital tools allowed him to rework the chase scene in *E.T.*, experimenting with this pivotal scene to make it seem less frightening to children (or at least to the parents of those young children). Similarly, the confrontation with Greedo becomes a pivotal point for interpreting the character of Han Solo. Digital effects can also bring old characters or actors back to life, as we have seen with the digital reconstructions of characters such as Princess Leia after the death of actress Carrie Fisher, a technique that led to a number of important philosophical and ethical questions. This use of motion capture could, in fact, allow filmmakers to create scenes depicting actors doing things they would not do normally. In fact, the ability to use motion capture has led some actors to stipulate in their wills how (or even if) their likenesses can be used posthumously. A second motivation is that digital effects can contribute to textual novelty, allowing filmmakers to revise and repurpose existing content in order to resell the same media properties multiple times, as with the twentieth- and thirtieth-anniversary releases of *E.T.*, or to draw in new audiences as those media franchises begin to age.

# DISTRIBUTION

Digital technologies have also significantly affected how movies are delivered to audiences, promoting the myth that these tools will allow filmmakers to transcend the physical barriers imposed by film's weighty materiality. Once again, this myth largely serves the media conglomerates that have come to dominate film distribution. Digital delivery provides studios with significantly greater flexibility to release movies in order to maximize profits. Because distributors no longer have to rely upon delivering bulky film prints around the globe, they can impose significantly greater control over where and when movies are shown. In addition, moviemakers now have the capacity to deliver movies directly to people's devices, whether that entails smart TVs, laptops, or even cell phones. While Netflix continues to be a dominant player, by 2020 each of the major studios had streaming services allowing them to deliver content straight to consumers, bypassing theaters altogether or, in some cases, making content available just weeks, rather than months, after its theatrical debut. Disney, in particular, has used its streaming service, Disney+, to manage the ways in which audiences access their movies. The so-called theatrical window, that is, the span of time when theaters had the exclusive right to show new movies, has now shrunk from six months to just a few weeks, and in some cases studios have begun experimenting with premiering movies directly to their streaming services, a process that Ted Striphas has associated with the logic of "controlled consumption" (2009, 180–82). But even while audiences can access content from the comfort of their couches, the idea of movies—and the practice of moviegoing—continues to be an important cultural idea. As Amanda Lotz has argued, "Media cannot be killed. . . . The distribution systems used to circulate media, however, evolve with considerable regularity" (2017, 1). Thus, even if the models used to distribute content change, the idea of moviegoing—the habits and social protocols that inform how we think about texts, especially sf texts—continues to be informed by the theatrical experience.

While digital delivery has opened up new options for studios to control the consumption of their movies—and to vastly expand the textual universes they oversee—digital tools also allow independent and low-budget filmmakers to distribute and promote films directly to target audiences. Since digital copies are far cheaper than film, the *Iron Sky* filmmakers could distribute to theaters at a much lower cost, while also selling their film to streaming services such as Netflix or Amazon Prime or even directly to a specific customer base through digital downloads and DVDs. But while digital delivery may seem to provide consumers significantly greater choice and flexibility, this idea of unlimited choice has not necessarily proven to be the case. One of the foundational myths of digital delivery is the idea of the "long tail," a term coined by *Wired* magazine writer Chris Anderson to describe the idea that low-cost storage can make it possible for content providers to make everything available to consumers at the click of a mouse (2006, 20–23). I describe this concept as a "myth" not to imply that it is false but to describe a deeply embedded belief that informs how we think about an idea such as cyberspace. Digital delivery, in its initial phase, seemed to hold out the promise that we could bridge geographic borders, historical limits, and even political or economic concerns, accessing previously unavailable movies. But just a few years after the streaming era had begun, these portals were becoming gatekeepers that operated to limit access. Digital geoblocking tools stop films from crossing national borders, and if a film does not provide a studio with sufficient value, it may not be available through streaming video at all. While Netflix initially secured contracts with several major studios, including Disney, to stream their films and television series, those same studios eventually created their own streaming services. Netflix and Disney+ have now locked down content, making it more difficult for films to circulate freely and requiring consumers to pay for several different monthly subscriptions. More significantly, digital delivery also enables another science fictional circumstance, new forms of surveillance, as companies like Netflix and Disney collect vast amounts of data on their users. Thus, as so many sf texts have demonstrated, the original technological fantasies of connectivity and personalized viewing experiences become mired in the challenges of maintaining that utopian ideal. Instead, as in many sf dystopias, these technological "solutions" to perceived challenges—the limitations of film as a medium—create new problems in which audiences are monitored and targeted with personalized content.

# Exhibition

The sf film has also played a vital role in ushering in the use of digital projection in theaters. In many ways, exhibition overlaps inextricably with distribution. Theater owners and studios constantly renegotiate the technologies used to show movies, whether that entails adding digital projectors or new sound systems or incorporating new rules about the theatrical window. Because of this focus on technological change and its relationship to industrial concerns, digital projection has played a vital role in reshaping not just the sf genre but also the film industry. In fact, the promise of sf spectacle played a crucial role in enticing theaters to convert to digital projection. Cameron's *Avatar* (2009) married digital special effects with 3D projection in a stunning visual world that, many felt, had to be experienced in theaters in order to be truly appreciated. Thus Cameron tirelessly (and successfully) promoted the idea that film viewers needed to "upgrade" their screening experience, stating that viewers should see *Avatar* in 3D or, more preferably, on an Imax screen (Tryon 2013, 82). *Avatar's* status as a major cinematic event provided theaters with a clear incentive to convert from conventional film projectors to digital ones. While prior to *Avatar's* release, only sixteen thousand screens worldwide were digital, by 2010, that number had jumped to thirty-six thousand, leading David Bordwell (2012) to describe the revival of 3D films as digital projection's Trojan horse, a means to sneak in a shift away from traditional film projection.

Against this backdrop of digital projection looms the specter of studios delivering movies directly into the homes of consumers and bypassing theaters altogether. Questions about a possible transition from reliance on theatrical exhibition to distribution through streaming platforms gained immediacy with the COVID-19 pandemic. Beginning in March 2020, theaters across the world were required to shut down temporarily, forcing studios to make difficult choices about their distribution calendars. Furthermore, many moviegoers were reluctant to return to theaters, even when they were reopened, making the decision to release a film theatrically even more complex. Universal, for example, chose to distribute several of its films, including *The Invisible Man* (Whannel 2020), an updated adaptation of the H. G. Wells sf classic; *The Hunt* (Zobel 2020), a satirical thriller; and, most notably, *Trolls World Tour* (Dohrn 2020) digitally through video-on-demand (VOD). Universal made the movies available on a wide array of services for the price of $19.99, less than the cost of two movie tickets in most markets, for a forty-eight-hour rental window. Universal reported that *Trolls World Tour* collected more than $100 million during the first three weeks of this release and announced that, even after theaters reopened, it would release future films to theaters and to streaming devices simultaneously. In response, AMC and Regal, two of the world's largest movie theater chains, indicated that they would no longer screen Universal movies (McNary 2020). While these announcements were likely part of an ongoing negotiation between studios and theater owners, they illustrate the potential fragility of the business of theatrical exhibition, if not the social protocols that have guided moviegoing for decades.

The temporary closure of movie theaters also served as a reminder of their broader social and economic roles within a wider community. Theaters have provided sf fans especially with the opportunity to gather publicly and share in the pleasures of encountering new chapters in their favorite movie franchises. When new movies from major sf franchises come out, fans often come to screenings dressed in character or carrying props identified with the story world. In the best cases, these screenings can become communal experiences where fans share the experience of exploring the story world of their favorite characters more deeply. Moreover, movie theaters are not merely sites where tickets are sold and movies consumed, but economic spaces where concessions—food and drinks—are sold and where local businesses advertise their products and services to captive audiences. More crucially, theaters provide the most visible opportunity to launch a movie franchise or extend the life of an existing one. They play a vital role in promoting the spectacle associated with sf film and perpetuating images of fandom as a collective experience.

## RECEPTION AND FANDOM

Finally, the digital seems to hold out the promise that our media have become more participatory and democratic. In *Spreadable Media*, Henry Jenkins, Sam Ford, and Joshua Green argue that new models of digital delivery entail "a more participatory model of culture, one which sees the public not as simply consumers of preconstructed messages but as people who are shaping, sharing, reframing, and remixing media content in ways which might not have been previously imagined" (2015, 2). Jenkins, Ford, and Green highlight a number of such "participatory" practices, including activism structured around *Avatar*, which saw groups of Palestinians and Israelis dressed as Na'vi, wearing blue paint, protesting the Israeli occupation of the West Bank. Camcorder footage of the protesters was cross-cut with scenes from the movie, drawing parallels between the situation of the Palestinians and the mistreatment of the Na'vi by American industry. For Jenkins and many others, digital tools have seemed to introduce the possibility of a more active, engaged audience, one that could become a participant in the meaning-making process. As these questions about exhibition suggest, the very practice of watching movies—and not just sf—has changed. As Casetti argues, instead of a model of spectatorship in which a viewer passively consumes a movie, we are now in a situation where spectators actively construct the situation in which they watch. In theaters, this construction involves everything from booking tickets online to choosing seats. Meanwhile, nontheatrical screenings involve a process Casetti describes in rather science fictional terms as a "technological doing," wherein the spectator chooses what to watch and how, a process that involves a series of actions including navigating menus, subscribing to a service, and setting up a device or moving between devices (2015, 187). Our various portals now even provide the opportunity to *simulate* the practice of moviegoing. Just days after the United States' COVID stay-at-home orders began, Netflix introduced what they called "Netflix parties," a tool that allowed viewers to sync movies and to chat while the movies were taking place. And sf, in particular, lends itself well to the intense speculation and reflection associated with such fan activity. As the fan pages associated with the *Matrix* films illustrated, sf is often linked to the processes of sense-making, whether that entails grasping the possibilities of new technologies of vision or making sense of what it means to be human.

More recently, through the digital editing tools that are commonly available on personal laptops (or even on YouTube), fans have been able to cultivate their own visual and media literacy skills through mashup videos and other forms of videographic analysis. When these tools were first popularized with the dawn of Web 2.0, fan reedits of movies like *The Shining* (Kubrick 1980) and *Back to the Future* (Zemeckis 1985) seemed to suggest that fans could demonstrate their grasp of different genres, reediting the time-travel movie *Back to the Future* into a fake trailer in order to foreground the male friendship between Marty McFly (Michael J. Fox) and Doc Brown (Christopher Lloyd). Cinema had become part of what Nicholas Rombes has termed a "global digital literacy" (2009, 99). This form of participatory culture has raised important questions about the altered dynamics between producers and fans. As we noted with the fan communities of *The Matrix*, viewers can take advantage of social media tools such as blogs, wikis, Facebook, and Twitter to collaborate on interpreting texts. The creators of the *Iron Sky* universe tapped into these impulses, building a massive fan base that was enlisted to support the film financially well before it hit theaters.

However, while internet-based fandom can promote new forms of community built around shared tastes and values, it can also produce harmful behaviors. In the most extreme cases, groups of fans have used social media to engage in cruel forms of harassment against actors who play characters they dislike. *Star Wars* fans, unhappy with the plot of *The Last Jedi* (Johnson 2017), launched vicious, even racist and sexist attacks on Kelly Marie Tran, who played Rose Tico. Similarly, actor Jake Lloyd, who played young Anakin Skywalker during the prequels, completely left acting and destroyed his *Star Wars* memorabilia after social media hordes attacked him online, while Ahmed Best, who voice-acted the character Jar Jar Binks, has acknowledged that he considered suicide after being harassed. The abusive treatment of Tran even led some media journalists to speculate that her character Rose was "sidelined" in *Star Wars: The Rise of Skywalker* because Disney worried about the risk to their franchise (Di Placido 2019). As this example suggests, transmedia franchises continue to cater to what Jenkins, Ford, and Green (2013) refer to as "affirmational fans" who are most often perceived to be young and male (151). Thus, even while digital media may enable new modes of transformation through digital special effects, the promotional paratexts and fan activities that frame interpretations of these new texts can conspire to reinforce traditional gender and racial norms.

While sometimes accommodating such toxic forms of fandom, transmedia storytelling can also exploit fan activity. Mark Andrejevic (2009) has described how the new digital regime grants fans the opportunity to interact within specific online communities, with the effect that their work becomes easier to monitor and to exploit for financial gain. As Andrejevic explains, fans "both construct popular websites and submit to the forms of monitoring and experimentation that are becoming an integral component of the interactive economy" (419). But while it would be tempting to treat these forms of fan engagement as mere exploitation, interactions between media franchises and their fans are often far more complex than that, especially given the ways in which fans learn to negotiate cultural meaning within these story worlds.

In turn, fan participation, as Jenkins, Ford, and Green argue, can become a form of "skilled labor" as those audiences develop new forms of engagement with their favored texts (2013, 151).

In fact, many of these fan activities have become formalized in the practices of crowdsourcing and crowdfunding, especially in niche communities. Crowdsourcing is the process by which an author or project leader invites public contributions to solving a problem or completing a project requiring a significant amount of specialized labor, such as creating a movie. Among the more widely discussed early examples of crowdsourcing is a Netflix contest that challenged participants to improve the company's recommendation algorithm. Netflix gave participating teams access to anonymized data and offered a $1 million prize to the first group to improve recommendations by 10 percent or more (Buskirk 2009). In the case of film productions, crowdsourcing often involves fan volunteers offering technical assistance by creating complex special effects sequences or contributing their talents as musicians or set designers. For *Iron Sky*, fans were also invited to design movie posters for a scene set inside a movie theater. These forms of digital participation can help to give audiences a sense of involvement and investment in the larger movie project, in effect, to feel like they are part of the film.

Crowdfunding, by comparison, involves soliciting funds in order to support the production of a creative project. Typically in crowdfunded projects, a filmmaker offers perks or rewards, often including digital copies of the finished film or set memorabilia to entice donations. Larger donors might earn the opportunity to visit the set, meet one of the film's stars, or appear in the movie as an extra (with speaking parts requiring higher donations). *Iron Sky*'s sequel, *Iron Sky: The Coming Race*, leveraged the original's cult success, with the producers raising over $1 million from seven thousand fans living in seventy-five different countries. Many of those fans were drawn to the opportunity to appear in the sequel and donated between five hundred and two thousand dollars for the chance to play a Moonbase citizen or even to be eaten by a dinosaur (Rosser 2015). Such reports seem to position the *Iron Sky* universe as a self-sustaining media franchise. However, the second film was plagued by production issues, including a copyright dispute based on digital contributions—a disagreement over who owned the rights to visual effects like backgrounds and spaceship designs, with the original designers claiming that they had only granted usage rights for a single film (see figure 15.2). The new conditions of reception created by digital tools have dramatically changed the industry at every level from blockbuster franchises like the *Star Wars* films to low-budget cult movies like *Iron Sky*, which is, ironically, a parody of *Star Wars*.

# CONCLUSION

Digital technologies have enabled a radical reconceptualization of film as a medium, a point that shows most obviously in the sf genre due to its frequent reflexive emphasis. As

FIGURE 15.2: Low-cost digital spectacle: the Nazi moon base in *Iron Sky* (2014). Walt Disney Studios.

we have observed, media involve not only the technological properties involved in filmmaking, but also the social protocols, habits, and experiences that are involved. Digital effects have permeated the story worlds of sf franchises, allowing filmmakers to create imaginary settings, to bring new life to characters played by actors who are not even living, and to influence our attitudes toward those characters. No longer is film primarily invested in the indexical representation of reality as it unfolds in front of a camera. Today even micro-budgeted filmmakers can take advantage of green screens and other tools to create new and increasingly exotic—or space-y—locations.

At the same time, digital delivery has, quite literally, fashioned a whole new world of distribution practices. We have opened up Pandora's digital box, as David Bordwell (2012) famously put it, and that has led to new ways of engaging with film texts. On the one hand, digital delivery allows studios to gain more control over the distribution of their films, monitoring when—and how often—a theater shows their films, while online ticketing services such as Fandango change how we plan our moviegoing experiences. On the other hand, streaming platforms are also altering our relationships to films. Finally, social media tools are changing the nature of fandom, even intensifying the relationships between fans and sf texts. In this context, filmmakers have been able to create new forms of engagement with audiences that can help to expand the experience of a text. For example, in his time-travel film *Looper* (2012), director Rian Johnson used podcasting tools to create an audio commentary track that audiences could download for free and listen to while watching the film in theaters (although Johnson encouraged people to use headphones to avoid distracting others). Johnson instructed viewers on how to synchronize the commentary track with the film as it played and encouraged use by promising the track would be different from the usual DVD commentaries. While audio commentary tracks have long been a staple of DVD issues, Johnson used this feature not only to encourage repeat engagement with his text, but also to provide audiences with an expanded film experience (Cornish 2012)—all while begging the question of which experience is the "official" one.

Ultimately, these developments reflect a change in orientation between cinema and its audience, one that is characterized by an increasing fragmentation. As Charles Acland has argued, "The contemporary environment of mobile, handheld, and personalized technological trinkets" contributes to "an über-individualism" in which the film consumer is positioned as being in control of the viewing experience, even as studios increasingly control how, when, and where we watch a movie—or can access it at all (2008, 83). Yet even with these hyperindividualized modes of access, the habits and protocols of moviegoing still shape our experience of sf cinema. We can still imagine the thrill of being immersed in a spectacular, even alien world produced by digital effects and imagine the excitement of attending a movie with a packed opening-night crowd of similar genre fans. These shared experiences can include a variety of expressions of fandom, whether online or in the space of a movie theater. They can even include participating in the making of a film, as with the crowdsourcing of the *Iron Sky* films.

As these examples suggest, the underlying myths that define the digital hold enormous power, especially within sf cinema, a genre that is consistently concerned with both the representational capacity of visual media and with the use of narrative to make sense of our social and political world. *The Matrix*, for example, imagines a world in which rebellious individuals can resist the technocratic modes of control that limit the mobility of human bodies, while *Iron Sky* satirizes the similarly totalitarian impulses associated with Nazism, even as it illustrates the potential of a truly democratized mode of cinematic production, one that involves the participation of fans and others willing to contribute to such a liberating text. Both films engage the digital and science fictional myths that, as Mosco describes, "point to an intense longing for a promised community" that we have not yet fully achieved (2004, 15). In other cases, digital effects can appear to offer what Kristen Whissel (2010) calls "The End," whether of celluloid filmmaking (as we saw in *Avatar*) or of certain modes of cinematic production. SF cinema, with its apocalyptic narratives, provides one powerful venue for expressing those fantasies. But thinking through the digital, like the best sf cinema, allows us to engage with the myths of technological progress and the possibilities of community, even as we live in an age of media and cultural fragmentation.

Ultimately, sf—especially in its intersections with the digital—provides us with one of the most useful means of making sense of the world and imagining possible futures. Paraphrasing the oft-repeated axiom that it is easier to imagine the end of the world than to imagine the end of capitalism, Dan Hassler-Forest argues that "we desperately need fictions that not only offer possible alternatives but also involve us as active participants in their construction" (2016, 6). Digital cinema and sf sit precisely at this intersection between utopian fantasies and dystopian speculation, corporate control and fan participation. Digital sf allows us to imagine the world anew, even as it challenges us to think through the ways it fosters various forms of world-building. SF has always been a genre that lends itself to self-reflexive processes, and mapping the concept of the digital onto sf serves as a powerful reminder of the ways in which we both inhabit worlds that are defined by spectacle and control, much like Neo in *The Matrix* or Wade Watts (Tye Sheridan) in *Ready Player One* (Spielberg 2018), and work to make sense of them.

## WORKS CITED

Acland, Charles. "Theatrical Exhibition: Accelerated Cinema." In *The Contemporary Hollywood Film Industry*, edited by Paul McDonald and Janet Wasko, 83–105. Malden, MA: Blackwell, 2008.

Alexander, Julia. "George Lucas Changed Han Solo's Scene with Greedo in *Star Wars: A New Hope*, Disney Confirmed." *The Verge*, November 12, 2019. https://www.theverge.com/2019/11/12/20961173/disney-plus-star-wars-han-solo-greedo-new-hope-edit-george-lucas.

Anderson, Chris. *The Long Tail: Why the Future of Business Is Selling Less of More*. New York: Hachette, 2006.

Andrejevic, Mark. "Exploiting YouTube: Contradictions of User-Generated Labor." In *The YouTube Reader*, edited by Pelle Snickars and Patrick Vonderau, 406–21. Stockholm: National Library of Sweden, 2009.

Bordwell, David. "Pandora's Digital Box: From Films to Files." *David Bordwell's Website on Cinema: Observations on Film Art*, February 28, 2012. http://www.davidbordwell.net/blog/2012/02/28/pandoras-digital-box-from-films-to-files/.

Buskirk, Eliot van. "How the Netflix Prize Was Won." *Wired*, September 22, 2009. https://www.wired.com/2009/09/how-the-netflix-prize-was-won/.

Caldwell, John Thornton. *Production Culture: Industrial Reflexivity and Critical Practice in Film and Television*. Durham, NC: Duke UP, 2008.

Casetti, Francesco. *The Lumière Galaxy: Seven Keywords for the Cinema to Come*. New York: Columbia UP, 2015.

Cheshire, Godfrey. "The Death of Film / The Decay of Cinema." *New York Press*, July 1999. http://www.nypress.com/news/the-death-of-filmthe-decay-of-cinema-DFNP101999123031 2309999.

Cornish, David. "Looper Director Makes Downloadable Commentary for In-Theater Use." *Wired*, October 10, 2012. https://www.wired.com/2012/10/looper-downloadable-commentary/.

Couldry, Nick. *Media, Society, World: Social Theory and Digital Media Practice*. Boston: Polity Press, 2012.

Di Placido, Dani. "'Star Wars: The Rise of Skywalker,' and the Shameful Sidelining of Rose Tico." *Forbes*, December 30, 2019. https://www.forbes.com/sites/danidiplacido/2019/12/11/why-i-cant-wait-for-the-star-wars-skywalker-saga-to-end/#14f13a1c6882.

Energia Productions. "*Iron Sky* Signal E21—Creating the Digital Effects." *YouTube*, September 2, 2011. https://www.youtube.com/watch?v=czpYwqV22p4.

Gitelman, Lisa. *Always Already New: Media, History, and the Data of Culture*. Cambridge, MA: MIT Press, 2006.

Hassler-Forest, Dan. *Science Fiction, Fantasy, and Politics: Transmedia World-Building beyond Capitalism*. Lanham, MD: Rowman and Littlefield, 2016.

Jenkins, Henry. *Convergence Culture: Where Old and New Media Collide*. New York: New York UP, 2006.

Jenkins, Henry, Sam Ford, and Joshua Green. *Spreadable Media: Creating Meaning and Value in a Networked Culture*. New York: New York UP, 2013.

Leitch, Will. "Steven Spielberg Finally Admits the Walkie-Talkies Were a Mistake." *Yahoo News*, September 15, 2011. https://www.yahoo.com/entertainment/blogs/the-projector/steven-spielberg-finally-admits-walkie-talkies-were-mistake-142746809.html.

Lotz, Amanda. *Portals: A Treatise on Internet Distributed Television*. Ann Arbor: Michigan Publishing Services, 2017.

McNary, Dave. "AMC Theatres Won't Play Universal Movies in Wake of 'Trolls World Tour' Dispute." *Variety*, April 28, 2020. https://variety.com/2020/film/news/amc-theatres-trolls-world-tour-dispute-1234592445.

Mosco, Vincent. *The Digital Sublime: Myth, Power, and Cyberspace*. Cambridge, MA: MIT Press, 2004.

Revkin, Andrew. "Showing in Theaters: The Digital Revolution—Cinemas Test a Projector Prototype That Makes Spools of Film Obsolete." *New York Times*, July 3, 1999. https://www.nytimes.com/1999/07/03/nyregion/showing-theaters-digital-revolution-cinemas-test-projector-prototype-that-makes.html.

Rombes, Nicholas. *Cinema in the Digital Age*. New York: Columbia UP, 2009.

Rosen, Phillip. *Change Mummified: Cinema, Historicity, Theory*. Minneapolis: U of Minnesota P, 2001.

Rosser, Michael. "*Iron Sky 2* Raises $500k from Crowdfunding." *Screen Daily*, January 2, 2015. https://www.screendaily.com/news/iron-sky-2-raises-500k-from-crowdfunding/5081416.article.

Striphas, Ted. *The Late Age of Print: Everyday Book Culture from Consumerism to Control*. New York: Columbia UP, 2009.

Telotte, J. P. *Science Fiction Film*. Cambridge: Cambridge UP, 2001.

Tryon, Chuck. "*Iron Sky*'s War Bonds: Cult SF Cinema and Crowdsourcing." In *Science Fiction Double Feature: The Science Fiction Film as Cult Text*, edited by J. P. Telotte and Gerald Duchovnay, 115–29. Liverpool: Liverpool UP, 2015.

Tryon, Chuck. *On-Demand Culture: Digital Delivery and the Future of Movies*. New Brunswick, NJ: Rutgers UP, 2013.

Whissel, Kristen. "The Digital Multitude." *Cinema Journal* 49, no. 4 (2010): 90–110.

# CHAPTER 16

·········································································

# FEMINIST MATERIALISM

·········································································

## FRANCES MCDONALD

FROM the rippling columns of Greco-Roman architecture to the gleaming tip of Luke Skywalker's lightsaber, the iconography of the phallus—tall, long, smooth, and hard—has dominated the skyline of human culture in both classic and contemporary times. Part of the work of feminism has been to identify such symbols and show how each tells the same story of male penetrative power and domination over Nature, where "Nature" includes all that is deemed passive and inert—soil and sky, yes, but also animals, minoritized groups, and women. A prime example of such investigative work is a 1986 essay by the feminist science fiction (sf) writer Ursula K. Le Guin (1989) that takes aim at one of the most iconic scenes in sf cinema: the famous match-cut from Stanley Kubrick's *2001: A Space Odyssey* (1968) in which a spinning Paleolithic bone is replaced by an orbiting satellite. Le Guin describes the scene as follows:

> That wonderful, big, long, hard thing, a bone, I believe, that the Ape Man first bashed somebody with in the movie and then, grunting with ecstasy at having achieved the first proper murder, flung up into the sky, and whirling there it became a space ship thrusting its way into a cosmos to fertilize it and produce at the end of a movie a lovely fetus, a boy of course. (167)

Jumping several million years into the future in a single shot, Kubrick's film plays out the androcentric fantasy of the bone to its inevitable conclusion—a final image of a limpid male embryo drifting in outer space. Cut free from Earth, this "lovely fetus," as Le Guin archly points out, exists "without any womb, any matrix at all" (1989, 167).

Le Guin's essay does more than just critique Kubrick's film. It also offers a positive alternative to it. "We've all heard about the sticks and spears and swords, the things to bash and poke and hit with, the long hard things," she writes, "but we have not heard about the thing to put things in, the container for the thing contained. This is a new story" (1989, 167). Le Guin's "new story" finds its scientific basis in American anthropologist Elizabeth Fisher's "Carrier Bag" theory of human evolution, which had just a few years earlier argued that the earliest cultural invention was not a weapon to be wielded but a vessel to be filled. According to Fisher's account, it was "women's invention of the carrier bag [that] was the take-off point for the quantum advance which created the multiplier effect that led to humanity" (1979, 56). The substitution of a bag for a bone forces a mutation in our origin story. Kubrick's sleek match-cut gives way to a "quantum . . . multiplier effect," the expansive looseness of which is echoed in Le Guin's own writing. Take, for example, her meditation on the wild profusion of forms that a carrier-bag can take: "A leaf a gourd a shell a net a bag a sling a sack a bottle a pot a box a container. A holder. A recipient" (1989, 168). If the story of the bone bashes and pokes until there is only one man left standing (or drifting in space), then the story of the carrier-bag collects, cups, and cradles multitudes without end.

Fisher's and Le Guin's feminist retellings of the story of human evolution anticipate feminist materialism, which firmed up as a scholarly field in the early 2000s. Broadly speaking, "feminist materialism" names various efforts among feminist scholars to intervene in another origin story produced and circulated by the sciences—the story of separable matter. In the seventeenth century, René Descartes developed a theory of matter (*res extensa*) as radically separate from mind (*res cognita*). The universe was thus split in two: on one hand, there was brute matter—passive, predictable, and unchanging; on the other, there were human beings whose powers of reason and deduction allowed them to "manipulate and reconfigure matter on an unprecedented scale" (Coole and Frost 2010, 8).[1] Since at least the 1970s, feminists have sought to expose the gender bias installed at the heart of this worldview. In France, Luce Irigaray argued that a Cartesian system not only presents mind *over* matter, but also culture *over* nature, subject *over* object, rationality *over* corporeality, and, last but by no means least, Man *over* Woman. In a patriarchal society, the two poles of the binary are explicitly gendered, with women fatally linked to the degraded second term. As Irigaray writes, "There is one pole of the opposition (the masculine) which constitutes the limit of the system and which plays with the other pole (the feminine) according to its needs" (1988, 160). In her experimental poetic work "Sorties," Hélène Cixous elaborated on the violence this oppositional system metes out to women, who must either accept their abasement or face annihilation. "Either woman is passive or she does not exist," Cixous writes. "What is left of her is unthinkable, unthought" (2001, 64).

While French feminists investigated the philosophical implications of Cartesian dualism, American feminist scholars asked how Descartes's dream of radical separability had infected and inflected scientific knowledge practices. In *Reflections on Science and Gender*, Evelyn Fox Keller presented modern science as an unabashedly masculinist project, where masculine "connotes, as it so often does, autonomy, separation, and distance. It connotes a radical rejection of any commingling of subject and object, which are, it now appears, quite consistently identified as male and female" (1995, 79). Emboldened by Descartes's neat bifurcation of the world into knower and known, scientists presumed themselves to occupy a hypothetical vantage point that was situated far above and outside of the material world. For Donna Haraway, an important figure in both feminist science studies and feminist materialism, this "conquering gaze from nowhere" was just one of many "God tricks" that characterizes the masculinized, militarized project of technoscience: "This gaze signifies the unmarked positions of Man and White, one of the many nasty tones of the word 'objectivity' to feminist ears in scientific and technological, late-industrial, militarized, racist, and male-dominant societies" (1988, 581). As a blueprint for world-building, Descartes's story of matter produces "an array of interlocking agonistic fields, where practice is modeled as military combat, sexual domination, security maintenance and market strategy" (Haraway 1994, 60–61). We can add to that list the state of ecological crisis in which we now live. Geologists call this the Anthropocene, a new geological epoch defined by the devastating effects of human beings on the Earth's ecosystems.

Feminist materialists believe that a reconstrual of matter as lively and dynamic, rather than as passive and inert, might allow us to repair our relationships with the world and each other. As Jane Bennett writes in *Vibrant Matter*,

> Why advocate from the vitality of matter? Because my hunch is that the image of dead or thoroughly instrumentalized matter feeds human hubris and our earth-destroying fantasies of conquest and consumption. It does so by preventing us from detecting (seeing, hearing, smelling, tasting, feeling) a fuller range of the nonhuman powers circulating around and within human bodies. (2010, ix)

Like Elizabeth Fisher's quantum theory of evolution, Bennett's story of vibrant matter reveals the multitudinous ways in which human beings are entangled in the material world. To detect the "fuller range" of forces that pass through and around us, we must resist the urge to separate humans from nonhumans (another iteration of Descartes's separation of "mind" from "matter"). Instead, Bennett asks that we understand "nonhuman powers" as agents in the ongoing, collaborative work of world-building. Recognizing nonhuman agency causes the protocols of knowledge production to shift—the cold detachment of the Archimedean point gives way to situated knowledges that comprehend the world through proximity and sensation, through "seeing, hearing, smelling, tasting, feeling" (ix). Acknowledging matter's vitality allows Bennett to map out three new styles of being-in-the-world: radical entanglement, nonhuman agency, and a hermeneutics of touch. These are the building blocks, too, of a strain of sf film that emerges at the same time and in the same spirit as feminist materialism. Before we turn to this cinema of feminist materialism, though, we must cast our eye back to take in the longer philosophical history of these ideas. Feminist materialism does not supervene from nowhere, one fine day.[2] Nor does it exist only in negative relation to Cartesian philosophy. Rather, it emerges out of and belongs to the rich but minoritized philosophical tradition of radical materialism, which understands the whole world and everything in it, including the human mind, as resolutely material.

Predating Descartes by a good millennium, radical materialism is rooted in the ancient atomism of Epicurus (341–271 BC). In Epicurus's schema, atoms have the capacity to swerve at "absolutely unpredictable times and places" and in so doing "annul the decrees of destiny and prevent the existence of an endless chain of causation" (Lucretius 1969, 216, 254).[3] The idea that atoms accord not to human laws of measurability but to "a strange logic of turbulence," as Jane Bennett so beautifully puts it, has animated the work of several important philosophers, including Spinoza, Nietzsche, and Deleuze and Guattari; however, these experiments in radical materialism have long been pushed to the margins by a philosophical tradition gridlocked in dualisms (Bennett 2010, xi). In the twentieth century, though, the sciences underwent a paradigm shift that loosened the Cartesian worldview's grip on modern thought. In 1913, the Danish physicist Niels Bohr proposed the first quantum model of the atom, which served as the structural seed for chaos theory, complexity theory, particle physics, and quantum field theory. In Bohr's model, matter is neither inanimate nor predictable—rather, atoms are buzzing balls of electrons that have the capacity to "jump" in discontinuous and indeterminate patterns. One of Bohr's best interlocutors is Karen Barad, a leading figure in feminist materialism who, like Haraway, received her doctorate in the sciences (Barad's in theoretical physics, Haraway's in experimental biology). In her influential work *Meeting the Universe Halfway*, Barad explains how Bohr's discoveries "[called into question] the Cartesian belief in the inherent distinction between subject and object, and knower and known" (2007, 138). When scientists believed themselves to be measuring a discrete object, they were in fact recording their own entanglement with the material world. As Barad explains, "What we observe in any experiment is a *phenomenon* or entanglement or the inseparability of the apparatus and the observed object" (2012a, 61).

Barad is particularly interested in how Bohr's dream of radical entanglement troubles our traditional notions of identity and causality. We are used to thinking of the world as made up of pre-formed and separate entities that can be categorized as either subject or object—yet Bohr's discoveries suggest that at the bottom of everything there are not discrete *things*, but dynamic and productive *processes*. To come to terms with this mind-boggling idea, Barad coins the term "intra-action":

> The neologism "intra-action" *signifies the mutual constitution of entangled entities.* That is, in contrast to the usual "interaction," which assumes that there are separate individual agencies that precede their interaction, the notion of intra-action recognizes that distinct agencies do not precede, but rather emerge through their intra-action. (2007, 33)

The prefix "intra" means "on the inside" or "within." Barad's model of intra-action, then, suggests that there *is* no exterior point from which we can regard the material world. Rather, everything that exists emerges out of dynamic, differentiating patterns of entanglement between material, discursive, human, and nonhuman entities. Barad calls this "the lively dance of mattering" (2007, 37). Feminist materialists manifest this in their own work by mingling terms such as mind, matter, human, nonhuman, nature, and culture with such fervor that thinking of them as binary oppositions becomes impossible. Haraway in particular is expert at this: her writing is shot through with hybrid terms such as "material-semiotic," "material-discursive," and "naturecultures" that invite us to recognize all the ways in which language and reality are folded in with one another. As Haraway argues, "There is no border where evolution ends and history begins, where genes stop and environment takes up, where culture rules and nature submits, or vice versa. Instead, there are turtles upon turtles of naturecultures all the way down" (2004a, 2).[4]

With the binary system exploded into stardust, feminist materialists considered the world with new eyes. What they saw was a churning field in which human and nonhuman forces entwine in "myriad unfinished configurations of places, times, matters, meanings" that Haraway in her most recent book calls "the muddle" (2016, 1). Once privileged as the only "active agent" in the known universe, humans now find themselves "becoming *with*" a vast array of nonhuman entities that each exert their own force on and in the world (1). Consider, for example, this patterning from Barad:

> electrons, molecules, brittlestars, jellyfish, coral reefs, dogs, rocks, icebergs, plans, asteroids, snowflakes, and bees stray from all calculable paths, making leaps here and there, or rather, making here and therefrom leaps, shifting familiarly patterned practices, testing the waters of what might yet be/have been/could still have been, doing thought experiments with their very being. (2012b, 207–8)

As well as collecting and curating a vast assemblage of animate forces, Barad also offers a meditation here on how quantum indeterminacy reworks causality. A temporal model based on intra- rather than inter-activity understands past, present, and future as entangled with one another. Cause and effect are no longer reckoned as "billiard balls following one upon another," to borrow one of Barad's images, but as involved in an open process of intra-action that blooms out in so many directions (2012a, 54). A sort of folding of time becomes possible—as Barad explains it, "The past, like the future, is not closed" (67).

Lively and entangled matter. A bristling field of nonhuman forces. Weird rewirings of time and space. If this all sounds like a list of ingredients for an sf story, that's because it *is*. In important ways, feminist materialists understand themselves as doing science fictional work: they use the "novum" of matter's intra-activity to imagine a world loosed from Cartesian binaries. For Bennett, Barad, and especially Haraway, sf does not so much illustrate critical ideas in feminist materialism as it is a crucial method of doing feminist materialism. As far back as 1984, Haraway asserted that "the boundary between science fiction and social reality is an optical illusion" and insisted that we invent "cyborg stories" that "reverse and displace the hierarchized dualisms of naturalized identities" (2004b, 8, 33). Karen Barad invents a practice of "diffractive reading" that weaves together science, history, and fiction into what she calls "entangled tales" (2018, 61). Bennett, too, draws on the motif of storytelling to describe her scholarship as a "speculative onto-story [that] hazards an account of materiality, even though it is both too alien and too close to see clearly" (2010, 4). These efforts remind us of where we started, with Ursula Le Guin's elaboration of Elizabeth Fisher's "new story" of human evolution. Le Guin's, Bennett's, Barad's, and Haraway's shared argument is as simple as it is urgent: if we and our companion species are to survive on an increasingly endangered planet, we must learn how to tell new, science fictional stories that "propose and enact patterns for participants to inhabit, somehow, a vulnerable and wounded earth" (Haraway 2016, 10).

So, words build worlds, and the stories that we tell matter. As readers rather than tellers of sf, the trick becomes learning how to distinguish between the "survivable stories" and the "killer stories" (Haraway 2008, 160; Le Guin 1989, 168). We've already encountered many of the key tropes and figures of each. We know to distrust any story— be it science, fiction, or sf—that operates like a sorting machine, carefully separating and subordinating the "nonhuman" (a category that, as we have seen, easily billows out to include women, people of color, and other minorized groups) to the "human." Not all sf is as obvious in this intent as Kubrick's 2001. Many films imagine hybrid forms and entangled encounters but nevertheless retain the power dynamics of the binary, from James Cameron's *Avatar* (2009) to Alex Garland's *Annihilation* (2018). Both *Avatar* and *Annihilation* are interesting examples because, while both stage elaborate encounters between humans, animals, aliens, nature, and biotechnologies, they also work furiously behind the scenes to sift mind from matter, humans from nonhumans, nature from culture, ad infinitum. The climactic scene in each film is the same: the hero (human, of course) makes intimate contact with an alien species. In both films, the alien is made to function as a mirrored surface—in the case of *Annihilation*, literally—in which our heroes can only ever see their own reflection cast back at them, *mise en abyme*.

Leaving these deadly stories behind, we might hunt instead for tales of entanglement that seriously unravel the binary form by swirling humans in among the lively play of nonhuman entities without hope of extraction.[5] Bruno Latour calls such stories "geostories," which he defines as "a form of narration inside which all the former props and passive agents become active without, for that, being part of a giant plot written by some overseeing entity" (qtd. in Haraway 2008, 41). Donna Haraway provides us with an example of such a narrative in "The Camille Stories," an sf story that she cowrote as part of a 2013 workshop run by Isabelle Stengers, another advocate for the speculative interplay of science and stories.[6] "The Camille Stories" follows five generations of a speculative kinship model as it unfolds on Earth. In important ways, it is the reverse of Kubrick's 2001: instead of leaving the Earth in order to preserve (and purify) the human, Haraway's tale of worldly survival depends on the intimate, earthly commingling of humans and nonhumans.

Adding to her ever-growing collection of sfs, Haraway describes her "The Camille Stories" as a "sym-fiction" because it imagines a mode of world-building based in sympoietic (collectively producing) systems rather than autopoietic (self-producing) ones.[7] The goal of such a model of world-building is not to reproduce the same thing endlessly, but to create emergent patterns of difference that pave the way for a more survivable Earth. In "Camille," the various practices of symbiosis create permanent entanglements between humans, animals, plants, and other nonhuman actors; as Haraway notes, "Symbiogenesis results in novel sorts of organization, not just novel critters. [It] opens up the palette (and palate) of possible collaborative living" (2008, 218). For obvious reasons, sym-fictions do not feature "heroes" in the traditional sense— if the Camille stories have a hero at all it is Earth itself, which is slowly reshaped and revived over the span of generations. Such stories are hopeful, but they are not utopian: a utopia after all is just one more God Trick that takes us "no place" and so relieves us from the responsibility of ecological and political recuperation. That we are vitally embodied and wholly embedded in the material world is and has always been Haraway's insistence. The question for feminist materialists, then, is not how to get out of an increasingly damaged world intact, but how to live in, with, and through it. As Haraway writes in "A Game of Cat's Cradle," "The most interesting question is, what forms of life survive and flourish in those dense, imploded zones?" (1994, 62).

One such "zone of implosion" is American filmmaker Shane Carruth's experimental sf film *Upstream Color*, which premiered at the 2013 Sundance Film Festival to critical acclaim. In this final section, I propose *Upstream Color* as an example of feminist materialist science fiction, which is to say, an example of feminist materialism. It, like Haraway, Bennett, and Barad before it, dissolves the binary system by reimagining nonhuman nature as an agential force that participates in the world's making in energetic and unpredictable ways. The film opens with a close-up of a thick paper chain curling out of a garbage bag. We don't know it yet, but if we were to unfold the paper chain, we would find passages from Henry David Thoreau's *Walden*—an earlier tale of worldly survival. Slowly, we become aware of movement in the frame—the bag shivers. Sunlight flares the camera's lens; our pupils become pinpoints. Music composed by Carruth adds sonic texture to the scene and is itself a hybrid, a mix of synth, samples, and recorded instruments that is structured as a series of surging suspensions. The title of the song—"Leaves Expanded May Be Prevailing Blue Mixed with the Yellow of the Sand"—gestures to *Upstream Color*'s interest in both scalar distortion and the rubbing together of different visual elements. The opening few frames of the film thus materialize the idea of entanglement in several ways: the way the grain of the film grazes our retinas, the way Thoreau's book is bodied forth as a series of thickly looping folds, the way the music tangles different types and times of sound, the suggestion of a bleeding color palette. This hybridity is mirrored too in the film's own generic instability; as one critic described it, *Upstream Color* is a "genre film wrapped in a love story dipped in the avant-garde" (Renshaw 2013).

As one would expect from a film "dipped in the avant-garde," the plot of *Upstream Color* is difficult to gloss. Most reviewers understand it as a love story that centers on two damaged people—Kris (Amy Seimetz) and Jeff (Shane Carruth)—who find themselves drawn to one another for reasons they cannot ascertain. For *New Yorker* critic Caleb Crain (2013), though, the "hero is the parasitic worm" that threads through the bodies and lives of the film's human characters. In an early sequence, a mysterious man known only as The Thief (Thiego Martins) plucks worms from the roots and soil of an orchid plant. At home, he distills from them a blue powder that has psychoactive properties. Later, in a back alley behind a bar, The Thief drugs Kris and forces her to ingest an untreated worm, which causes her to enter a trancelike state of compliance. Much later, another unnamed man who is listed in the credits as The Sampler (Andrew Sensenig) extracts the worm from Kris's body and implants it into the cellular tissue of a live pig. This field of crossings and contaminations makes up the complicated plot of the film.

With Haraway's Camille stories in mind, we might consider *Upstream Color* as having no hero—human, worm, or otherwise. In interviews, Carruth takes pains to remind us that the worm is only one agent in an assemblage that he calls the "worm-pig-orchid cycle" (Anders 2013). We might notice Carruth's glaring omission of "human" from this patterning. Although human agents participate in the worm-pig-orchid cycle, either as transmitters (The Thief, The Sampler) or as hosts (Kris, Jeff), Carruth asks us to train our eye instead on the lively behaviors of what Latour terms "all the former props and passive agents" (qtd. in Haraway 2016, 41).

The film's novum is the "blue" that is produced by the worm-pig-orchid cycle. While in other sf films this alien object might be explained as originating either from outer space or from a scientific experiment gone awry, Carruth's film is resolutely bound to Earth and to nature. "I wanted [the blue to be] embedded in nature," he explained to one interviewer.

> I want it to feel like it's permanent, and that it's been here as long as we have, and that it is just outside our normal experience, but nothing strange or alien, as far as an alien presence or whatever. It needed to be cyclical, and it needed to continue on its own volition—nothing conspiratorial, not somebody managing the process, but something that would just keep going. (Anders 2013)

We might say that the film resists the alien narrative so that it can construct an alien form of narration. For what Carruth gives us is not an anthropocentric (or androcentric) fantasy of first contact, but a new narrative form that makes possible the telling of "earthly stories of survival," to paraphrase Haraway.[8] What follows briefly examines three of these formal strategies, which we can add to our feminist materialist toolbox: the first and second, Carruth's use of cinemicroscopy and experimental formal composition to stage scenes of entanglement; the third, his use of nonnarrative editing to suggest the recursivity of time.

As in feminist materialism, in *Upstream Color* entanglement is both a motif and a formal method. Bodies are not discrete entities, but porous fields that enter into all types of intimate contact with one another. This contact includes the elemental bodies of water and light that are the thickened media through which Carruth's characters (both human and nonhuman) move. In one of the film's final scenes, Kris discovers submerged in a swimming pool a yellow orchid in full bloom. Carruth composes an underwater shot that places the orchid in the foreground so that it eclipses our view of Kris, who hangs suspended behind it. A strong backlight carves her figure into a crescent shape—the outline of her body deforming and reforming with the undulation of the water. The composition of the shot literally backgrounds the human as it asks us to bear witness to the thickly entangled co-presence of water, light, human, orchid—we could say that Kris is becoming-orchid.[9] Carruth uses the technique of cinemicroscopy to similar effect (see figure 16.1). Microscopic scenes show the worm burrowing into Kris's cellular tissue and washing through her bloodstream before being flushed into the pig's circulatory system by The Sampler. As with her orchid encounter, Kris is permanently altered through

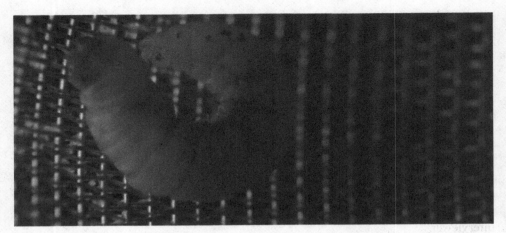

FIGURE 16.1: The use of cinemicroscopy in *Upstream Color* (2013): the mysterious worm examined. VHX.

her material relations with the worm and the pig: these scenes of bio-entanglement are also scenes of bio-transformation. The hope of building a truly symbiotic relationship between the human and nonhuman world is played out in the final image of the film, which sees Kris cradling a piglet as if it were her own child.

Time, too, is given material texture by Carruth. The nonnarrative sequencing of the film formalizes the intra-activity of past, present, and future as described by Karen Barad. The audience comes to understand what has happened / is happening / will happen through Carruth's heavy use of temporal jump-cuts that loop together sounds, scenes, and images in intricate patterns. The thick co-presence of past/present/future helps to establish the field of intersubjectivity that is growing between Kris, Jeff, and the two pigs that are now hosts to their respective worms. Memories, perceptions, and feelings flow freely between these four actors—Jeff's and Kris's childhoods begin to blur, and when the pigs are separated from their piglets their trauma is felt by Kris and Jeff, who curl up in a bathtub in a scene of acute, uncomprehending mourning. This "being affected at a distance" is a key motif of the film that creates an uncertainty about what (or who), precisely, the active agent in any given moment is, or, in Carruth's terms, "a confusion about what's leading what [where]" (Anders 2013). Because *Upstream Color* focuses on intra-active processes rather than interactive entities, this indeterminacy would persist even if we were to rearrange the film's events in chronological order.

All three of the techniques described above—the nonnarrative braiding of time, the use of cinemicroscopy, and Carruth's formal composition of shots—invite the viewer to practice a mode of reading film that is based in touch and feeling. Carruth achieves this "aesthetic of tactility," as he describes it, through intensive layering and looping of times, colors, sounds, and figures, which require that the viewer give up the Archimedean point in favor of her immersion in the rich material texture of the film (Pearis 2013). Confronted with the film's looped temporal structure, for example, the viewer gains

FIGURE 16.2: Human intimate "entanglement" (Shane Carruth and Amy Seimetz) in *Upstream Color* (2013). VHX.

an intuitive grasp on the plot's movement not by disentangling its timeline, but by apprehending it *all at the same time*, as one would a painting (see figure 16.2). This painterly sensibility is evident too in the vividly colored cinemicroscopy scenes, which bathe the viewer's face in tableaus of reds, oranges, blues, and yellows to mesmeric effect. It is hard not to feel physically *touched* by the film, which also thematizes touch and feeling in its narrative. A close-up of a hand brushing lightly across a textured surface repeats often, and in a crucial sequence toward the film's end Kris and Jeff, unsure of which direction they should be pointing themselves in, wander down an open country road. Kris turns to Jeff. "I *feel* like you know," she tells him.

Despite its emphasis on entanglement, *Upstream Color* is not a utopian film. Predatory forces intervene in the worm-pig-orchid life-cycle in the figures of The Thief and The Sampler, both of whom understand matter (inclusive of women and animals) as a raw resource to be plundered. Both men use their powers of extraction to play "God Tricks" on other characters in the film. We know that The Thief extracts worms from the soil of orchid plants and then through a chemical process distills the "blue" from their flesh. Having administered the psychoactive "blue" to Kris in a violent scene that codes strongly as rape, The Thief becomes to Kris a type of God—an ever-present but never fully visible entity whose orders she obeys to the letter. A very similar set of behaviors is evident in The Sampler, who extracts and separates from nature various sounds—a leaf rustling, a brook babbling, a pig crying—that he then remixes as *musique concrète* compositions. Through this technology, The Sampler is able to "listen in" on the lives of all the people from whom he has extracted worms—Kris and Jeff included, while

246     FRANCES MCDONALD

he himself remains invisible. In an extended montage, we see him stand unnervingly close to people who, like Kris and Jeff, are victims of The Thief. A woman distractedly pushes food around a plate, a man drives with glassy eyes fixed straight ahead, a woman stares at a mannequin in a shop window for a little too long. And everywhere they are, The Sampler can be too. It is by listening to one of his albums (perhaps the one titled "Extractions," although there are many others) that Kris is able finally to locate him.[10] When she does, she shoots him dead.

Exemplifying a feminist materialist cinema, *Upstream Color* stages a confrontation between two stories of matter, and, in the end, allows radical entanglement to win out over radical separability. The final image of the film—that soft close-up of Kris cradling a piglet—imagines the strange intimacies that might flourish between humans and nonhumans when the forces of masculinist Science, here represented by The Thief and The Sampler, have been permanently disabled. Like Haraway, Bennett, and Barad before him, Carruth understands that sf has the power not only to illustrate but to *precipitate* such a disabling by inventing different narrative shapes, speeds, and textures that confound the oppositional thinking upon which masculinist Science depends. *Upstream Color*'s narrative complexity is designed to activate a feminist materialist sensibility in the viewer, who is required to pay attention not only to the film's content but also to its form. For as Ursula Le Guin (1989) teaches us, directors can play God Tricks, too. Remember that what Le Guin wants us to see in her analysis of Kubrick's *2001* is how the film expresses violence in its very form by way of a match-cut that annihilates three million years in the blink of an eye. By way of comparison, in *Upstream Color* match-cuts abound, but they work to layer and loop together different times, places, and actors—they work, in other words, in service of connection rather than separability. Like the suspended chords that characterize the film's soundtrack, these layers and loops do not resolve neatly, if they resolve at all. There are no God Tricks here, only the insistence that we learn how to coexist with a lively world, the movements of which we can never fully know or predict.

## NOTES

1. In a letter to Richard Bentley, Isaac Newton (2017) describes the world as composed from "inanimate, brute matter."
2. This phrasing is lifted from Jacques Derrida, who, in his *Mémoires: For Paul de Man*, writes that "deconstruction is not an operation that supervenes afterwards, from the outside, one fine day. It is always already at work in the work" (1989, 74).
3. The Epicurean "swerve" is taken up for feminist purposes by Sandra Gilbert and Susan Gubar in their pathfinding work *The Madwoman in the Attic* (1979). They use the term to describe a specifically female mode of writing that "swerves" from patriarchal codes to carve out a radical new space for female self-definition.
4. Other scholars perform similar gestures. Consider, for example, Myra Hind's (2003) examination of what she calls the "culture of matter" and the "matter of culture," or Elizabeth Grosz's exploration of "the entwinement of the orders of ideality and materiality" (2017, 5).

FEYMINIST MATERIALISM    247

5. An example of a survivable story is Octavia Butler's *Xenogenesis* trilogy (2012), in which an alien species called the Oankali, human beings, biotechnologies, and the natural environment become entangled with one another (at times literally, thanks to the dense drift of tentacles that constitute the Oankali's sensory system) to create a new hybrid species that is literally unthinkable in binary terms.

6. In *Staying with the Trouble*, Haraway acknowledges the cowriters of "The Camille Stories" as "the filmmaker Fabrizio Terranova and psychologist, philosopher, and ethologist Vinciane Despret" (2008, 134).

7. Included among her roster are: science fiction, speculative fabulation, string figuring, and speculative feminisms.

8. A documentary detailing some of Haraway's contributions to modern thought is *Donna Haraway: Story Telling for Earthly Survival* (2016). The director is Fabrizio Terranova, with whom Haraway conceptualized and cowrote the Camille stories.

9. The orchid is an important bit-player in one of Gilles Deleuze and Felix Guattari's most-cited examples of becoming in *A Thousand Plateaus*. As Jon Roffe and Hannah Stark write, "[Deleuze and Guattari] are fascinated by the way certain orchids display the physical and sensory characteristics of female wasps in order to attract male wasps into a transspecies courtship dance. . . . As these wasps move from flower to flower, desperately trying to copulate with them, so too does the pollen which has been transferred to their bodies. Through this seduction the wasps are unsuspectingly co-opted into the orchid's reproductive apparatus. This is a signal example of what Deleuze and Guattari call a *becoming*: the wasp, enlisted into the reproductive cycle of the orchid, engages in a becoming-orchid. This effect is not, they stress, an act of imitation, but a genuine incorporation of the body of the wasp into the orchid's reproduction. The same is true in turn for the orchid, which engages in a becoming-wasp, not by copying the female wasp, but by crossing over into the zone of indiscernibility between it and the wasp in a series of de- and re-territorializations" (2015, 1).

10. We see Jeff leave a record store with an armful of CDs by The Sampler. Some of the titles include: *Repetico, Extractions, Artifacts*, and *Echo Trilogy: Part 2*.

## WORKS CITED

Anders, Charlie Jane. "How Shane Carruth's *Upstream Color* Explains Your Dysfunctional Relationships." *Gizmodo*, April 2, 2013. https://io9.gizmodo.com/how-shane-carruths-upstream-color-explains-your-dysfun-465799671.

Barad, Karen. "Interview with Karen Barad." In *New Materialism: Interviews and Cartographies*, edited by Rick Dolphijn and Iris van der Tuin, 48–71. Ann Arbor, MI: Open Humanities, 2012.

Barad, Karen. *Meeting the Universe Halfway: Quantum Physics and the Entanglement of Matter and Meaning*. Durham, NC: Duke UP, 2007.

Barad, Karen. "On Touching—The Inhuman That Therefore I Am." *differences* 23, no. 3 (2012): 206–23.

Barad, Karen. "Troubling Time/s and Ecologies of Nothingness: Re-turning, Re-membering, and Facing the Incalculable." *Posthuman Temporalities*, special issue of *New Formations: A Journal of Culture/Theory/Politics* 92 (2018): 56–86.

Bennett, Jane. *Vibrant Matter: A Political Ecology of Things*. Durham, NC: Duke UP, 2010.

Butler, Octavia. *Lilith's Brood: The Complete Xenogenesis Trilogy*. New York: Open Road Media, 2012.

Cixous, Hélène. "Sorties: Out and Out: Attacks/Ways Out/Forays." In *The Newly Born Woman*, by Hélène Cixous and Catherine Clement. Trans. Betsy Wing, 63–134. Minneapolis: U Minnesota P, 2001.

Coole, Diane, and Samantha Frost. *New Materialisms: Ontology, Agency, and Politics*. Durham, NC: Duke UP, 2010.

Crain, Caleb. "The Thoreau Poison." *New Yorker*, May 9, 2013. https://www.newyorker.com/books/page-turner/the-thoreau-poison.

Derrida, Jacques. *Mémoires: For Paul de Man*. New York: Columbia UP, 1989.

Fisher, Elizabeth. *Women's Creation: Sexual Evolution and the Shaping of Society*. New York: McGraw-Hill, 1979.

Fox Keller, Evelyn. *Reflections on Gender and Science*. New Haven, CT: Yale UP, 1995.

Gilbert, Sandra, and Susan Gubar. *The Madwoman in the Attic: The Woman Writer and the Nineteenth-Century Literary Imagination*. New Haven, CT: Yale UP, 1979.

Grosz, Elizabeth. *The Incorporeal: Ontology, Ethics, and the Limits of Materialism*. New York: Columbia UP, 2017.

Haraway, Donna. "A Game of Cat's Cradle: Science Studies, Feminist Theory, Cultural Studies." *Configurations* 2, no. 1 (1994): 59–71.

Haraway, Donna. "Introduction: A Kinship of Feminist Figurations." In *The Haraway Reader*, 1–6. New York: Routledge, 2004.

Haraway, Donna. "A Manifesto for Cyborgs: Science, Technology, and Socialist-Feminism in the 1980s." In *The Haraway Reader*, 7–47. New York: Routledge, 2004.

Haraway, Donna. "Otherworldly Conversations: Terran Topics, Local Terms." In *Material Feminisms*, edited by Stacy Alaimo and Susan Hekman, 157–87. Bloomington: Indiana UP, 2008. 157–87.

Haraway, Donna. "Situated Knowledges: The Science Question in Feminism and the Privilege of Partial Perspective." *Feminist Studies* 14, no. 3 (1988): 575–99.

Haraway, Donna. *Staying with the Trouble: Making Kin in the Chthulucene*. Durham, NC: Duke UP, 2016.

Hind, Myra. "From the Culture of Matter to the Matter of Culture: Feminist Explorations of Nature and Science." *Sociological Research Online* 8, no. 1 (2003): 92–103.

Irigaray, Luce. "Luce Irigaray, Paris, Summer 1980." In *Women Analyze Women: In France, England, and the United States*, edited by Elaine Baruch, Elaine Hoffman Baruch, and Lucienne J. Serrano, 149–66. New York: New York UP, 1988.

Latour, Bruno. *Facing Gaia: Eight Lectures on the New Climatic Regime*. London: Polity, 2017.

Le Guin, Ursula K. "The Carrier Bag Theory of Fiction." In *Dancing at the Edge of the World: Thoughts on Words, Women, Places*, 165–70. New York: Grove Press, 1989.

Lucretius. *On the Nature of Things (Book II)*. Trans. Martin Ferguson Smith. Cambridge: Hackett Press, 1969.

Newton, Issac. "Original Letter from Isaac Newton to Richard Bentley." *The Newton Project*, October 2017. http://www.newtonproject.ox.ac.uk/view/texts/normalized/THEM00258.

Pearis, Bill. "An Interview with Shane Carruth." *Brooklyn Vegan*, February 20, 2013. http://www.brooklynvegan.com/an-interview-wi-81/.

Renshaw, Scott. "Upstream Color." *Salt Lake City Weekly*, April 19, 2013. https://www.citywee kly.net/utah/upstream-color/Content?oid=2287412.

Roffe, Jon, and Hannah Stark. "Introduction." In *Deleuze and the Non/Human*, edited by Jon Roffe and Hannah Stark, 1–16. London: Palgrave Macmillan, 2015.

CHAPTER 17

# OBJECT-ORIENTED ONTOLOGY AND SCIENCE FICTION CINEMA

## LEVI R. BRYANT

WHAT might the philosophical school of object-oriented ontology (OOO) have to teach us about science fiction (sf) cinema? Object-oriented ontology is a realist ontology premised on the deceptively simple thesis that all of being is composed of objects. As theorist Ian Bogost straightforwardly explains, a thing or object is any *unit* of being or existence (2008, 5). While objects differ in powers and capacities among themselves, they are nonetheless all objects, or what Aristotle referred to as substances. We might consider a few random examples of objects: a quark, an institution, a class, a flea, a planet, a solar system, a smartphone, a person, a tree, a work of art, and the sun. Clearly all of these things are quite different in nature, and a thorough ontology would have to specify what is unique and what is distinct about each of these *types* of objects, yet OOO is committed to the thesis that at the most general level of abstraction or generality, they are nonetheless objects—or as Bogost underscores, "all things equally exist, *yet they do not exist equally*" (2010). Thus, while no object is more or less an object than any other, objects nonetheless differ in dignity or value within an assemblage, or in the degree to which they contribute to any assemblage. One of the primary reasons OOO can be of value to our thinking about sf cinema is that such cinema is unique in the interest it takes in objects of all kinds. Indeed, it would not be amiss to argue that sf cinema is the genre for which the object is the primary point of study, its central character.

As a *realist* ontology, OOO is committed to the thesis that objects exist independently in their own right. In other words, they are not, as idealists and social constructivists would have it, simply the constructions of mind, society, language, or signs. While there are notable exceptions, philosophy since Kant has been dominated by what Quentin Meillassoux refers to as correlationism, that is, "the idea according to which we only ever have access to the correlation between thinking and being, and never to either term considered apart from the other" (2008, 5). In this context, we might recall the basic aim and ambition of ontology: to articulate the being of being qua being. In other words, what are the minimal features something must have, no matter what it might be, to be a being or entity? If, ontologically, there is a problem with correlationism, it is that it privileges a single form of relation, the human-world relation, such that, as Graham Harman, one of OOO's founding figures, has noted, we begin with the premise that there are humans and then everything else (2020). In short, rather than articulating the being of beings, correlationism restricts itself to articulating what beings are for humans, society, or language. Yet should ontology not strive to articulate the being of beings in their being, regardless of whether humans exist or encounter them?

The tendency of this sort of approach is, as Stacy Alaimo (2010) notes, to reduce things to mere vehicles for human meanings, or constructions of human minds and society, thereby overlooking any contribution that things in their "thingliness" might make (1). The being in itself of objects, as well as their agency, is therefore lost. In this context we might also recall Jean Baudrillard's observation in his *System of Objects*: "The arrangement of furniture offers a faithful image of the familial and social structures of a period" (2006, 13). Under his analysis, the furniture that makes up a living room becomes reduced to a text to be interpreted and deciphered, thereby erasing the agency of the things themselves *as* things. Such is the nature of correlationism. But OOO calls for a more promiscuous, more pluralistic ontology that refuses the gesture of erasing or reducing beings to mind, social forces, or linguistic constructions so that we might preserve the agency of things as things (Harman 2016, 7–13).

For the purposes of this essay, it would take us too far afield to give a detailed discussion of the intricate ontology proposed in OOO, but I do want to focus on two general claims that are at the heart of this theoretical approach and that offer a useful vantage for considering sf. First and foremost, as Harman articulates it, all objects might be described as "withdrawn" (2011, 36). To say that objects are withdrawn is to say that they are so thoroughly independent of their relations that real objects never appear or manifest themselves as they are in themselves. Rather, in their interactions with one another, objects always distort each other as if seen through the bottom of a bottle, such that the nature of the object as it is in itself is never directly encountered. One object "translates" the other object, presenting it as other than it itself is. We might think of a marble falling on a wooden table. Upon impact the table "translates" the marble, producing the resonant sound as the impact reverberates through the wood. Likewise, something similar happens within the marble as it "translates" the table. The marble does not encounter the table *as* a table, nor does the table encounter the marble *as* a marble. Rather, both objects are distorted for one another and they never directly encounter one another. For this reason, Harman will elsewhere offer a correlative account, describing all objects as "vacuums," since no object ever fully deploys itself in the world (2005, 82). If all objects are vacuums, it is because no object is ever fully present. It is not simply that objects are absent to *us* because, as phenomenological philosophy has argued, they can never be presented all at once; rather, it is that they are absent to all other objects in the universe, regardless of whether humans regard them or not. We might here recognize the Kantian idea of the noumenon or thing-in-itself (what Harman calls "real objects") that is forever beyond its phenomenal manifestations (sensual objects). The twist that Harman adds, however, is that while Kant saw this inaccessibility of the thing-in-itself as restricted to how *humans* apprehend beings, OOO views this inaccessibility or withdrawal of objects as true for *all* relations between objects, regardless of whether humans are involved. No two objects ever encounter or relate to one another as they are *as real objects*. It is for this reason that, as Harman puts it, "an object is a box of surprises" (2005, 78). Insofar as OOO argues that objects are withdrawn, it follows that this approach does not constitute a traditional *epistemological* realism, but rather what we might term an *ontological* realism. The claim of OOO's realism is not that we can grasp, know, or represent objects as they truly are—which is impossible because all objects are absolutely withdrawn, and because Kant was right: the thing-in-itself is unknowable and can only be thought without being known. Instead, the claim is that objects are real beings existing in their own right, independent of mind, society, and all other objects. Consequently, OOO, as Harman acknowledges, represents a rather "weird realism" (2012, 17).

A second claim of OOO is that objects can be viewed either as objects or as assemblages of relations. Thus Harman suggests that we think of the universe as "made up of objects wrapped in objects wrapped in objects wrapped in objects" (2005, 85). Unlike Aristotle and Leibniz, who tend only to grant the status of substance to living beings that possess what might be described as an absolute unity, OOO argues that assemblages of objects are themselves objects, and that objects are infinitely decomposable. In this claim we again encounter the "weirdness" of this sort of realism, for insofar as objects are withdrawn from one another in such a way that they never directly touch or encounter one another—and as an analogy we might think of black hole stars whose interior we can never see or know—the interiors of real objects are withdrawn from themselves and are themselves bubbling with surprises, lying in wait to be unleashed upon the objects they compose.

From the foregoing account, it becomes clear as to why OOO should be of great interest to the theorist, critic, and even the creator of sf cinema. The sf cinema is one of those genres that foregrounds the object, and in such a way that the thing, such as a spaceship, becomes a veritable character among characters within the film. Within the worlds of sf we routinely encounter universes teeming with mysterious objects of all kinds. Often these mysterious objects, these boxes of surprise, disrupt assemblages within which people live, requiring the reconfiguration of these assemblages and potentially even transforming the nature of life. Consider, for example, Frank Darabont's 1998 film, *The Mist*. At the film's start, we are presented with a small, bucolic town where people more or less get along. To be sure, there are tensions here and there, but they are of the sort that make up the regular fabric of day-to-day life. Following the appearance of a mysterious mist and of a variety of terrifying objects or creatures that emerge from it, the social fabric of the town is fundamentally transformed. In response to these unruly objects, the people of the town who are trapped in the local supermarket shift from being an "us," to dividing into groups along the lines of race, class, religion, and education. Of course, we can argue that these divisions were always virtually there and that would not be wrong; however, these divisions had not yet actualized themselves. This situation is illustrative of how it often is with sf: a new object emerges on the scene, forcing a reconfiguration of how we act, live, and form the fabric of society—or as Darko Suvin puts it, resulting in a pattern of "estrangement and cognition" (1979, 7).

On other occasions people find themselves trapped within networks of objects, so the film narrative becomes a question of whether it is possible to achieve that sf commonplace: escape velocity. Here we might think of Steven Spielberg's 2002 film *Minority Report*. As Tom Cruise's character, John Anderton, flees the authorities chasing him for "precrime," he finds himself trapped in a world of objects that make evasion nearly impossible. This futuristic world is filled with networks of devices that show people-targeted advertising holograms aimed at them and them alone, activated by their eyes being scanned as they walk down streets and through buildings. These holographic devices thus are advanced versions of the algorithms that govern news feeds on platforms like Facebook and Twitter. In the world of the movie, they can, of course, be used as surveillance and tracking devices. So long as Anderton retains his eyes, it is impossible for him to evade the authorities chasing him. It is only when he removes his eyes and replaces them with another set of eyes—thus revealing that he himself is both an object and assemblage of objects—that he is able to achieve escape velocity from the network in which he is trapped and move throughout the world. Let us now turn to a few other examples of sf cinema to see how OOO might allow us too to read them with fresh eyes.

# Jan Švankmajer and Distributed Agency

Jan Švankmajer's 1968 short film *Byt* or *The Flat*, while often classified as a surrealist work, takes on a very different character when read through the eyes of OOO; it begins to seem more appropriately classified as sf. Containing no dialogue and at a mere twelve minutes and thirty-five seconds in length, the film begins with a man tumbling into a mysterious room. Behind him threads stitch up about the door through which he entered, preventing any egress. As he explores the small flat, he encounters increasingly disturbing curiosities. First, he goes to strike a match and light a lamp, only to have water mysteriously bubble up from the lamp and extinguish the match. Next, he notices that a picture hanging on the wall is askew, but when he goes to right it, the picture above it becomes askew. Because the upper picture is out of reach, he grabs a chair to stand on so as to adjust it, only to have the chair shrink to the floor. Later he notices a table bearing bread, soup, beer, and an egg. As he goes to eat the soup, all of the liquid flows through the spoon. When he tries to drink the beer the glass continuously changes shape and size, such that he gets less than a thimble full to drink. When he places the mug back on the table, it returns to its normal size. As he goes to crack the egg with his spoon, it is the spoon that breaks in half. He next tries to crack the egg on the table, only to have it punch a hole in the table and fall painfully on his foot. Frustrated, he throws the egg across the room at the wall, leading it to disappear into the wall as if it had passed through. When he attempts to punch through the wall to retrieve the egg, the wall absorbs his hand. Upon retrieving the egg, he finds it now crushed in his hand. Later he attempts to lay down in a bed in the corner of the flat to take a nap, only to have it decay into a pile of dust once he closes his eyes. Backing away from the bed in dismay as he wipes the debris from his body, he finds himself against a wall, only to have nails come out entangling his pants. As he rushes to remove his trousers to escape, he falls to the floor, where he finds his hands pierced by more nails that suddenly slide out. The film ends with him carving his name into the plaster of a wall filled with the etched names of countless others who have apparently also been trapped in this strange place. As the film concludes we get the distinct sense that suicide might not be far in the future, since it might offer the only escape from such a place of despair. However, who can know if suicide is even possible in such a place?

Given this strange sequence of events, we can easily see why this film is typically classified as a surrealist work. Its reality is obviously not consistent with our own and seems to follow unknown laws. And yet, it is precisely *because* of this sequence of events that the film might just as well be thought of as science fictional. Through the *presence* of these events, the film draws attention to the *absence* of those laws of the universe upon which our agency conventionally depends. In this way, we learn that our human agency is, as Jane Bennett puts it, something that we might think of as "distributed" (2010, viii–ix). In a normally functioning universe, the tendency is to think of agency as something that resides in us alone. I walk and I find myself across the room; I choose to sit and I sit. However, the object-oriented claim that agency is distributed constitutes a claim that agency never resides solely in the individual, but rather it is the result of an assemblage of the individual plus features of the universe and the objects through which he or she executes any action.

In this context we might also recall a passage from Deleuze and Guattari's "Treatise on Nomadology," where they propose that "the stirrup . . . occasioned a new figure of the man-horse assemblage, entailing a new type of lance and new weapons" (1987, 399). Deleuze and Guattari's point is that the fearsome agency of the warrior does not reside in the warrior alone, but rather that agency is distributed throughout a complex assemblage composed of the warrior + horse + lance + stirrup. If the stirrup is particularly crucial to this assemblage, it is largely because we know from Newton that for each action there is an equal and opposite reaction. Prior to the advent of the stirrup the use of weapons like the sword and lance on horseback were practically unthinkable because the warrior would have been violently thrown from the horse upon his lance's impact. Thus we should begin to see a very basic physics at work in this assemblage. With the birth of the stirrup, the awe-inspiring power of the horse could now be transferred to the lance and the sword, with the result a terrifying experience of being run down by a mounted and stirrup-equipped warrior. This power of assemblages of objects to confer quite different types of agency is ubiquitous throughout every aspect of our experience, yet it is generally invisible to us as we are so integrated with the objects we use and as the physics of our environment is largely stable, veiling the withdrawn surprises that objects harbor. In this connection we might recall when the Apollo astronauts took their first steps on the moon.[1] The moon is about 1.2 percent of the earth's mass. As the astronauts took their first steps, they very quickly discovered that it is *impossible* to walk conventionally on the moon. As we watch the footage of this event, we see them falling over and spinning about. Very quickly they discovered that they must invent an entirely new form of movement that resembles a sort of bunny hop. Ordinarily we think of ourselves as walking *on* the earth, yet this moon footage reveals otherwise—that we walk *with* the earth. Even something as simple as our agency in the case of walking is fundamentally distributed between ourselves and the earth.

There is a curious way in which the world that makes up the ubiquitous backdrop for all of our actions is nonetheless invisible. The brilliance of Svankmajer's short film is that it performs what Suvin famously described as a sort of "cognitive estrangement" (1979, 7), an effect that, he argues, is fundamental to the sf genre. It is in this way, as noted earlier, that the film readily stakes a claim for sf kinship. However, Suvin's concept of the cognitive estrangement performed by the typical sf situation essentially refers to how we become aware of other *political* possibilities of life in and through the genre's effects by envisioning alternative possible worlds that allow us both to see the structures governing ours and to imagine other possibilities—and here one might think of John Carpenter's 1988 film *They Live*, which presents us with an allegory of how ideology functions and how seeing correctly can help us to rise up against it. Svankmajer's cognitive estrangement, in contrast, is one in which we are made aware of the way in which our agency depends both on certain laws of physics and on objects of all kinds that afford and constrain our possibilities of action. By depicting the suspension of these laws, affordances, and constraints, Svankmajer allows us to discern how they function and enable our action all the more clearly.

His film thus provides us with a template for reading the role of other extraordinary technologies and circumstances that are distributed throughout the world of sf cinema. While depictions of technologies in this cinema indeed allow us to envision futuristic worlds—and one wonders whether some technologies such as smartphones and tablets would have come into being had it not been for sf cinema—these technologies also allow us to discern the nature of the present, whether by magnifying and exaggerating contemporary technologies, or by depicting technologies that disrupt contemporary ways of living and relating to one another. Here, for example, we might think of Jonathan Mastow's film *Surrogates* (2009) that took internet life to its logical extreme by portraying a world where people can purchase humanlike, robot surrogates through which they can live their day-to-day lives while lying at home in bed (see figure 17.1). A man, for example, might live the life of a woman through such a robotic surrogate, while never himself leaving his home. What *Surrogates* allows us to discern is how, at the height of the internet chat room's popularity, many people were already living their lives in this fundamentally science fictional way. The film thus brought into relief invisible features of the world in which we were already living.

FIGURE 17.1: Detective Greer (Bruce Willis) and robotic FBI agents in *Surrogates* (2009). Walt Disney Studios.

## GRAVITY, ESCAPE VELOCITY, AND *TOTAL RECALL*

If objects afford and constrain our agency, then the power they exercise might be described as a sort of gravity. Within the Newtonian paradigm, gravity is conceived as forces of attraction and repulsion. Somehow the earth simultaneously attracts and repels the moon. However, this paradigm changed with Einstein's general theory of relativity, which saw gravity not as a force, but rather as a curvature of space-time produced by the mass of an object. In curving space-time, the object produces a *path* along which another object moves. The mass of the earth thus produces a curvature of space-time, or path, along which the moon then moves. Without assuming all of Einstein's theory of gravity, we can think of all objects as exercising a sort of gravity path generator, and these paths can, in turn, be thought of as affordances and constraints. As an analog, we might consider a marriage. I choose this example advisedly to contrast the concept of gravity as OOO conceives it with Einstein's notion of gravity.[2] Obviously a marriage does not generate gravity by virtue of mass, because a symbolic contract does not have mass. On the one hand, there is the one object, the marriage, within which, on the other hand, two other objects, the couple, are entangled. Hopefully this entanglement is by choice, but as the institution of arranged marriage indicates, it is not always so. As an object, the marriage exercises gravity, creating space-time paths along which those who are married and others who relate to them move. As a consequence of being married, a person's tax status changes in the United States. Likewise, others might relate to the married couple differently than they did before, now treating them, after the fashion of the government, as a unit, as a sort of two-in-one person, rather than as discrete individuals. In fact, all sorts of things change in the lives of these individuals as a result of the gravity of the object in whose orbit they have been caught. It would thus be a mistake to conceive of a marriage as a mere subjective pact on the part of two people, for a marriage has a reality that goes beyond that of the subjective agreement of those who enter into it, because it is a socio-symbolic reality with a legal standing. While a marriage does not have the same sort of substantiality as a rock, it nonetheless has a substantiality independent of what is merely thought, and as a result, some of the gravitational transformations produced by it are affordances, while others are constraints.

We can classify objects in terms of the sort of gravity they exercise or have exercised upon them within assemblages. Perhaps foremost among these are what we might term "bright objects"—like stars—that exercise immense gravity, capturing other objects in their orbit. The COVID-19 pandemic, for example, is a bright object by virtue of the way in which it has captured all of us the globe over in its orbit one way or another, requiring us to modify how we live, interact with one another, teach, work, and travel, while thoroughly disrupting the world economy. The pandemic has effectively restructured the paths along which we act and move. Of course, if there are bright objects that capture other objects in their orbit, then those other objects caught in the orbit of a bright object might be thought of as satellites. Thus we have all become satellites of the pandemic. In addition to bright objects and satellites, there are also what we might call dim objects. Dim objects exist within a situation, assemblage, or world, but they only dimly appear insofar as they exercise very little gravity. In this case, for example, we might think of the homeless who are there all about us, but who often go unnoticed and who have very little power or representation. There are also terrifying objects such as black holes that are hopefully rare. A black hole is an object that exercises such powerful gravity that escape is impossible. Examples of familiar—or cultural—black holes might be crushing student loan debt, severe drug addictions, and terminal illnesses. Some might even see human-caused climate change as a black hole. To round out our gravity-based classifications, we might also include versions of astronomy's rogue objects and dark objects. Astronomy has discovered rogue planets, stars, and perhaps even black holes that are not attached to any particular system in galaxies, but that wander aimlessly throughout the galaxies to which they belong. We might similarly conceive of rogue objects that appear, as if out of nowhere, disrupting and reconfiguring relations between objects within an assemblage. For common examples of such rogue objects, we might consider revolutions, new technologies, or, once again, the pandemic. And then there are the dark objects that are so thoroughly withdrawn from our perceptions that they don't manifest themselves or seem to appear at all. In fact, whether they actually exist becomes a purely speculative question, for in their absolute withdrawal they offer no evidence of their presence. However, we might be surrounded by them in all sorts of ways and, were we to disturb

OBJECT-ORIENTED ONTOLOGY AND SCIENCE FICTION CINEMA    257

them in just the right way, they might erupt into manifestation, impacting our lives and the world around us in a variety of ways.

We can read many sf films within this context, as exploring the gravity of hypothetical worlds and situations, and often investigating how people might get away from or achieve escape velocity in such circumstances. Escape velocity is the speed an object must reach in order to escape the gravity of another object. For example, an object must reach about thirty-three times the speed of sound to escape the gravity of earth, or 11.2 kilometers per second. But even apart from such astrophysical circumstances and scientific measurements, we all can find ourselves, in one way or another, enmeshed in assemblages or worlds from which we would like to escape, and by mapping the way gravity is structured in these situations, we might devise ways of achieving the sort of escape velocity we need or desire.

One especially effective exploration of this notion of escape velocity can be seen in Paul Verhoeven's *Total Recall* (1990) starring Arnold Schwarzenegger. Because Schwarzenegger has two identities within the film, most discussions of *Total Recall* have focused on questions of personal identity and how we might distinguish between fantasy and reality. It is a favorite reference text for philosophy professors who are teaching the work of Descartes and trying to illustrate his method of radical doubt. On the one hand, the film poses significant questions about how it is possible to distinguish between simulation and reality. On the other hand, it raises very different questions about the nature of who we are and the reliability of our memories in the constitution of the self or subject. While not discounting analyses that approach these thematic elements, an OOO perspective would suggest that the film is also concerned with how it might be possible to achieve a vitally necessary escape velocity.

Verhoeven's film opens with Schwarzenegger's double-identity character, Quaid/ Hauser, walking along the side of a Martian cliff in a spacesuit, accompanied by a woman. While walking, he slips on the loose Martian soil, tumbles down the side of the cliff, and on landing cracks his face mask on a rock. There he lies on the ground, his eyes, tongue, and veins grotesquely bulging as he gasps for oxygen. Already in this short sequence of images we encounter one of the primary, if unseen characters in the film: oxygen. Oxygen becomes the bright object around which the film's main events all unfold, for it is, on Mars, a key or possibly "bright" object that captures the people of Mars—especially the "dim objects" in the form of the mutants—in its orbit, determining the paths along which they must travel and live. The central question of the film will be whether oxygen is indeed a bright object from which it is possible to achieve escape velocity, or whether oxygen is a black hole, forever trapping the dim objects or mutants within its gravity.

Rather than suffocating, Schwarzenegger's character wakes up gasping, for the scene was merely a dream. His wife, Lori (Sharon Stone), asks him if he was dreaming of Mars (and "that" woman) again, and he expresses his desire for them to move to Mars. She enumerates her reasons as to why they cannot and should not move there, and Quaid/ Hauser responds by explaining that he feels he is meant for greater things than his mundane life as a construction worker on Earth, and that he wants his life to count for something. Like so many of us, he is trapped within the gravity of an unsatisfying life from which he wishes to escape. The discussion also suggests that he is not only dissatisfied with his career as a construction worker, but that he is also alienated in his marriage. Unable to physically escape his circumstances, he decides to escape into the imaginary by having memories of a trip to Mars implanted in his mind by the Rekall company, which provides its customers with memories of an imaginary vacation. He chooses an adventurous Martian trip on which he will have the identity of a deep undercover secret agent whose goal is to save Mars. If he cannot "be someone" and "make a difference" in his real life, Quaid/Hauser reasons that he can at least have artificial memories of having done so.

However, when Quaid/Hauser is undergoing the procedure, the Rekall staff discovers that he already has memories of Mars that had been partially erased or suppressed. This situation opens onto one of the central conceits of the film. The question that it presents to viewers—and that the film never quite resolves—is whether these erased/suppressed memories of being a secret agent are a part of the artificial memories being implanted in him as part of his virtual vacation, or whether they are in fact *real* memories and that his life as a construction worker and married man (as Doug Quaid) were all an artificial fabrication as a part of his cover as a secret agent (as Hauser). For the sake of this analysis, we might simply bracket this never-answered question so that we can bring other elements of the film into relief.

Previously I referred to the gravity of oxygen as structuring life on Mars. In *Total Recall* the planet has been developed as a mining colony for a fictional substance called "turbinium." While we are never told the purpose of this valuable, precious substance, the film hints that it has some sort of energy or military application. The governor of Mars, Vilos Cohaagen (Ronny Cox), rules with an iron fist, even requiring the people of Mars to pay for their oxygen. Because of poor oxygen filtration and inadequate domes capable of filtering out highly charged cosmic particles that cause mutations, the first generation of Martian turbinium miners suffered severe mutations that not only transformed the human bodies of their offspring, but also gave some of them psychic powers. These mutants might be seen as dim objects within the Martian assemblage, living a life of deep poverty where they must resort to prostitution and gambling to make a living. While they attempt to rise up against the despotic regime in charge, it is to no avail; as long as Cohaagen controls the oxygen, he controls the people, since they need it in order to live. At one particularly crucial juncture in the film, in the midst of a mutant revolt spurred by Quaid, Cohaagen shuts off oxygen to one region of the colony. After Quaid has been apprehended, one of Cohaagen's staff asks if he should turn the oxygen back on, but the governor answers in the negative, stating that the rebels' deaths will serve as a lesson to the people of the other zones should they contemplate revolting against his regime. Thus, we see how the lives of all of these people are structured by the bright object of oxygen—which is in scarce supply, has to be manufactured, and is controlled by the mining outfit—to the extent that they have been reduced to the status of satellites trapped in the orbit of oxygen. They can, of course, try to rise up again and again, but so long as Cohaagen controls the oxygen, their struggles will be in vain.

The main question of the film, then, becomes whether it is at all possible to achieve escape velocity from the bright object of oxygen. At this point in the narrative we encounter the rogue object of a recently unearthed alien technology discovered by the Martian miners. We learn that an alien species had lived on Mars a million years ago and had left behind, hidden within one of the mountains—perhaps Olympus Mons—an enormous and mysterious machinery made entirely of turbinium, composed of giant rods situated above a glacier that makes up the core of Mars. The governor has sought to suppress this discovery because he is terrified that if activated it will set off a chain turbinium reaction, killing everyone on Mars. But in the film's final sequence, Quaid activates the gigantic machine, leading the rods to heat up and slide down into the glacier, causing a chemical reaction that releases immense amounts of oxygen and water vapor, creating an atmosphere for Mars, turning the sky blue, and filling it with clouds (see figure 17.2). Because of a prior explosion that had destroyed part of the inhabitants' domed living quarters, Quaid and his female companion Melina (Rachel Ticotin) have been sucked out on to the surface of Mars. As the reaction begins, they lie there gasping for air, their eyes bulging from their heads—just as we saw in the opening sequence—apparently on their way to a horrific death. Yet as the turbinium reaction continues, powerful waves of oxygen roll across the planet, shattering the windows and domes of the colony, bringing air to the people. In the film's closing scenes, we see the people standing in the shattered windows of the colony, freely breathing air on the surface of Mars. Thanks to the rogue object of the alien technology, the gravity of the Martian situation has now been utterly transformed. The governor—or the governor's replacement, since Cohaagen seemingly dies on the Martian surface in these scenes—no longer controls the oxygen and therefore no longer controls the people. While there might be

FIGURE 17.2: The "bright object" of oxygen suddenly spews forth in *Total Recall* (1990). Tri-Star Pictures.

further armed conflict, the dynamics of the situation have changed dramatically with oxygen no longer functioning as a bright object reducing the people of Mars to helpless satellites.

To understand the significance of this sort of reading of *Total Recall*, we might compare it to the original *Star Wars* trilogy (1977; 1980; 1983). While the original *Star Wars* films are pervaded by spaceships, light sabers, laser guns, and aliens, they might not be called sf cinema simply because the films do not foreground technologies, science, or the laws of physics as characters or agencies within their narratives. Rather, these films, as enjoyable as they are, are something of a mash-up of fantasy, Samurai stories (for example, *The Hidden Fortress* [Kurosawa 1958]), and Westerns. The most dramatic difference between these films is in the nature of the struggle they depict between the despotic forces of the Empire and the rebels. In the *Star Wars* films, triumph over the Empire is largely a matter of the resolve of the revolutionaries in their struggle. The element of cultural transformation largely revolves around the agency of humans. What we saw in the discussion of *Total Recall* is that often our oppression is not simply the result of *people*, but rather it is built into environments themselves and how their gravity is structured. To be sure, the governor and his security police are major forces of repression in *Total Recall*, but the core of the problem is the scarcity of oxygen. People's ability to overcome the gravity of their situation entails overcoming this scarcity.

# Unruly Objects, Strange Mereologies, and *Cube*

At the beginning of this essay I mentioned that we can view objects both as units and as assemblages. Objects are themselves irreducible units of being, but they also comprise objects that are themselves irreducible units of being. These are therefore strange and unruly objects that always risk falling apart from within as the objects that compose the larger-scale object clamor to go their own way. For this reason, OOO considers scale to be irrelevant as to whether something counts as an object—for example, whether we are talking about the tiniest subatomic particle or about entire galaxies and galactic clusters. In the case of *Total Recall*, there is, on the one hand, all of those objects that we encounter in the story—the technologies, the mutants, the corporation, the alien technology and artifacts, Mars, Earth—but on the other, there is the object comprising all of these objects that we might refer to as "the situation." The situation is an uneasy object, something that attempts to maintain a certain balance of power and organization arising out of all these parts, where the parts threaten the object from within, exposing it to the danger of falling apart. Here we encounter the dimension of "mereology" that is important to OOO thinking. Mereology explores the relationship between parts and wholes. "The situation" is a whole, an object in its own right, composed of all these parts. Among OOO's central claims is the thesis that the parts of an object are themselves objects. As a result, there is a sort of uneasiness at the heart of all objects since the objects that compose any object are themselves withdrawn or unperceived and threaten to disrupt the larger-scale object. Often sf explores these fraught part/whole relationships and the startling results that arise from them. In this connection we might think of Bong Joon-ho's 2013 film *Snowpiercer*, a narrative about a train that, following a failed effort at climate engineering, travels constantly across a frozen Earth landscape. The train itself is populated by all sorts of objects: the different classes of people that live on the train, children, insects, food, explosives, the parts of the train, soldiers, and so on; but the train itself is a whole constituted by these other parts or objects. These parts and the whole are always in an uneasy relationship with one another—a relationship that perpetually threatens to fly apart, as revolutions and repressions occur. It is this friction that propels the film narrative, as it forms a drama of how a whole both forms itself and how it struggles to maintain itself. By examining the nature of this "friction," we might gain a better vantage not only on the different needs of the various objects, but also on the issue of injustice that might be involved in the maintenance of the whole.

In a similar vein, we might further this discussion of parts and their frictions in sf cinema by considering Vincenzo Natali's 1997 sf/horror film *Cube*, a narrative about disparate individuals trapped in a cube-shaped room. Earlier we saw that we can simultaneously view objects as units that exist in their own right *and* as assemblages of other objects. We might recall that since each object is a unit in its own right, it is irreducible to either the subject regarding it or the other objects that compose it. Moreover, each object is what we have termed a box of surprises, waiting to be unleashed on the world in all sorts of unexpected ways. Here again we find ourselves within that branch of ontology referred to as mereology, and particularly a strange mereology in which objects are simultaneously units in themselves yet made up of additional units or objects that are independent of the larger scale objects that they compose. This situation implies that every object contains within it an element of chaos or instability, since it is never fully able to integrate the components of which it is composed. No matter how well integrated these components might be, they nonetheless "plot" against the larger-scale object as they pursue "adventures" of their own.

*Cube* effectively illustrates this notion. In the film five strangers mysteriously awake in a room made of what looks like stainless steel. They have no idea why they are in the room or how they came to be there. In the middle of each of the walls are square hatches that lead to other, apparently similar chambers. However, the strangers quickly discover that if they go through the wrong portal they can die a horrible death. Some paths engulf people in intensely hot flames. Other paths activate grids of lasers that, working like a potato cutter, reduce people to cubes of flesh. Anyone who follows the wrong path will die horribly. But the strangers also discover that on the ports behind the portals between the rooms there are numeric codes, and one character, Leaven (Nicole de Boer), who is a student of mathematics, uses her skill in order to unlock the codes. While her observations allow them to begin making their way through the labyrinth, it turns out that her skills are somewhat inconsistent.

After one of the strangers dies a particularly gruesome death, another character—a dark, pessimistic, and quiet man named Worth (David Hewlett)—recognizes the structures and believes that he might have worked on the project, particularly in designing the outer shell of the massive cube that contains the various rooms. At this point, the character Holloway (Nicky Guadagni), a woman who tends toward conspiracy theories, exclaims that she knew that it was all a conspiracy and that the cube is obviously a dark government or corporate experiment designed to gather information about how people behave. But at this point Worth launches into an extraordinary monologue about the nature of the cube, emphatically denying that anyone created the structure as part of a dark plan or even that there was some sort of central oversight. Rather, he says, the cube was an *emergent* effect of countless government agencies that had little or no communication with one another, but all which worked under the constraint of having to find ways of spending funding that had been allocated to them. Each part of the "project" had been completed independently of the others, and in a way that followed no unified *plan*. Somehow all of these elements had come together through a sort of self-organizing principle. In short, the cube was a product of design-without-a-designer: the result of an acephalous subject, a collective not organized around any common cause or goal, wherein each element of the collective was unaware of even being an element in the collective. The monstrous technology of the cube is something that came into being without any unifying intention.

As a commentary on this situation, we might consider one of the key aphorisms of the French psychoanalyst and theorist Jacques Lacan, that "the big Other does not exist" (2007, 700). In a number of ways, this thesis carries a crushing implication. We would like to think that behind governments, institutions, bureaucracies, corporations, city designs, and so on, there are unifying, purposeful intentions that govern, guide, and design things. When Lacan claims that the big "Other" is nonexistent, he is, among other things, also claiming that some such comforting agencies do exist. And make no mistake; even in the case of conspiracy theories that explain the madness and irrationality of our world by discerning a rational plan behind its *apparent* disorder, and even when that plan suggests dark and nefarious motives, we still take some comfort in knowing that there is order and purpose hidden within the apparent chaos. This cold comfort is what makes conspiracy theories so appealing. But to claim that the big "Other" does not exist is to claim that there is no plan behind the world, behind this larger assemblage.

If the character Worth is to be believed, it is precisely this circumstance that *Cube* depicts. The film presents us with a world in which objects, units of being, come into existence without a plan or order behind them. In doing so, the film draws our attention to the way in which the world we inhabit is often structured in an aleatory, unintentional way as well. In our day-to-day lives—in our workplaces, the government agencies we must contend with, the institutions we belong to, and even many of the technologies we must use and that seem like extensions of our world—we repeatedly experience vexing situations that leave us wondering how anyone could have been so daft as to design things in a certain way. What *Cube* suggests about these objects is that *no one* designed them in this way, but rather that they are emergent effects of countless small and independent decisions, coupled with heterogeneous features of objects brought to-gether to form larger-scale objects, producing what, in effect, are things as a sort of com-promise. *Cube* very simply strikes to the core of much sf, reminding us of how the genre endlessly and in countless ways explores the manner in which our technologies, our inventions, our imaginings are designed for a specific purpose or aim, but nonetheless have a life of their own that can thwart our intentions and lead to monstrous—as well as utopian—outcomes. Our highest mastery, that found in the sciences and technology, becomes that which is often most unruly. The doctors might find a vaccine for cancer, but in so doing everyone might become a zombie.

# Conclusion

In the way it invites us to approach sf cinema, OOO demonstrates that we can do more than read our sf films as allegory, approach them through ideological critiques, or try to uncover how they might help us to imagine alternative futures and ways of living at the level of the political. It can prompt us to attend to the role that objects inevitably play within these narratives and in our lives. If we approach sf cinema as a medium that foregrounds the object and its properties, science, and the laws of physics, we will see it shedding new light on the world we inhabit and on ourselves as its inhabitants—a light that might allow us to discover new paths to our own escape velocities. By mapping the hitherto invisible ways in which we find ourselves trapped in the world and bound by its frictions, it can help us learn how to dismantle these impediments to our own agency.

## Notes

1. For an illustration of this concept, see the footage of "Walking on the Moon," YouTube, May 1, 2008, https://www.youtube.com/watch?v=aQX9KOCS7MA.
2. A detailed discussion of the concept of gravity within OOO can be found in my *Onto-Cartography* (Bryant 2014).

## Works Cited

Alaimo, Stacy. *Bodily Natures: Science, Environment, and the Material Self*. Bloomington: Indiana UP, 2010.

Baudrillard, Jean. *System of Objects*. Trans. James Benedict. London: Verso, 2006.

Bennett, Jane. *Vibrant Matter: A Political Ecology of Things*. Durham, NC: Duke UP, 2010.

Bogost, Ian. "Materialisms: The Stuff of Things Is Many." *Ian Bogost*, February 21, 2010. http://bogost.com/blog/materialisms/.

Bogost, Ian. *Unit Operations: An Approach to Video Game Criticism*. Cambridge, MA: MIT Press, 2008.

Bryant, Levi. *Onto-Cartography: An Ontology of Machines and Media*. Edinburgh: Edinburgh UP, 2014.

Deleuze, Gilles, and Felix Guattari. *A Thousand Plateaus: Capitalism and Schizophrenia*. Minneapolis: U of Minnesota P, 1987.

Harman, Graham. "The Battle of Objects and Subjects: Concerning Sbriglia and Zizek's *Subject Lessons* Anthology." *Open Philosophy* 3, no. 1 (2020). https://www.degruyter.com/view/journals/opphil/3/1/article-p314.xml?language=en.

Harman, Graham. *Guerrilla Metaphysics: Phenomenology and the Carpentry of Things*. Peru, IL: Open Court, 2005.

Harman, Graham. *Immaterialism: Objects and Social Theory*. Cambridge: Polity Press, 2016.

Harman, Graham. *The Quadruple Object*. Winchester, UK: Zero Books, 2011.

Harman, Graham. *Weird Realism: Lovecraft and Philosophy*. Winchester, UK: Zero Books, 2012.

Lacan, Jacques. "Subversion of the Subject and the Dialectic of Desire." In *Ecrits: The First Complete Edition*. Trans. Bruce Fink, 671–702. New York: Norton, 2007.

Meillassoux, Quentin. *After Finitude: An Essay on the Necessity of Contingency*. London: Continuum, 2008.

Suvin, Darko. *Metamorphoses of Science Fiction: On the Poetics and History of a Literary Genre*. New Haven, CT: Yale UP, 1979.

# CHAPTER 18

........................................................................................................

# POSTHUMANISM

........................................................................................................

## J. P. TELOTTE

JACK Arnold's *The Incredible Shrinking Man* (1957) is hardly a title that immediately comes to mind when thinking about the various "new" science fiction (sf) cinemas. It is, after all, part of that great surge of sf films in the 1950s when the genre was first staking its claim on the cultural imagination. However, it is also one that forecasts a number of the key impulses that are being worked out in posthumanist thinking and in the contemporary sf films that this thinking has increasingly informed. The movie tells the story of Scott Carey (Grant Williams) who, after accidentally being exposed to radioactive dust, begins shrinking, learns that he can never return to his original size—or his original and quite conventional suburban life—and so must come to terms with his freakish condition and eventually his infinitesimal size. As he transitions from man to a doll-sized figure and then to, as he puts it, something "smaller than the smallest," Carey wonders about the changes he faces and, ultimately, what he has become: "What was I, still a human being or was I the man of the future?"; and he finally comes to accept his "posthuman" place in what he can only describe as the mysterious continuum of "the unbelievably small and the unbelievably vast." While fantastically confronting viewers with the possibility of radical change in the human species, these multiple turns in the narrative and the multiple roles into which Carey finds himself cast—as a distortion of the human, stared at by others; as someone who is presumed dead, when he is thought killed by the family cat; and as a transcendent explorer of the universe's immeasurable "vastness"—can help illustrate the primary trajectories of posthuman conception and the developing posthuman narrative that increasingly informs contemporary sf cinema.

As a concept, posthumanism, much like postmodernism, admits of no simple description nor single definition, although as *The Incredible Shrinking Man* illustrates, we might isolate three principal lines of development. First, and most obviously, the notion of the posthuman advances a new, and generally non-anthropocentric view, one that examines inevitable changes that are transpiring in the human condition, thanks especially to developments in biotechnology and information science, and that have produced what, from a contemporary vantage, might at times seem like freakish variations or differences in the human. Second, it speculates on the death and disappearance of the conventional human (whether eaten by cats, self-created catastrophes, or the cosmos—perhaps a suddenly appearing Black Hole), and on what the aftermath of that disappearance might be like. And third, it considers what we might become in a universe that is simply characterized by constant change, including changes over time that come to all life forms, perhaps allowing us to transcend a recognizably human state, to become "smaller than the smallest" or, as the central character in Luc Besson's *Lucy* (2014) demonstrates, to not only change her body in various ways, but transform into a figure beyond all size and situation, becoming, as the film offers, "everywhere." These are all strategies that serve to cast an eye to both the near and far future—a future whose outlines we are already seeing in instances of genetic manipulation, of boundary crossing between humans and technology, and of efforts to both confront and, hopefully, forestall the possibilities of complete human extinction.

These broad posthuman considerations offer us useful strategies for conducting the sort of thought experiments, in this case about our own nature, that have always been the stock-in-trade of sf. But they are also assays in better understanding and critiquing the current state and immediate prospects of the human: in the face of climate crises, global pandemics, systems of mutually assured destruction, and a possible technological/AI singularity that could well suddenly replace the human, or at least confront us with a self-imposed challenge to our place on the evolutionary scale and a new competition for survival with the artificial intelligences we have created. In that element of current critique we can readily see a political dimension that is practically fundamental to the posthumanist position, which can help explain why it often seems to have close connections to—and is sometimes confused with—various other critical/philosophical vantages that have emerged from the postmodern turn, including object-oriented ontology, postgender study, the Anthropocene, and a feminist materialism (all explored in more depth and given clearer distinction elsewhere in this volume). At some common level, these are all vantages that have sought to address what key posthuman theorist Francesca Ferrando describes as the multiple "exclusionary practices" that are found in all cultures and that have at various times in history been used to demarcate, divide, subjugate, and ultimately even define the human. As she offers, "sexism, racism, classism, ageism, homophobia, and ableism"—among other such -isms—are the sorts of categories that have typically "informed the written and unwritten laws of recognition as to who was to be considered human" and who might be excluded from the "family" (Ferrando 2019, 4). Thus, like many cultural studies practices, the posthuman vantage advances by first recognizing and then reevaluating those conventional "laws of recognition," while also underscoring their constantly shifting nature and a correspondingly hopeful potential—that those seemingly natural "laws" are also subject to change.

Yet posthumanism represents more than just a strategy for making visible and focusing attention on the various cultural determinations of the human. Its larger aim is to challenge the many "limits and symbolic borders" (Ferrando 2019, 5) that characterize the big human picture. In fact, it is this process of border violation, drawing on Michel Foucault's notion that border crossings of every sort are simply inevitable, especially when dealing with the various "subjugated knowledges" of human experience (1980, 81), that might be seen as the most fundamental component of the posthumanist project. Thus, while acknowledging that cultural critique is both a starting point for posthumanist thinking and an important value that comes from its application, Ferrando emphasizes that we need to frame its theoretical activity more broadly. Describing posthumanism as "the philosophy of our time" (Ferrando 2019, 1), she holds that in its most valuable application posthumanism—much in the vein of sf itself—represents "an invitation to investigate perspectives we usually leave aside . . . an intellectual exercise towards a posthuman future which will radically stretch the boundaries of human comprehension" (2012, 14). In effect, she sees it as an ambitious, fundamental, and ultimately necessary reorientation—or stretching—of how we conceive the human trajectory. As such, it represents, on the one hand, a counter to traditional humanist attitudes, and especially to the use of the human and human reason as fundamental measures of all value, which should lead to an examination of the values almost invisibly embedded in a conventionally anthropocentric worldview. On the other hand, posthumanism involves a rethinking of how such fundamental concepts as consciousness and embodiment fit into any vision of human nature, as the human becomes not a single conception, but part of what Bruno Latour, another major contributor to the posthumanist turn, has identified as part of the "material network" (2005, 229) of being—a network we see modeled in the *Lucy* figure's transformation to "everywhere." (see figure 18.1).

This approach to the human, seeing "it" not simply as the coherent, embodied individual, but as an element in a broad "network" or range of sentient possibilities, opens a door onto a wide array of connections and of a sort that have been particularly productive for the sf imagination: acknowledged links between humans and other forms of sentient life, openness to technological and other sorts of enhancements or modifications to the human, and recognition that the current state of human existence is but one stage among many. We can see those recognized links being played out in a variety of contemporary situations. One familiar example is the different forms of animal rights activism, which find a pointedly sf parallel in the effort to gain official "rights" for robots—a move

**FIGURE 18.1:** The title character (Scarlett Johanson) transforms into a networked being in *Lucy* (2014). Universal Pictures.

that provoked much debate when, in 2017, Saudi Arabia granted citizenship to a humanoid robot named Sophia (Sigfusson 2017). Another instance is the development of various sorts of human enhancements, whether technological or genetic—such as artificial retinas, computer-chip implants, laboratory-grown skin, blood, and appendages, genetic splicing—that has prompted both excited commentary about miracle cures for various handicaps, along with protests about radical changes that might lead "to people who are no longer either physically or psychologically human" (Masci 2016). And recent advances in computer technology, including the idea of uploading the human mind to a computer (or to the disembodied "cloud") have led to predictions of a next stage in human development that will constitute a kind of technological immortality, along with a literal disappearance of the visually recognizable or "embodied" human. All of these possibilities spring from that posthuman tendency to challenge the boundaries of how we think about the human—or even about thinking—or separate the concept of humanity from how we see other elements of life.

Just as this emerging field of posthumanist studies seems to turn in many directions, thereby demonstrating what Ferrando describes as its "dynamic openness" (2012, 16), so too does it find its origins in a variety of fields and major theoretical thrusts. As the work of such key figures as Donna J. Haraway (1989), Rosi Braidotti (2013), and N. Katherine Hayles (1999) has illustrated, cybernetics, biotechnology, philosophy, cultural studies, and gender studies have provided some of the most important contributions to the posthuman concept and particularly to its attempt to move away from the notion of a human fixity—that is, a single, stable conception of the self. For Haraway this shift is famously bound up in what she refers to as the "myth of the cyborg"—the image of an organic/technological creation that might help free us from much that constitutes our traditional sense of the human. In its very constructed-ness, she suggests, the cyborg figure allows for a way out of traditional notions of gender and the contestations attached to it (and thus for a postgendered vision of the human), as well as for a kind of liberation from much else in culture that has generally been conceived as "natural" rather than constructed. Thus Haraway argues that our new and constantly evolving relationship to machines—a relationship that has made possible the cyborg as an empowering myth, as a combination of human and technology that is free from the past—will make "thoroughly ambiguous the difference between natural and artificial, mind and body, self-developing and externally-designed, and many other distinctions that used to apply to organisms and machines" (1989, 152), while also opening up promising pathways for human exploration and development.

In Haraway's rejection of a broad array of dualisms that have long characterized Western thought and culture we can see an additional tenet of much posthumanist thought, one that has proven particularly central to the work of another major theorist, Rosi Braidotti (2013). She observes that "the common denominator for the posthuman condition" should be the assumption that a "non-naturalistic" structure undergirds not only our conventional sense of the human but of "all living matter" (2), and she allows, following Haraway, that this assumption has led to productive cultural analyses based on "the binary opposition between the given and the constructed" (3). However, she also sees that assumption as just a starting point, and one that eventually falters in light of contemporary scientific and technological advances, such as the very ones that Haraway heralds. For these same technological advances, Braidotti (2013) suggests, have now "blurred" easy distinctions between oppositional elements (3) and created a detached subjectivity, an unmoored sense of self, leading her to argue the need for "recasting subjectivity" (186), so that we can begin to see the self in a "nomadic" and "relational" context (189). As a result, she defines posthumanism in this unbounded, nonoppositional way, as "a process of redefining one's sense of attachment and connection to a shared world, a territorial space: urban, social, psychic, ecological, planetary" (192). We might find a simpler translation of this vision in Ferrando's (2019) description of posthumanism as a kind of methodology, "a philosophy of mediation," one that consistently reduces dualisms and hierarchies into ongoing processes and shifting positions, while taking the possibility of hybrid forms of all sorts as a point of departure (3).

We can find further echoes of Haraway's translation of the self as cyborg and Braidotti's "process-oriented vison of the subject" (2013, 190) in N. Katherine Hayles's work, particularly as she distinguishes between a human and a posthuman subjectivity. She has described how, spurred by the development of information technology, "a historically specific construction called *the human is giving way to a different construction*" (1999, 2)—one that is no longer bound by notions of body, brain, or even the organic. In the place of that earlier, "natural" construction, which had privileged "informational pattern over material installation" (3), mind over body, she has argued for a series of key propositions: for a view of consciousness not as primary, as Descartes posited, but rather as a kind of way station in human evolution; for the importance of revisioning "embodiment," that is, seeing the self as "contextual, enmeshed within the specifics of place, time, physiology, and culture" (Hayles 1999, 196); and for a merging of "bodily existence and computer simulation, cybernetic mechanism and biological organism, robot teleology and human goals" (3). Much like Haraway, Hayles foresees—or already discerns (as is implicit in her description of the human as "giving way")—a dissolution of the traditional construction of the self as conscious body, gendered identity, and "free" individual into a new techno-organic possibility that draws on "a distributed cognition" (1999, 3)—or Latour's "material network" (2005)—and bulks beyond all notions of individuality.

While hardly a thorough accounting of the posthumanist position, these three major theorists and the closely allied stances they take can provide us with appropriate touchstones for considering how sf cinema has both echoed and built upon the developments of posthumanist thinking. All draw upon a recognized and ongoing destabilization of the conventional self, one that finds itself enmeshed in an increasingly technologized and electronic reality that, as Scott Bukatman has offered, increasingly poses "a set of crucial ontological questions regarding the status *and power* of the human" (1993, 2). Probably the most obvious form that ontological challenge has taken is seen in the rise of various sorts of human-machine hybrids (Haraway's cyborgs) that possess skills, powers, or durabilities surpassing those of any conventional (or nonenhanced) humans. However, the challenge that is made explicit by such figures is not just a physical one, but oppositional and contrary to an anthropocentric way of thinking (thus Braidotti's sense of the "blurred" distinctions that are resulting), as it compels us to see the human in a "relational" way that dissolves traditional dualisms and hierarchies. The result of such thinking, as Hayles explains, is the emerging notion of a "contextual, enmeshed" self that is able to recognize its shared identity—a kind of transhumanism—with other life forms. And these, what we might term *phases* of posthumanist thought readily inform the key directions that, as we have noted, sf cinema's posthumanist narratives have commonly taken: in stories of emerging difference, of a possible human extinction, and of radical human change, even transcendence.

As we have already observed, while posthumanist theory and its critical application are relatively recent phenomena, posthumanist concepts actually have long informed both sf literature and films. Stories of robots, human-mechanical hybrids—such as the robot with an implanted human brain in *The Colossus of New York* (Lourie 1958)—and human medical experiments easily represent the most dominant strain in this development. In Fritz Lang's landmark film *Metropolis* (1927), for example, the scientist Rotwang (Rudolf Klein-Rogge) announces that he has created a robot engineered to be the perfect worker; it is a type that, he ominously boasts, will replace all human workers. A similar, albeit more benevolent impulse drives the later *Tobor the Great* (Sholem 1954), wherein a robot is designed to aid in the conquest of space by taking the place of human astronauts who are proving unable to endure the physical rigors of spaceflight. A similarly beneficent interest results in an experimental surgical procedure on the title character of *Charly* (Nelson 1968), a mentally handicapped adult. While that operation triples his intelligence, transforming him from a figure of public amusement to one of amazing mental abilities, the transformation proves only temporary, as he soon lapses into a childlike state and again becomes a mental outcast. The more recent *Limitless* (Burger 2011) offers a drug-induced version of such a human transformation with its protagonist (Bradley Cooper) consuming an experimental drug that vastly expands his mental powers, while also leading him down an amoral path of material acquisition and political power. In all of these cases we see science creating ways to augment or move beyond the human, producing figures that are at times amazing, frightening, even pitiful, but those efforts have most often served as warnings against our scientific efforts to challenge the given situation of humanity—in effect, warning against the difference that is at the very core of the posthumanist vision.

In many of the *Alien* films (Scott et al. 1979–), though, we can begin to see a more nuanced investigation of a posthuman identity and even destiny. For example, these films not only examine the human tendency toward speciesism, particularly in their basic narratives about capturing and exploiting other, even quite dangerous alien species for profit and military potential, but they also populate those stories with a succession of androids, demonstrably superior in many ways to their human creators, and at various times figures of curiosity, menace, but also support. The fourth film in the series, *Alien Resurrection* (Jeunet 1997), is particularly telling in the way it parallels these treatments of different "species" (both aliens and androids) with efforts to engineer another sort of being, as it depicts ongoing cloning experiments designed to create a hybrid by resurrecting the dead protagonist of the previous films, Ripley (Sigourney

Weaver), along with the alien egg with which, before her death, she had been fertilized. Using blood samples drawn prior to her death in *Alien 3* (Fincher 1992), scientists have thus cloned Ripley, whose DNA is mixed with that of an alien Xenomorph, but in one of their labs this new form of Ripley—Ripley 8—discovers the results of the scientists' previous failed attempts: seven various and monstrously deformed hybrids of Ripley and the Xenomorph. These freakish distortions of two quite different forms, most dead, but one still alive, in great pain, and begging Ripley to kill her, make for a grotesque display of one possible posthuman future—one of unpredictable, horrifying hybrids. While Ripley 8 euthanizes that pleading hybrid and later, drawing on the acid alien blood she has developed through the cloning process, kills another young hybrid, she lives on and returns to Earth as a living challenge to the human status quo. Although she looks human, Ripley is, as she says at film's end, ultimately "a stranger" to the human world, although one who has become better able to survive and perhaps to help others precisely because of her hybrid, posthuman nature.

Another, somewhat less common scenario has also depicted a posthumanist vision by opening up another avenue of analysis: tales that are literally about a *post* human situation, as they envision a possible human extinction and speculate on what might come after. Such apocalyptic visions might be framed through Earth destruction stories, as in the case of films such as *Deluge* (Feist 1933) and *When Worlds Collide* (Matè 1951), the former following a man and woman who, after a great global flood, believe themselves the last man and woman alive, and the latter depicting an elaborate rocket-born effort to leave Earth prior to a life-extinguishing collision with a rogue star system. But in both instances the narratives pull back from their near-human-extinction scenarios, as *Deluge* reveals that other pockets of civilization have survived and the flood waters have begun to recede; and *When Worlds Collide* provides a safe landing on another planet for the ark-survivors, now ready to seed human life—just as it existed on Earth— elsewhere in the universe. At least in these early efforts, the boundary line of human elimination—and the contemplation of what might follow—is never really crossed, and it is only tentatively challenged in such later films as *The Time Machine* (Pal 1960; Wells 2002), *The Omega Man* (Sagal 1971), and the original *Planet of the Apes* (Schaffner 1968). Both versions of H. G. Wells's time-travel story *The Time Machine* (1895) present a far-distant future in which a barely surviving humanity has developed into two strains, the passive and peaceful Eloi and the largely brutish, predatory Morlocks, who prey upon the Eloi for food; *The Omega Man* (also see other versions of this story: *The Last Man on Earth* [1964] and *I Am Legend* [2007]) depicts a group of nocturnal mutants largely ruling a ravaged Earth following global biological warfare, with the leader of the mutant "family" describing the last normal man as "obsolete," "dead," and "the refuse of the earth"; and each entry in the *Planet of the Apes* series envisions humanity mostly destroyed through atomic warfare with the result that an advanced ape society is now ascendant. All of these films (and their remakes and sequels) evoke what Ferrando refers to as common culture's "laws of recognition" (2019, 4)—for example, *The Omega Man*'s albino, sore-covered mutants look ghastly, so they must be bad, their rule feared—but in doing so the films also allow viewers to interrogate what seem like natural "laws" or human boundary markers, precisely the sort of "binary oppositions" that Braidotti explicitly challenges. The resultant glimpses of other possible societies, other forms of life, whether evolutionary developments, mutants, or other ruling species, suggest an increasing tendency to think beyond the extinction boundary and begin to address some of posthumanism's basic concerns.

The potential for human transcendence—that is, for evolving into another state, even if just the sort of shift in dimensional thinking described in *The Incredible Shrinking Man*—has proven to be one of the more difficult phenomena to depict and to interrogate. Is it possible to think outside of speciesism, to envision an absence of those laws of recognition, to *be* not as an individual but as part of a matrix? Would hybrid versions of the human still be a factor in such an equation or would all conventional notions of the human become little more than nostalgic traces of the sort depicted in Steven Spielberg's *A.I. Artificial Intelligence* (2001)? In that film's conclusion, the "child" android David (Haley Joel Osment) is awakened from a two-thousand-year slumber to a radically changed Earth, but he is still driven by the memories contained in his databanks. The "child" is a figure programmed to love, although the human objects of that love, and indeed the rest of humanity, have now all vanished, leaving his longing for his "parents" their sole trace. These are the sorts of issues that posthumanism and especially a posthumanist cinema wrestle with in trying to forecast and visualize what might even constitute a human transcendence or a radical version of Hayles's "distributed cognition."

Arguably the most successful effort at meeting this challenge is Stanley Kubrick's *2001: A Space Odyssey* (1968), a work set precisely in the year when Spielberg with *A.I.* was offering his own speculations about such a situation. *2001* famously concludes with its image of the "star child," a large, human-seeming embryo floating in space, but what it will eventually look like, how it will grow and evolve, and what it will do, we are left to speculate. That final image leaves even dimension impossible to calculate. Perhaps recognizing the power of those "symbolic borders" Ferrando describes (2019, 5), Kubrick simply stops his narrative at this point, thereby challenging his audience to imagine what might come next. However, the film does lay a foundation for that imagining: by providing evidence of a transcendent intelligence at work through the mysterious monoliths that, at various points in human history, prompt evolutionary change; by suggesting one possible postbiological future through its incarnation of the HAL 9000 computer that decides to eliminate the interfering humans on the spaceship *Discovery* as it voyages to Jupiter; and, as Marcia Landy offers, by "invoking images of faces, eyes, interiority and exteriority" that "place spectators outside the traditional anthropocentric context" and force them into "a rethinking of the relations between mind and body, affect and thought, sight and sound" (2006, 88). The key part of that "rethinking" is, in fact, a reimagining of the very relationship between time, space, and the human, as Kubrick, prior to the "birth" of the star child, depicts his astronaut protagonist Dave Bowman (Keir Dullea) in a most unusual/posthuman situation: transported to an ornate eighteenth-century suite (that is, still bound within the human species' Age of Reason), where in a series of point-of-view shots, Bowman sees himself in different positions in the room and in different stages of life, as if he were able, simply by glancing, to produce his own future image occupying, in each shifting glance, a different space. Less an image of transcendence than a potential demonstration of its process, this scene, as posthumanist theory might explain, calls into question those seemingly most restrictive of borders—of time and space—as it imagines other possibilities that might open up to the human.

For a more developed application of posthumanist thinking, though, we might turn to a much more recent film, and one that has proved to be a touchstone for several other contributions to this volume, Denis Villeneuve's *Arrival* (2016). It is a work that, while not depicting any of the common posthuman tropes we have described, draws on many of the key ideas explored by figures like Haraway, Braidotti, Hayles, and Ferrando. Based on Ted Chiang's novella "Story of Your Life" (1998), the narrative involves no physical hybrids, envisions no human extinction event, and imagines no transcendent human evolution, and yet it readily opens onto a variety of the concerns of the posthumanist project as it depicts the consequences of a sudden and highly visible alien "arrival" on Earth. But as its title might hint, *Arrival* is also about another sort of "arrival," that of humanity—in what we might think of as a moment of species singularity—as it attains a point of multiple new awarenesses: that it is not alone in the universe, that it shares a destiny with other species, including these alien visitors, and that it is not bound by the familiar and seemingly inviolable "laws" of space and time.

Of course, all alien encounter or alien invasion narratives inevitably invoke some posthuman implications by virtue of their central premise, that there are other intelligent species in the universe. That premise, along with the advanced technologies typically employed by those alien visitors, confronts us with the fact that the various hierarchies of intelligence and ability we have embraced—hierarchies that typically place the human at an evolutionary apex—must be rethought or abandoned. But *Arrival* also compounds that situation by deploying the unexpected alien appearance to challenge other sorts of common human hierarchies, such as those involving different countries, nationalities, and races. Thus, rather than appear to just one of the more "advanced" Earth cultures, the aliens here land twelve ships in twelve locations scattered around the globe, and they attempt to communicate different parts of a larger message to each of the cultures involved. The implication is that each of these seemingly quite different cultures is part of a single human species that can—or should—work together for the common good and even speak a single, uniting language, such as the complex one the aliens employ.

If speciesism poses barriers among Earth's populations, often prompting us to recoil in fear, apprehension, or a sense of the abject from both other races and other life forms, it is no less the case when faced with this film's particular alien presence. *Arrival* confronts those who must communicate with the visitors with a disturbing vision—of cephalopod-like creatures, dubbed "heptapods" because of their seven arms, who stay separate from humans in murky, smoky, atmosphere-controlled capsules, and who communicate by emitting inky substances that form intricate circular logograms. They are, quite simply, difficult to look at and produce a clear recoil in those tasked with first communications, such as the linguist Louise Banks (Amy Adams) and the astrophysicist Ian Donnelly (Jeremy Renner). Moreover, the circumstances in which these encounters take place—wearing hazmat suits, in a trying atmosphere, with gravity affected, and with outside communications inconsistent—further problematize the physical encounter, allowing Louise and Ian to work with the aliens for only brief periods. Our very human physicality, as well as the psychological boundaries it forces on us, the film suggests, poses one of the most fundamental challenges to a properly posthumanist vantage and behavior.

Yet as we noted earlier, the challenge that nonhuman figures like the heptapods pose is not just physical, not just the sort of fearful response endemic to most horror/sf hybrid films, but it is oppositional in the way it questions a conventionally anthropocentric way of thinking (see figure 18.2). In *Arrival*'s case those heptapods present Louise, Ian, and the other representatives elsewhere on Earth with a language that can only be deciphered by abandoning traditional dualisms and thinking rather in a "relational" way. Obviously, the presentation of the aliens' message in parts, given to the people in each of the twelve cultures where they land, requires one sort of reorientation—working together to decipher and correlate the various pieces, despite long histories of cultural and racial mistrust. As Braidotti, in a broader gesture, has offered, such efforts demand "a qualitative shift in our thinking about what exactly is the basic unit of common reference for our species, our polity and our relationship to the other inhabitants of this planet" (2013, 2). But in an imaginative turn on Marshall McLuhan's notion that "the medium is the message" (1964), the language of the alien message proves just as significant and as daunting as finding human commonality. The alien logogram is repetitive yet with subtle differences, insistently circular in both design and function, and admits of multiple interpretations—thus nearly bringing disaster when a Chinese communicant interprets a key word as "weapon" while Louise translates it as "tool." But the language's strangely circular and incrementally repetitive form is central to its meaning, as a device

FIGURE 18.2: At the boundary layer between human and alien in *Arrival* (2016). Paramount Pictures.

for seeing through time and thus for bringing present and future together—a *gift* offered by the aliens because they have seen that in approximately three thousand years they will need the help of a developed humanity.

That evolved humanity will have to be a kind of mutant or cyborg species, a condition that Louise, through a series of premonitions, is already anticipating. Thanks to her intense efforts to understand the alien language, her brain is beginning to be reprogrammed, and as a result, she starts to experience flashforwards wherein she sees scenes of a daughter, Hannah, she does not yet have, and she views these future events as if they were memories of a past she has not experienced. By literally, as Ferrando would put it, expanding the "boundaries of human comprehension" (2019, 4), the heptapods have effectively rendered Louise a kind of cyborg, one of those "future" humans who will be available to help them—just as she does by using that same newfound ability to see through time, locate a private message between the Chinese leader (Tzi Ma) and his dying wife, and use it to forestall an attack on the aliens. That mutant capacity also helps explain why her marriage to Ian, which also has not yet occurred, eventually fails. For while she can foresee the birth, life, and happy moments with their daughter Hannah, Louise also has foreknowledge of Hannah's death, which she withholds from him, knowing that he might not want to have the child if it means enduring the tremendous emotional pain that would accompany her eventual suffering and demise. With that withheld knowledge, a posthumanist vantage would argue, we see a sure sign of the difference between these figures, with one—Ian—trapped in the present, bound to a conventional subjectivity, while the other—Louise—lives with a different sense of both time and space, lives, in effect, as a cyborg already possessing a posthuman subjectivity.

It is in this sense especially that we might read *Arrival* as a fundamentally posthumanist narrative. For the film is clearly not just another alien invasion/encounter story, but rather one about the evolutionary results such an encounter might produce, that is, not just life-changing results but being-altering effects. Obviously, it offers the sort of cultural commentary that, as we have noted, is an implicit dimension of much posthumanist thinking, as humans around the Earth are forced to see themselves in common ways that defy and ultimately subvert racial, national, and ideological boundaries. But until *Arrival* is read through a posthumanist lens—indeed, until it is seen as a series of challenges to humanity to transcend our current state culturally, psychologically, and even physiologically—it mainly seems like a puzzle picture, a story about overcoming the difficulties of communicating with aliens that is set in parallel to the same difficulties we continually have with other humans. Its greater significance, though, is for another sort of boundary thinking, that involved in the boundaries of how we conceive of our world and ourselves, and how we might open ourselves to other possibilities of life. Thus we need to think of the "arrival" that this film chronicles not just as that of aliens on our planet, but as our own coming into a fundamentally different mode of being—as exemplified by Louise—one that marks us as also quite a bit alien from our former lives, but able to look back at what we were while also looking forward to all that is open to us.

As a final gloss on such posthumanist thinking and its potential for a productive reading of sf cinema, we might consider three films from very different periods: a 1950s sequel to *Creature from the Black Lagoon* (Arnold 1954), *The Creature Walks among Us* (Sherwood 1956); the previously cited entry from the *Alien* series, *Alien Resurrection*; and the recent android/AI effort *Ex Machina* (Garland 2016). All of these posit the same sort of "arrival" question as does *Arrival*, while challenging viewers to answer that question in various ways. Against its background story about failed evolutionary branches of life on Earth, the first of these films tells of scientists recapturing the escaped and badly burned "gill man" of the previous *Creature* films (1954; 1955), performing a surgical procedure that gives him some normative human appearance, and helping him breathe through lungs rather than gills. Yet after his multiple unhappy encounters with humans and even efforts to, as the title suggests, "walk among us," he finally abandons the human world and returns to the sea, not as an evolutionary dead end, but as another, coexisting species. As we have already chronicled about *Alien Resurrection*, it too offers us a kind of "new," scientifically altered being, the clone human-Xenomorph-hybrid Ripley 8 who, at film's end, prepares to return to Earth—a "home" she has never actually experienced—and as she anticipates entering the human population, Ripley confesses with some trepidation, "I'm a stranger here myself." It also seems an appropriate description of the escaped android/AI Ava of *Ex Machina*. Curious about the human world, and indeed about life itself, she kills her imprisoning creator; dons artificial skin, makeup, and a wig; and, at film's end, lands by helicopter in a crowded city. While a stranger there, she is one who, with an appropriate appearance and expanded intelligence, easily passes for human and quickly disappears among the crowds. As the posthumanist perspective might suggest, these films, and in more recent times others like them, often develop their narratives through such momentous moments of contact. While acting out the point made by the 1956 *Creature* title, that a new sort of creature already "walks among us," they find their thematic thrust in revealing our own estrangement and posing the challenge of just who—or what—that "us" really is.

## WORKS CITED

Braidotti, Rosi. *The Posthuman*. London: Polity, 2013.

Bukatman, Scott. *Terminal Identity: The Virtual Subject in Postmodern Science Fiction*. Durham, NC: Duke UP, 1993.

Chiang, Ted. "The Story of Your Life." 1998. In *Stories of Your Life and Others*. New York: Tor, 2002. 91–146.

Ferrando, Francesca. *Philosophical Posthumanism*. London: Bloomsbury, 2019.

Ferrando, Francesca. "Towards a Posthumanist Methodology: A Statement." *Frame* 25, no. 1 (2012): 9–18.

Foucault, Michel. "Two Lectures." In *Power/Knowledge: Selected Interviews and Other Writings, 1972–1977*, edited by Colin Gordon, 78–108. New York: Pantheon, 1980.

Haraway, Donna J. "A Cyborg Manifesto: Science, Technology, and Socialist-Feminism in the 1980s." In *Simians, Cyborgs, and Women*, 149–81. New York: Routledge, 1989.

Hayles, N. Katherine. *How We Became Posthuman: Virtual Bodies in Cybernetics, Literature, and Informatics*. Chicago: U of Chicago P, 1999.

Landy, Marcia. "The Cinematographic Brain in *2001: A Space Odyssey*." In *Stanley Kubrick's 2001: A Space Odyssey: New Essays*, edited by Robert Kolker, 87–104. Oxford: Oxford UP, 2006.

Latour, Bruno. *Reassembling the Social: An Introduction to Actor-Network-Theory*. Oxford: Oxford UP, 2005.

Masci, David. "Human Enhancement: The Scientific and Ethical Dimensions of Striving for Perfection." *Pew Research Center*, July 26, 2016. https://www.pewresearch.org.

McLuhan, Marshall. *Understanding Media: The Extensions of Man*. New York: McGraw-Hill, 1964.

Sigfusson, Lauren. "Do Robots Deserve Human Rights?" *Discover*, December 5, 2017. https://www.discovermagazine.com.

# CHAPTER 19

## QUEER CINEMA

### CATHERINE CONSTABLE AND MATT DENNY

This chapter demonstrates the usefulness of queer theory for exploring the differing constructions of the artificial person in sf cinema, focusing specifically on films in the *Alien* franchise (1979–). It draws on Judith Butler's (1990) influential reconceptualization of subjectivity, wherein she argues for the concept of gender performativity, emphasizing that "gender is a *doing* and not a *being*" (qtd. in McCann and Monaghan 2020, 124). Our approach combines Butler's theory with Aylish Wood's (2020) comments on technoscience in film to show how a queer perspective can illuminate the artificial person's potential to act as a site of denaturalization and critique, deconstructing both normative gender roles and traditional definitions of "humanness." The chapter offers detailed readings of several android characters from the *Alien* franchise, notably Call (Winona Ryder) from *Alien Resurrection* (Jeunet 1997) and David (Michael Fassbender) from *Prometheus* (Scott 2012) and *Alien: Covenant* (Scott 2017). The analyses demonstrate the value of bringing an alternative perspective to texts that have lacked the general critical acclaim of the first two films in the franchise, showing how an askance, queer approach can illuminate their complex subversion of gendered norms and human values.

As McCann and Monaghan (2020) note, much queer theory is characterized by an awareness of ambiguity and plurality that expresses a broader political commitment to social destabilization and subversion (1). Butler's deconstructive approach to the allegedly fixed categories of sex, gender, and sexuality both pluralizes and destabilizes them. Her argument that such categories are "performative" (1990), that is, inscribed rather than given, profoundly denaturalizes them, thereby offering a way of countering repressive conceptualizations of queer desire and identity that, following Freud, have typically pathologized homosexuality as a bad copy of heterosexuality. The notion of performativity also challenges the naturalness of traditional hierarchical binaries of male/female, heterosexual/homosexual, original/copy, while not simply erasing these categories. Butler's approach provides a crucial means of interrogating them by charting the ways in which inscriptions of gender and sexuality serve to sustain or destabilize normative power structures.

# FROM BAD COPY TO FELLOW IMITATOR

Before considering a queer reading of the *Alien* films, we explore two influential accounts of subject formation, the Cartesian model of the autonomous subject and Butler's performative analysis, showing how each offers a distinctly different approach to the analysis of film androids. The first account presents automata as (bad) copies of the human, while the second views gender and "humanness" as a series of acts performed by both human and technological beings.

Despina Kakoudaki (2014) argues that Descartes's writings mark a key shift in thinking about the artificial person from a premodern scale in which the difference between human and artificial was measured by motion or liveliness, to a Romantic/ Gothic scale of imitation and verisimilitude (17–18). Famously, Descartes's *Meditations on First Philosophy* (1996) begins by advocating a process of radical doubt as a means of finding a solid and secure foundation for all knowledge and truth. Descartes's narrator doubts the validity of his sense perceptions—what he can taste, touch, and see—the last exemplified by viewing a group of men crossing the town square: "I normally say that I see the men themselves. . . . Yet do I see any more than hats and coats which could conceal automatons?" (21). Kakoudaki links such radical Cartesian skepticism to "paranoid trends . . . in popular culture . . . in which what look like people turn out to be robots, automata or aliens" (2014, 181). The *Alien* series is an exemplary instance—as figures thought to be humans are repeatedly revealed to be androids or hosts for alien life forms.

Descartes's project in *Meditations* is to establish that which cannot be doubted. The process of radical doubt is a journey inward, leading to the discovery of one foundational point or thought: "this alone is inseparable from me. I am, I exist—that is certain" (Descartes 1996, 18). This moment of self-knowledge takes place in complete isolation; it has no relations to others or even to the thinker's own body. The journey to "the cogito" establishes an influential model of a completely autonomous self that takes the form of an interior core. Rational consciousness is conceptualized as a kind of invisible interiority, forming an important, if elusive, marker of humanness.

The lasting legacy of such thought is a series of binary distinctions that have dominated much of Western philosophy: mind/matter, consciousness/inert materiality, animate/inanimate. In viewing automata as imitations that might be mistaken for men, Descartes puts in play three other influential binaries: human/artificial, original/ copy, and authentic/inauthentic. While the key marker of humanness—the possession of rational consciousness—supports such clear divisions, Kakoudaki (2014) notes that it undermines the premodern notion wherein obvious bodily markers, such as disjointed, robotic movement, enabled easy identification of an artificial person (18). This conflict sets up a major problem, namely, what signals the crucial difference between human and artificial beings? The Cartesian model thus marks the beginning of a scale of imitation, while also presenting verisimilitude as troubling—a problem that must be solved in order to secure the binary distinctions on which it is founded.

Kakoudaki explains it is imitation that "forms the dominant understanding as to why we would want to imagine artificial people . . . how they would look physically . . . what we could imagine doing with them . . . and why they would succeed or fail (they imitate too well or not well enough)" (2014, 19). The last point recalls Romantic and Gothic narratives in which successful imitation is transgressive—imitating *too* well—and therefore linked with a fatalistic sense of inevitable failure, since the imitation can never become the original. *Frankenstein* is one of many such stories of "overreaching scientists and their constructions" wherein "the aspiring human does not become a god and the artificial person does not really become a person" (19). If, as the quote suggests, the scientist and the artificial person offer mirroring instances of overreaching and transgression, their failure reinscribes key binaries, cementing the difference between creator and created, human and artificial, original and copy.

This traditional model of an interior, core self has been subjected to sustained critique, most notably by Michel Foucault and Jacques Lacan. Butler (1990) draws on both to explain the construction of gendered identities. She argues that the illusion of a coherent, unchanging self is produced by the positing of a necessary connection between sex, gender, and desire, constructing a "natural" equation of male/masculine/desire for the female and female/feminine/desire for the male. The terms are welded together by the presupposition of "a causal relation among sex, gender, and desire" suggesting "that desire reflects or expresses gender and that gender reflects or expresses desire" (31). Butler thus argues that it is desire, in the form of heterosexual desire for the opposite gender, which is taken for the true expression of the unity of the three terms, underpinning the oppositional organization of the two other categories into the familiar binaries of male/female, masculine/feminine. The regulation of sex, gender, and desire into such binaries is part of the "institution of compulsory and naturalized sexuality" accomplished by the hegemonic power structures of the heterosexual matrix (31).

Butler also attacks the "authentic-expressive paradigm" of the self, arguing that the conventional expressions of gender—typically masculine or feminine actions or behaviors—do not arise from an internal core: "There is no gendered identity behind the expressions of gender; that identity is performatively constituted by the very expressions that are said to be its results" (1990, 31, 34). This formulation is a radical rethinking of subjectivity itself, defining it as a process in which the self is continually constructed through repetition. Masculine or feminine behaviors are thus performative—a series of act/ions through which gendered identity is constantly inscribed and reinscribed—and it is not possible to opt out of creating a gendered identity. This necessity, coupled with the normative values of the heterosexual matrix, means gender performativity is "a strategy of survival within compulsory systems . . . with clearly punitive consequences. Discrete genders are part of what 'humanizes' individuals within contemporary culture; indeed, we regularly punish those who fail to do their gender right" (190).

Butler also draws attention to cultural practices that foreground gender performativity, thereby highlighting the constructed status of gender and offering potential sites of subversion. For example, she notes that drag performances by male artists disrupt the heteronormative unity of sex/gender/desire by displaying "three contingent dimensions of significant corporeality: anatomical sex, gender identity and gender performance" (1990, 187). Thus, the male drag artist who is gendered masculine and performs femininity is said to offer a proliferation of incompatible gender enactments that cannot be causally anchored in his anatomical sex. Moreover, the drag act reveals the mechanism through which all gendered identities are continually being formed and reformed: "*In imitating gender, drag implicitly reveals the imitative structure of gender itself—as well as its contingency*" (187).

In foregrounding the structures of imitation underlying all gender identity, drag also serves as a mode of gender parody undermining the opposition between original and copy. As Butler notes, "The parody is *of* the very notion of an original; . . . gender parody reveals that the original identity after which gender fashions itself is an imitation without an origin" (1990, 188). In revealing that gender is inscribed from the outside in, via the mechanism of imitation, drag shatters "the illusion of a primary and interior gendered self" (188), undermining both the notion of an interior point of origin and the status of the heteronormative as the original, proper standard of sexuality, gender, and desire. Gender parody thus has the potential to challenge Freudian and post-Freudian models that classify lesbian and gay identities as bad copies of the heterosexual norm. Importantly, in so doing, gender parody also opens up "performative possibilities for proliferating gender configurations outside the restricting frames of masculinist domination and compulsory heterosexuality" (193).

Butler's model of gendered identity as a process of inscription that occurs from the outside in has provided Aylish Wood with a key means of thinking about the ways in which artificial persons in sf cinema attain human identity. The privileged "terms of humanness" in the films that Wood analyzes are "self-reflexive consciousness, a capacity for communication, caring, a rationality balanced by emotions, freedom of choice and the need for community" (2002, 182). Importantly, the first of these fundamentally reworks the Cartesian marker of humanness—possession of rational consciousness understood as an interior core. Once consciousness itself becomes a question of behavior—an aggregate of exterior actions—being human is simply "to behave in ways which are recognised as human" (118). Paralleling Butler's notion of drag's denaturalization of gender, Wood argues that in sf films "where humanness is a state acquired by or attributed to a technological being, the element of performance is especially foregrounded" (2002, 12). Such enactments shatter a traditional causal logic in which the qualities of humanness are viewed as natural expressions of specific biological bodies. Moreover, once a technological being "can perform as though it were human to the extent that its difference is either not apparent or is of little consequence, then any notion of an authentic humanness becomes redundant" (120). In this way, performativity breaks down the binaries of authentic/inauthentic, original/copy, positing an ontological equivalence in which all performers of humanness become fellow imitators.

For Wood, the key questions to address when considering the presentation of technological beings in sf are "which categories of humanness are at play" and "how are they validated or invalidated?" (2002, 120). Wood's identification of categories of humanness splinters a monolithic concept of "the human" into a range of types and styles, enabling the analysis of key differences, including multiple variants of gender roles and sexualities, as well as vital issues of race and class. Following Butler, Wood's attention to a film's validation or invalidation of different categories of humanness involves linking the performative with normative power structures and traditions of thinking. The film readings that follow take up this approach to identify gendered types and their inter-section with aspects of race, class, and nationality where appropriate. They explore the linking of gender performativity and sexualities—specifically whether the texts open up or close down a play of erotic possibilities beyond the heteronormative. As we shall see, the android characters Call and David offer complex inscriptions of humanness, and the readings trace the ways in which each denaturalizes normative gender roles, thereby facilitating a critique of associated human values.

# ASH

Call and David are presented in ways that evoke the android of the first *Alien* film (Scott 1979), Ash (Ian Holm). There are narrative parallels in the presentation of Call and Ash: both "pass" for human until violence reveals their artificiality and both have a hidden agenda concerning the alien. In contrast, characterization links Ash with David's perfor-mative construction as a white British male. Ash is initially presented as indistinguish-able from the human crew of the *Nostromo*—waking in a cryogenic sleeping pod, eating meals, and bickering with the commanding officers. His artificial status is a shocking revelation, conforming to Kakoudaki's (2014) conception of Cartesian trends in popular culture. The epistemological dangers of such successful imitation are resolved through the creation of a key sign of physical difference that informs the entire series, the white fluid of the android body. Ash's bodily fluid is revealed after Ripley angrily attacks him on discovering special order 937, which commands the science officer to return an alien specimen to earth and deems the rest of the crew expendable. His subsequent attack on Ripley displays another key difference, inhuman physical strength, and it takes the combined efforts of Parker and Lambert to rescue her.

The crew reanimates Ash's head alongside the wreckage of his body, the display foregrounding his nonhuman physicality. Ash's admiring description of the creature as "a perfect organism" puts him entirely at odds with the remaining human crew members. Indeed, his comments—"I admire its purity—a survivor, unclouded by conscience, re-morse or delusions of morality"—extol both the alien life form and himself by placing both firmly outside Wood's key criteria of humanness. The lone survivor has no need of community and associated values of empathy and ethics. This extolling of emotion-free, survivalist logic is also the moment when Ash's own rational consciousness is revealed to be programming. Obeying the Company's commands to the last—he provides no information as to how to destroy the alien—Ash epitomizes the remorseless drive and functionality of the machinic. Indeed, his lack of "conscience" also carries philosophical and theological implications in suggesting the absence of a soul. The scene thus forcibly asserts a series of binary divisions: human/artificial, human/nonhuman, consciousness/programming, constructing Ash as a bad copy. Yet his final, ironic expression of sympathy to the remaining crew is a rare moment of humor, suggesting a malicious enjoyment of their predicament, which disconcertingly crosses the scene's clear divisions between emotionally informed and emotion-free rationality. Holm's English accent and sardonic delivery of the last line reinscribes Ash as villainous in a different way, shifting the android from being a machinic tool of the Company to the familiar (human) character type of the British villain.

# CALL

While Call follows Ash in hiding her "synthetic" status and agenda for the alien, she differs from him in being the only female android within the *Alien* franchise thus far. If Ash's bickering behavior is indistinguishable from the majority of the crew who are male, Call's femininity is defined in relation to a smaller group of female characters. *Alien Resurrection* begins with the introduction of the only other female crew member of the *Betty*, the pilot Hillard (Kim Flowers). She is presented alongside Captain Elgyn (Michael Winnicott) in a high-angle shot from behind as they pilot the ship. This shot is followed by a medium close-up of Hillard in the foreground but out of focus, with Elgyn in focus in the background. It is only after he remarks, "No matter how many times you see it, the sight of a woman all strapped up in a chair like that just . . . ," and moves toward her that Hillard comes into focus. She is dressed in a form-fitting, sleeveless costume that pays tribute to both pulp sf and 1990s "bodycon" high fashion. Cinematography, script, and costume thus combine to introduce Hillard through a lens of male heterosexual desire, and her response—ready acquiescence—sets up an initial paradigm of heteronormative gendered behavior from which both Call and Ripley will depart.

Call is first presented via a series of shots from multiple angles, situating her within the space of the engine room and providing a holistic rather than fetishized conception of her body. The sequence opens with a high-angle shot in which Call is framed centrally and shown dressed in a practical jumpsuit operating a winch. This introduction is followed by a tighter low-angle shot from below, and finally a shot aligned with Vriess (Dominique Pinon), showing Call from behind. This shot travels the length of Vriess's disabled, male body, which is strapped in a chair, a de-eroticized echo of Hillard, and leads to a reverse shot of Vriess turning to look at Call and smiling.[1] The editing emphasizes Call's activity and professionalism, aligning the audience gaze with a male character whose body is on display. The lack of dialogue ensures Vriess's gaze is open to a range of readings—sexual desire, paternal indulgence, platonic affection, or collegial recognition—in a way that Elgyn's is not.

The different formal strategies used to introduce Hillard and Call construct two differing representations of femininity. The former is aligned with the heteronormative paradigm while the latter is distanced from a long tradition of female "sexbots" in sf film, from the ersatz Maria (Brigette Helm) of *Metropolis* (Lang 1927) to Ava in *Ex Machina* (Garland 2014). Intertextually Call is inscribed as human through references to Ripley, via her costuming (the practical jumpsuit worn in *Alien*) and employment (specifically the job of power-loader operator in *Aliens*). These references also generate a sense of gender ambiguity. Weaver's most famous role has been read in terms of "musculinity," the enactment of traditionally masculine signifiers of strength by the female action heroine (Tasker 2013, 148–49). By comparison, Ryder's diminutive stature and gamine styling emphasize the youthfulness of her version of androgynous femininity. Both the similarities and age differences suggest the possibility of succession, presenting Call as "the Ripley" of this installment in the franchise.

The first scene with Call, Vriess, and Johner (Ron Perlman) establishes her potential as both a protagonist and locus of ethical values. Positioned above the others in the cargo bay, Johner appears, mimicking the sounds and movements of an ape, before aiming his knife so that it impales Vriess's thigh, the animalism preceding the casual violence presenting both as primitive. Call reacts angrily before Vriess notices, saying, "What is wrong with you?" Johner justifies his use of a fellow crew member for "target practice" by taunting Vriess for his physical limitations—his inability to feel the pain of the injury. The close-up of Vriess's face as he turns away, registering distress at the taunt, is immediately followed by a close-up of Call, whose quivering lower lip indicates her empathetic response. The brief exchange contrasts Johner's action—a primitive assertion of physical superiority enforced by his placing and Perlman's stature—with Call's civilized condemnation of it.

Call's reactions—instant condemnation and emotional sympathy—inscribe the character with key aspects of humanness and human values, namely, caring for and about others. The enactment of empathy enables Call to pass for human and also feminizes her by drawing on traditional gender roles in which femininity is associated with nurturing. Johner's obliviousness to the ethical issues raised by his action is reflected by his immediate demand for the return of his knife, and Call pauses with the weapon before snapping the blade off against the ship's superstructure. Her gesture connotes castration, punishing Johner's action and rejecting his attempt to assert masculine superiority via physical strength. The confrontation thus uses enactments of gender roles to present a clash between different systems of value, pitting the ethics of caring against "brute force."

The confrontation rhymes with a later rebuffing of Johner's heterosexual pretensions when the *Betty*'s crew are introduced to Ripley 8 (a clone of Ellen Ripley and the Alien Queen). Viewing her training with a basketball, Johner comments that he "can't leave off the tall ones." Ripley's all-brown costuming has a sleeveless top with buckle fastenings at the back, evoking fetish wear and punk aesthetics, while offering a queer parody of Hillard's feminine bodycon clothing. Johner's attempts at flirtation are mockingly mirrored by Ripley 8, who rests her arms on his shoulders, bringing her face inches from his, as she continues to pass the ball from hand to hand. Despite Johner's acknowledgment that Ripley 8's height and athleticism are the sources of his attraction, he is clearly unsettled when she refuses a properly feminine show of submission. Instead, she outperforms him, fending off his attempts to steal the ball before using it to attack him in another symbolic castration. Ripley 8's actions showcase the physical strength and agility bequeathed by her alien DNA, while also offering a subversion of Hillard's properly feminine behavior, deploying nonhuman otherness to parody and disrupt heteronormative gender roles.

Call's covert agenda is revealed when she sets out alone to find Ripley 8, believing her to be incubating the alien and planning to kill them both. On discovering that the alien has already been removed, Call offers to kill Ripley 8 to "make it all stop," reframing her intended action as a release from a life that is not worth living. Ripley refuses the offer, asking, "What makes you think I would let you do that?" while impaling her hand by driving it onto the knife Call is holding. The self-penetration is an ambiguous gesture, at once aggressive and yielding, nonhuman and human, as her apparent inability to feel pain both demonstrates her alien genealogy and links her with another human character, Vriess. The parallel foregrounds the limits of Call's conception of caring, which extends to Vriess as human, but not to Ripley, whose nonhuman life is not worth living. In thwarting Call, Ripley 8 refuses to be viewed as an object of pity and asserts her status as a self-determining individual—albeit one who does not know who she is or what she might become.

The moment of penetration also eroticizes the encounter between the "naïve terrorist" and her target (Wood 2002, 140), turning it into "a sadomasochistic seduction scene in which the two shift roles and power" (Stacey 2003, 270). Ripley 8 takes Call's hand and holds it, tracing the contours of her own face while describing her connection to the alien queen. Kneeling upward to loom over her, "back into dominatrix mode, Ripley grabs Call by the throat, holding her face roughly in her large hands. . . . In a pose which imitates the conventional heterosexual prelude to a first kiss (male above, looking down, seizing the female's face in his hands as she gazes up into his eyes), Call submits once again to Ripley's superior strength" (Stacey 2003, 272). Ripley follows this gesture by laying her cheek gently against Call's, showing their generational difference, and juxtaposing the violently erotic with a more maternal tactility (see figure 19.1). Stacey notes that the film plays with a "narcissistic aesthetic of duplication" (272) associated with homoerotic desire by emphasizing the stars' faces: "the similar bone structure (high cheek bones) and proportioned features (big eyes, small noses and fine lips) echo each other in conformity to cultural norms of white beauty" (269). The complex scene offers almost delirious levels of imitation: Ripley 8's rapid take-up of masculine and feminine roles interlaced with animalistic postures indicative of her alien genealogy (Constable 1999, 193), Call as Ripley's successor/lover, and both characters as differential copies of Ellen Ripley. Importantly, this play of varied, multilayered, performative roles creates a plethora of erotic possibilities outside the heteronormative—homoerotic, cross-species, bio-technological couplings—that continue throughout the film.

Call's change of attitude toward Ripley 8 occurs in a later scene where the survivors come across the lab containing versions 1 through 7 of the Ripley / Alien Queen clones. While the majority are preserved in specimen jars, the last is still alive and clearly in pain. She is connected to a number of pipes and inadequately covered by a flimsy sheet

FIGURE 19.1: *Alien Resurrection* (1997) juxtaposes the erotic with maternal tactility, as Ripley (Sigourney Weaver) embraces the android Call (Winona Ryder). 20th Century-Fox.

that Ripley 8 gently adjusts, affording her some measure of privacy and dignity. This small gesture and her distressed expression on hearing the clone's stuttered appeal, "Kill me," marks Ripley 8's first display of empathy. The clone's request recalls Call's offer of death to Ripley 8, changing its terms. In this case it is not the intruder who decides that the clone's life is not worth living; instead, the experimental subject appeals to a fellow experiment for death. Call wordlessly hands Ripley 8 the flame-thrower, which she uses to torch the lab, putting all her fellow clones beyond further experimentation. Ripley's tears and ragged breathing attest to the emotional cost of this act and differ strongly from her previous detachment, bringing her closer to the Ellen Ripley of the previous films. Call's understanding, expressed through her willing provision of weaponry, contrasts with Johner's uncomprehending dismissal of Ripley's action as "a waste of ammo" and "a chick thing," reinforcing the film's initial gendering of empathy, compassion, and caring as feminine. The scene thus both humanizes and feminizes Ripley 8, while expanding the parameters of Call's caring to include the human/nonhuman clones.

Later in the film, it is Ripley who extends compassion to Call when she is "outed" as a synthetic, having, apparently miraculously, survived being shot in the chest by Wren (J. E. Freeman). In a gesture that reverses her covering up of the clone, Ripley 8 gently pulls aside Call's jacket to expose her injury, an extreme close-up showing the white bodily fluid coating her fingers as she probes the wound. The bodily penetration parallels the pair's earlier intimacy, while the subjective shot and close-ups of Call's pleading, distressed expression as she looks at Ripley present this as a romantic moment in which a lover reveals a terrible secret. While the narrative placing of this late reveal resembles that of *Alien*, Call's wounded, humanoid body does not offer the visceral horror of Ash's dismembered, malfunctioning physique. The editing places the moment within the development of a relationship, contrasting with the special effects–driven display of Ash's body as a nonhuman object of visual spectacle. Within *Alien Resurrection*, it is only the reactions of the remaining human survivors that assert the exclusionary binary logic of human versus nonhuman, objectifying Call as a rare specimen or malfunctioning machine. The editing serves to align the spectator with Call, offering a critique of her exclusion from the human community.

Ripley's initially incredulous response differentiates her from the others: "I should have known; no human being is that humane!" The comment explicitly divorces the values associated with Call from the possession of a biologically human body, foregrounding the performativity of caring. The reveal breaks the naturalized linking of caring and biological femaleness that is traditionally posited on the basis of reproductive sex roles wherein the female nurtures the young. A later conversation between the pair in the chapel returns to this issue. Call reveals her discovery of the government's covert operations, her failed endeavor to thwart them by assassinating Ripley, and her aim to save life on earth, including the crew of the *Betty*. On being asked, "Why do you care what happens to them?" Call replies that she is programmed to do so, an excuse that Ripley 8's mocking rejoinder rejects as inadequate: "You're the new asshole model they're putting out these days? Come on!" In drawing attention to the gulf between her aspirations and her actions, Call presents caring and compassion as human ideals that she endeavors to instantiate rather than programming, thus differentiating her behavior from the fulfillment of order 937 that marks Ash's actions as machinic. Within *Alien Resurrection*, Ash's remorseless, logical, emotion-free rationality is presented by the head scientist, Wren, who has no qualms about developing weaponry by experimenting on sentient creatures—from clones to human beings. In contrast, Call's caring involves "considering the outcome of experiments on people and the world . . . an amalgamation between rationality and emotions," making her a key figure of human/humane rationality (Wood 2002, 144).

The late revelation of Call's artificial status ensures that all aspects of her humanness—her emotions, burgeoning relationship, rationality, and ethical ideals—are denaturalized and recognized as performative. This effect allows the film to offer a more general critique of "the state of humanness, that is, humans do not care enough anymore" (Wood 2002, 143). While the rendering of the android as "more human than human" might be seen as simply re-inscribing humanist values, the film's pairing of compassionate ideals with advanced technology presents both as artificial constructions that exceed their human creators. Importantly, the focus on the performative provides a key way of seeing how the film utilizes complex inscriptions of gender and humanness to effect critique and open up erotic and ethical possibilities. This is crucial to appreciating how the film repeatedly gestures beyond the heteronormative and the human-centric in its proliferation of biotechnological and cross-species couplings and its expansion of a human-centered conception of care to include the posthuman subjects of biological and technological experimentation.

# DAVID

*Prometheus* is a prequel to the *Alien* series, chronicling a scientific expedition to trace and potentially contact the ancient alien "Engineers" apparently responsible for life on earth. The film indicates David's android status from the start, opening with shots of him monitoring and maintaining the ship, thereby distinguishing him from the human passengers who require cryostasis. David's operational tasks are interleaved with less functional actions: learning and practicing languages, watching *Lawrence of Arabia* (Lean 1962), dying and styling his hair in imitation of Lawrence, and rehearsing Peter O'Toole's clipped delivery of "the trick, William Potter, is not minding that it hurts." David practices his imitation in front of a mirror, the reflection adding yet another imitator to the succession of performers (see figure 19.2). This staging also echoes a key theme of *Lawrence of Arabia*, the impossibility of knowing the "authentic" T. E. Lawrence. Importantly, both David's language lesson and his rehearsal involve repeating the lines spoken by a person appearing on a screen, presenting perfect mimicry as a means of attaining fluency in both languages and humanness.

While David's appropriation of humanness involves adopting the accent and mannerisms of an Oxford-educated British officer during the First World War, his performance of a historicized Britishness differs from Ash's. In *Alien*, both Ash and Kane are played by British actors, Holm and John Hurt. Both actors deliver their lines using English Received Pronunciation, common at the time, enabling the former to pass for human by suggesting a shared education and class background. In contrast, David's

**FIGURE 19.2:** The android David (Michael Fassbender) adopts the look of T. E. Lawrence (*Prometheus*, 2012). 20th Century-Fox.

adoption of a historicized British masculinity, indicated by formal courtesy and upper-class English accent, distinguishes him from the informality of the largely American crew. In imitating the Lawrence of the 1962 film—the failed white savior who defied his military and political superiors by attempting to secure an Arab homeland—David also constructs himself as an individualist who is not bound by the rules governing the human crew. His character is thus associated with a particular model of white, masculine individualism in which uniqueness and privilege are expressed by *not* conforming to social norms.

*Prometheus* sets up a rival sibling relationship between Weyland's (Guy Pearce) biological daughter Vickers (Charlize Theron) and his creation David, who, again recalling T. E. Lawrence, is a kind of illegitimate son. Physically similar, both Vickers and David are tall, blonde, and athletically slim—their resemblance is emphasized during a brief confrontation in the ship's corridor. A close-up of the pair in profile accentuates their angular faces, drawing attention to Theron's and Fassbender's striking bone structures, while the blue lighting allies such white beauty with unnatural coldness. While David's reserved and formal behavior aligns with his gender and assumed nationality to function as markers of individuality, Vickers's similarly cool and distant enactments of patriarchal authority sit uneasily alongside her cultural situation as an American woman, prompting Janek (Idris Elba) to ask if she is a robot. The insult equates Vickers's abrasive, impersonal authority with a lack of emotion that renders her unfeminine and inhuman, which she counters by agreeing to have sex with Janek. While the exchange recalls the heteronormative gender disciplining of female scientists in films of the 1950s (Bould 2012, 50–54), the racial and power dynamics are very different.

The late revelation that Weyland is alive and on board the *Prometheus* reprises the sibling rivalry between David and Vickers, using rhyming scenes in which each is placed at the father's feet. A postoperative Shaw (Noomi Rapace) bursts in to discover David preparing Weyland's aging body for the final expedition. He kneels before his creator to wash his gnarled feet, the posture suggestive of the swearing of fealty to a sovereign, combining feminine caring with suitable subservience. In contrast, Vickers is unsympathetic to her father's quest for immortality. Arriving in his chamber, she carefully positions herself lower than his seated figure, saying, "A king has his reign, then he dies. It's inevitable," the blunt comment mitigated by the intimate rubbing of her cheek against his hand. Her father's clenching of his fist and turning away indicate rejection, and the following close-up shows Vickers's cold fury. Her brief enactment of subservient and seductive femininity is in marked contrast to her habitual demeanor. These scenes denaturalize the siblings' traditional gender roles by swapping them around—David is the obedient, caring daughter and Vickers is the ambitious, patrilineal heir.

While David's performative characterization fluctuates between gendered roles, from Lawrence to dutiful daughter, the former intersects with his position as co-protagonist in the film's narrative arc of scientific exploration and discovery. The first outing on LV 223 returns two items to the ship for further analysis, Shaw bringing back the helmet-encased, severed head of a long-dead engineer, and David smuggling in a mysterious black canister. The objects are associated with two different modes of scientific inquiry. The ship's scientists' collaborative endeavor is unsuccessful—the examination of the head results in its complete disintegration. In contrast, David's lone, surreptitious investigation leads to the extraction of a particle of the alien life form, whose lethal properties he establishes through experimentation on the unwitting Holloway (Logan Marshall-Green). At this point, David's Lawrentian individualism, being outside the rules that govern the human community, is aligned with a long-standing model of scientific inquiry as a lone pursuit involving callous detachment from objects of experimentation and a disregard for collective social consequences, famously emblematized by Victor Frankenstein.

The negative connotations of the experiment on Holloway contrast with a later scene in which David reawakens the engineers' derelict spacecraft and his scientific curiosity becomes a sense of wonder. He clearly delights in being first to access the craft's control room: settling into the seat at the ship's control console, pressing buttons, smiling, and looking up, as though he were an on-screen avatar of the audience anticipating the scene's spectacular climax. Holographic projections of the engineers appear, one taking David's place at the controls, and the music builds triumphantly from the single fluting notes that unlock the controls to a full orchestral score as the ship springs to life. A long shot gives the full view of the projected orrery with David framed centrally and dwarfed by the multiplanar, holographic imagery. Cutting between wider and tighter framings emphasizes David's small stature within this spectacle, underscoring the childlike spontaneity of his reactions, as he leans back and joyfully turns on the spot with arms outstretched, before walking into the vast, whirling spaces of the luminescent cosmos. David's demonstrated emotional range here aligns with the spectator's response to the beauty and complexity of the CGI spectacle, strongly humanizing him.

This demonstration of David's wonder and delight invites a reading of the scene as revealing true, inner depths, as it reverses the play of surfaces that marked his adoption of the Lawrence persona and displays the depth logic of what Butler terms the "authentic-expressive paradigm" of the self (1990, 31). His response also recalls Shaw's awestruck wonder on discovering the cave paintings on Skye, briefly paralleling him with her reverence for archaeological investigation. The film thus draws on a long history in which the capacity for wonder and awe formed the link between the human and the divine, inscribing David with an inner self that strongly resembles the pre-Cartesian Christian concept of the soul, and contradicting Weyland's assessment of him as perfect but soulless. It is worth noting that in aligning the soul with a white European male, albeit in synthetic form, the film reinforces traditional gendered and racial hierarchies. However, the positive inscription of David's inner depths is not sustained across the narrative; it forms another contradictory addition to a series of surface appearances, rendering the character profoundly ambiguous.

*Alien: Covenant* departs from the linear chronology of *Prometheus*, juxtaposing moments across David's life and inviting the spectator to compare them. The opening prologue shows a conversation between David and Weyland in which the former identifies works of art in the room and takes his name from Michelangelo's famous statue. Weyland positions himself as David's progenitor, saying "I am your father" and drawing attention to their contrasting roles as creator/created, artist/artwork. The staccato rhythm of David's responses and the similarly stilted nature of his movements across the room add another opposition, that of human/machine. Directing David to play the piano, Weyland situates him within a long tradition of what Kakoudaki (2014) terms "mechanical spectacles," specifically that of the "lady musician" automaton whose intricate movements imitated the feminine accomplishments of eighteenth-century bourgeois ladies (20–21). The performance of such automata attests to their creator's talent, just as the accomplishments of bourgeois wives and daughters reflected the household patriarch's status, and both suggest a feminization of David as an object of display. However, Weyland does not praise David's rendition of "The Entry of the Gods into Valhalla" from Wagner's *Das Rheingold*, but dismisses it as "a little anemic without the orchestra," thereby characterizing both the music and the android as pale imitations.

This brief depiction of an earlier stage in David's life encourages a re-reading of his later enactment of the Lawrence persona as part of a trajectory toward more convincing humanness. The appearance of the younger Weyland reminds us of his ultimate failure to achieve his ambition, his search for the engineers ending in death rather than immortality. Weyland's fate follows the Romantic model of imitation as transgression, offering a venture-capitalist variant of the overreaching scientist who fails to become a god. This fatalistic model suggests that David's reduction to a severed head at the end of *Prometheus* might be seen as punishment for his transgressive attainment of humanness. However, David's prescient comment to the younger Weyland—"I will serve you, yet you are human. You will die, I will not"—distinguishes human mortality from the android's godlike longevity, suggesting the possibility of horizons of aspiration beyond becoming human.

While the prologue augments the multiple iterations of David across his lifespan, the next scenes on the colonization vessel, *Covenant*, complicate this further by adding his double, the synthetic Walter, also played by Michael Fassbender. Walter acts in accordance with "Mother's" commands, tending the ship, its human crew, and the cargo (2,000 colonists and 1,140 embryos), without indulging in nonessential activities like watching films or dying his hair. His American accent and informal dress make him vocally and visually indistinguishable from the other crew members, who are coming to terms with the sudden death of the captain, Jake Branson (James Franco). The grieving spouse Daniels (Katherine Waterston) tells Walter about Jake's dream of the couple as pioneers, building a log cabin beside a lake in a new world, saying, "Now I wonder why I bother." Walter gently responds, "Because you promised to build a log cabin on a lake." While Walter's attentive expression and gentle, quiet intonation comfort Daniels to the point of laughing through her tears, her emotions highlight the restrained nature of his response. His reply focuses on the literal construction of the cabin rather than its symbolic significance for a "pioneer" community. Walter's enactment of the role of safe, asexual confidant thus reinforces his android status, while the film positions him as a queer outsider in a traditional, heteronormative community.

Walter's alignment with a reserved Americanness affects Fassbender's performance of David's Britishness in *Alien: Covenant*, shifting away from restrained formality to a theatrical literariness evident in his delivery of quotation-laden dialogue. The contrast underscores David's individuality—a lone prototype discontinued for being "too human, too idiosyncratic"—presenting his successor as a muted copy. David's placement as an experienced elder is shown in his endeavors to teach Walter to play the flute. The resulting homoerotic scene draws on a Platonic conception of tutelage in male-male relationships, while also exaggerating a standard "mise-en-scene of excessive sameness" through Fassbender's dual roles (Stacey 2003, 273). Warmly lit and placed either side of the central flute, Walter forms an embouchure and exhales in rhythm, while David guides his fingers, transforming breath into music in an act of physical intimacy. David's praise for Walter, "You have symphonies in you, brother!" evokes a familiar scenario in which the seducer awakens latent potentialities in the seduced, paralleling sexuality and artistic creativity. While this scene plays with succession and imitation in a way that recalls the erotic encounter between Ripley 8 and Call, the heteronormative framing of the later film forecloses any exploration of erotic possibilities, simply aligning David's queerness with a "disturbing" deviance and villainy.

After this music lesson, David takes Walter outside onto a high platform overlooking the desolate cityscape, placing the pair on a level with the towering monuments, above a landscape studded with blackened corpses. He recites lines from Shelley's (2017) famous poem "Ozymandias" about the tyrant Rameses II—"My name is Ozymandias, king of kings / Look on my works, ye Mighty, and despair!"—and there follows a flashback to David's arrival on the planet and his releasing of a pathogen that rains down onto the terrified populace like volcanic ash, killing and embalming men, women, and children. The dark tendrils of the CGI pathogen resemble a hoard of carnivorous locusts as they actively penetrate and puncture bodies, with this destruction of an entire civilization evoking the wrath of the Old Testament God (Alpert 2017–18). The flashback ends with a zoom-in to a close-up of David's white, windswept face, his blue eyes wet with tears as he gazes down at the devastation, and a voiceover of his ecstatic repetition of Shelley's second line forms a sound bridge to the present. This shift is immediately followed by Walter's accurate and dispassionate completion of the poem: "Nothing beside remains. Round the decay / Of that colossal wreck, boundless and bare, / The lone and level sands stretch far away." David's omission of the poem's hubristic framing fundamentally reworks the quote as a celebration of Ozymandias's arrogant pretensions, paralleling the tyrant's monumental "works" with his own attainment of godlike powers through the genocide of the engineers. In thus destroying Weyland's "gods," the created obliterates his creator's creator, exaggerating the Romantic model of overreaching and transgression.

As Robert Alpert notes, David's actions recall both the warfare of the *Blitzkreig* and Nazi concentration camp experimentation on prisoners. He argues that such acts demonstrate the android's machinic nature: "David is morally and emotionally repulsive . . . intelligence without consciousness, reason without emotion, and unbounded ego without limitation" (2017–18). While Alpert's reading parallels David with Ash as similarly lacking a conscience, both soulless imitations of humanity, it overlooks the ways *Alien: Covenant* insistently emphasizes David's individuality. The flashback ends with his expressive face and voice, strongly suggesting that his most evil actions arise from his successful adoption of the worst of humanness—arrogance pushed to a tyrannical egocentrism. While Walter reads David's misattribution of Shelley's poem to Byron as evidence of machine malfunction, the reference highlights the human figure at the heart of the Romantic linking of individuality and creativity with transgression. David thus evokes a set of figures who illustrate the clash between individual aspiration and the values of community: Lawrence as privileged outsider, Frankenstein's antisocial scientism, and Byron's aristocratic rejection of social norms and hierarchies. His performative humanness enacts and displays the problems inherent within a long tradition of white, Western, masculine individualism.

The final confrontation between Walter and David pits the goodness of the former, demonstrated by his returning to rescue Daniels, against the evil of the latter, expressed in a line that misquotes Satan's speech from *Paradise Lost*: "It's your choice, them or me, to serve in Heaven or to reign in Hell."[2] The scene ends with both poised to strike a decisive blow and with Walter apparently victorious. Once back on board the *Covenant*, Daniels tends his wounded face, thanking him for saving her life—their interaction replaying an earlier conversation between the two androids in which David suggested that Walter's sacrifice of his hand for Daniels was an expression of love. Walter's disclaimer, "It's my duty," is followed by Daniels' subjective view of him gently touching his stump arm with his remaining hand, his gesture marking the inadequacy of the explanation and affirming a romantic motivation. The subjective shot conveys Daniels's response, traveling up the android's body as she views him for the first time as a physically attractive, potential partner. This brief interlude shifts Walter from asexual confidant to heterosexual partner—a connection later confirmed by Daniels's proposal that they build the cabin together—while it also demonstrates the only way an android could achieve full inclusion within the heteronormative community of the *Covenant*.

Approaching *Alien Resurrection*, *Prometheus*, and *Alien: Covenant* from a queer perspective brings out the ways in which the films use their central android characters to display and deconstruct gender roles and associated human values, creating a distance that enables them to act as a locus of critique. It constitutes a departure from reading Call as an example of "safe" or positive technology that is aligned with the human and therefore humanized (Wood 2002, 144–45; Alpert 2017–18), and David as bad technology whose villainy is purely machinic (Alpert 2017–18). *Alien Resurrection* uses the figure of the female android as a means of interrogating and expanding the ethics of care, ending with a burgeoning, cross-species, biotechnological community that includes Call, Ripley 8, Vreiss, and even Johner. The film deploys its nonhuman characters to confuse, undermine, and refuse binaries, repositioning human heterosexuality as just one option in a plethora of erotic possibilities. In contrast, *Alien: Covenant* forecloses its exploration of a technologized homoeroticism, ending with David's violent rejection of Daniels's misplaced proposal to pursue his project of scientific, cross-species experimentation, and positioning him as a threatening, queer outsider to the heteronormative community. Both of the later films foreground David's individualist aspirations so that his villainy becomes a demonstration of the toxic consequences of too perfect an imitation of particular forms of white, Western, masculine individualism. Focusing on the performative allows us to trace the ways in which both Call and David offer critiques that are grounded in the gendering of their distinctive enactments of humanness—the first a vital reminder of the ideals of caring and community, and the second a terrifying warning about the dangers of individualism.

The authors are indebted to Professor Richard Dyer for his inspiring comments on whiteness and colonialism in *Prometheus*.

## NOTES

1. In the 2003 alternate cut of the film produced for the Special Edition DVD, Vreiss attracts Call's attention to tell her a crude joke. While its omission from the theatrical version follows a standard and problematic de-eroticization of disabled characters within Hollywood cinema, it also makes Vreiss's attitude to Call much more ambiguous. For more on the relationship between queer theory, disability studies, and the emerging field of crip theory, see Kafer (2013) and McRuer (2006).
2. This phrasing reverses the order of Satan's famous declaration in book 1, line 263 of Milton's (2017) *Paradise Lost*: "Better to reign in hell, than serve in heaven."

## WORKS CITED

Alpert, Robert. "Ridley Scott's *Prometheus* and *Alien: Covenant*: The Contemporary Horror of AI." *Jump Cut* 58 (2017–18). http://www.ejumpcut.org/archive/jc58.2018/AlpertAlienPrequels/index.html.

Bould, Mark. *Science Fiction*. London: Routledge, 2012.

Butler, Judith. *Gender Trouble: Feminism and the Subversion of Identity*. London: Routledge, 1990.

Constable, Catherine. "Becoming the Monster's Mother: Morphologies of Identity in the *Alien* Series." In *Alien Zone II*, edited by Anette Kuhn, 173–202. London: Verso, 1999.

Descartes, René. *Discourse on the Method and Meditations on First Philosophy*. Trans. John Cottingham. Cambridge: Cambridge UP, 1996.

Kafer, Alison. *Feminist, Queer, Crip*. Bloomington: Indiana UP, 2013.

Kakoudaki, Despina. *Anatomy of a Robot: Literature, Cinema, and the Cultural Work of Artificial People*. New Brunswick, NJ: Rutgers UP, 2014.

McCann, Hannah, and Whitney Monaghan. *Queer Theory Now: From Foundations to Futures*. London: Red Globe, 2020.

McRuer, Robert. *Crip Theory: Cultural Signs of Queerness and Disability*. New York: New York UP, 2006.

Milton, John. *Paradise Lost*. Revised 2nd ed. Ed. Alastair Fowler. London: Routledge, 2007.

Shelley, Percy. "Ozymandias." In *Selected Poems and Prose*, edited by Jack Donovan and Cian Duffy. New York: Random House, 2017. 57.

Stacey, Jackie. "'She is not herself': The Deviant Relations of *Alien Resurrection*." *Screen* 44, no. 3 (2003): 257–76.

Tasker, Yvonne. *Spectacular Bodies: Gender, Genre, and the Action Cinema*. London: Routledge, 2013.

Wood, Aylish. *Technoscience in Contemporary American Film: Beyond Science Fiction*. Manchester: Manchester UP, 2002.

CHAPTER 20

# UTOPIANISM

## CAROLINE EDWARDS

THERE'S a curious oversight in scholarship on science fiction (sf) film that remains puzzling: an almost universal consensus that utopian film does not exist. Seen from a literary studies background, where discussions of utopian method, possibility, temporality, and genre often dominate studies of sf literature,[1] one is hard pressed to understand this scholarly lacuna. Robert Shelton (1993) addressed this question, asking why the wide interest in film genre studies had failed to extend into a scholarly discourse on utopian cinema: "Trying to find 'Utopian Film' in the various indexes of film is worse than the proverbial search for the needle in a haystack, because, it turns out, no one seems to have bothered to put the needle in the haystack in the first place" (22). Whether through deliberate avoidance and erasure or simple lack of interest in the matter, analyses of utopian sf cinema have limited themselves to a handful of texts: H. G. Wells's collaboration with William Cameron Menzies and Alex Korda, *Things to Come* (1936), Frank Capra's adaptation of James Hilton's vision of Shangri-La, *Lost Horizon* (1937), and John Boorman's mind-bendingly weird meditation upon the utopian enclave in *Zardoz* (1974). This small caucus grows somewhat if we include the utopias that have turned inexorably into dystopias—in films such as Fritz Lang's *Metropolis* (1926), Terry Gilliam's *Brazil* (1985), and Michael Anderson's adaptation of *Logan's Run* (1976). To mark the quincentenary of the publication of Thomas More's originary *Utopia* (1516), a 2017 special issue of the journal *Science Fiction Film and Television* returned to the question of utopia on screen, with Simon Spiegel coming to a similar conclusion as Shelton: "There is a wide agreement among scholars that a typical eutopia lacks some very basic elements of a typical narrative film: it neither features a conflict that drives the plot forward nor real characters with individual traits—both of which are required by feature films in the classical Hollywood tradition" (2017, 53). While Spiegel argues that utopian films in the Morean vein exist, he sees them not in the places where film scholars have been looking. Nonfiction films, he suggests, offer more fertile ground for discovering a nascent genre of utopian filmmaking: one that reflects upon the sociopolitical conditions of its contemporary moment from the perspective of an embryonic sense of possibility.

This chapter broaches the question of the unavailability, or erasure, of utopian film from a different perspective. For some years, utopian scholars have argued that utopia is better understood as an *analytical method* than as a distinct *literary genre*.[2] My own work in this area draws on the rich tradition of utopian studies—which encompasses sociology, political philosophy, cultural studies, literary criticism, architecture, urban planning, anthropology, and sound studies—to formulate a method of utopian thinking that excavates the repressed utopian content of narratives in order to demonstrate the ongoing usefulness, as well as ubiquity, of what the German philosopher Ernst Bloch called the utopian impulse or *Noch Nicht* (translated as "not yet" or "still not"). This chapter demonstrates this utopian hermeneutic. Where Simon Spiegel formulated a response to the paucity of discussions of utopia in film scholarship by noting that *we have been looking in the wrong places* and advocated that scholars turn to nonfiction film, I suggest that utopia has been in front of us all along but that *we have been looking for it in the wrong kind of way*.

Alongside Ernst Bloch's utopian philosophy, my guide here is the politicized aesthetics of Bloch's contemporary Herbert Marcuse. In uncovering utopian possibility within oppositional social movements that attempted to resist advanced industrial capitalism and American consumer society, Marcuse insisted that the formal properties of cultural texts were themselves proto-political gestures of what he termed "the Great Refusal." Aesthetic form, Marcuse writes in *The Aesthetic Dimension*, "is not opposed to content, not even dialectically. In the work of art, form becomes content and vice versa" (1978, 8). It is an important distinction to bear in mind when approaching sf cinema. The rich, enticing, and even sometimes overwhelming world-building that undergirds sf visual narratives can focus discussion around questions of spectacle, special effects, costume, and performance, as well as the obviously estranging content of such storytelling. However, as I hope to demonstrate, focusing exclusively on the content of sf film in utopian sociological and political readings (following the literary tradition of *utopia-as-genre*, of looking for representations of utopian groups and societies within filmic texts) overlooks an important ingredient of these texts' utopian *Vorschein* (anticipatory consciousness): their complex formal properties and innovative narrative organization.

# Toward a Filmic Utopianism

Although we might agree with Peter Fitting when he writes that the dialogic structure of classical utopian literature "has little affinity with film" (2010, 138), there is sufficient evidence to demonstrate that a *non- or postgeneric understanding of the utopian impulse* has informed various studies of cinematic history.[3] A critical attentiveness to what might be described as the utopian dimension of cinematic form has been considered by a small number of scholars. Richard Dyer offers a good example of this approach. In his 1977 article "Utopia and Entertainment," Dyer differentiates between models of utopian worlds (what I am referring to as sociological utopias) and "the utopianism contained in the feelings [entertainment] embodies," focusing on the musical as an anticipatory utopian form (2002, 20). This interest in the affective after-effects of utopian form on its viewers has also been considered by Malte Hagener, who has written more recently of the "possible utopia" of immersive spectatorship, reception, and exhibition that avant-garde films attempt as they strive to dissolve boundaries between "screen and auditorium, life and art, theory and practice, film, and spectator," thereby becoming "total cinema" (2007, 121).

In developing this kind of *filmic utopianism*, we should also bear in mind Ernst Bloch's writings about early cinema, which remain almost entirely ignored by film historians and utopian theorists alike. As Johan Siebers (2014) notes, with particular reference to Bloch's 1932 essay "Significant Change in Cinematic Fables," Bloch had "an aesthetical understanding of cinema which emphasises the way in which cinema can take apart the integrated experience of reality and distort, fragment or transform it by virtue of its technical affordances—zooming in and out, panning, slow motion and fast forward, but also the use of music" (Siebers 2014, 46). Although Bloch was frequently critical of Hollywood (he notoriously referred to it as a "dream factory" in the 1910s and 1920s that had been perverted into a "poison factory" by the 1930s) (Bloch 1986, 410), he was also aware of the revolutionary utopian potential that popular cinematic storytelling held. As he writes in *The Principle of Hope* (*Das Prinzip Hoffnung*):

> Ever since [D. W.] Griffith cut the heads of people into the action for the first time, since this employment of the close-up, the play of facial muscles has also appeared as revealed suffering, joy, hope. The spectator now learns from the close-up of a gigantically isolated head, much more visibly than from that of the speaking actor on the whole stage, what incarnate emotion itself looks like. (1986, 406–7)

Bloch's interest in cinematic techniques—close-ups, tracking shots, dissolves and fades, use of slow motion, and editing techniques that can either establish narrative continuity or disrupt it (as in montage)—leads him into a comparative reading of cinematic and dramatic performance that focuses on what he terms "expectant emotions." Unlike greed or envy (so-called filled emotions), the expectant emotions of hope and fear project something undeniably new into their differently imagined futures: a horizon of expectation that "becomes capable of a rapport with the objectively New" (108–9). In particular, the close-up (as I consider later in Lars von Trier's *Melancholia*), offers the utopian possibility of watching this "objectively New" flicker across an actor's magnified visage on the silver screen. Meanwhile, avant-garde editing techniques such as Eisenstein's and Pudovkin's experiments with montage powerfully demonstrated to Bloch "that a different society, indeed world, is both hindered and circulating in the present one" (411).

I want to make the case for a Blochian approach to identifying the utopian impulse within sf cinema at the aesthetic level. With their use of formal complexity, sophisticated generic discontinuity, and auteurist sensibility, the films I focus on in this chapter dismantle the integrated experience of our current reality, as Bloch would put it. In so doing, these sf films ask viewers to reflect upon the disintegration of human exceptionalism and the sovereign human subject at a time of ecocatastrophic crisis—a utopian project of the highest order, given the current climate emergency. I have selected three Hollywood sf films for analysis: Christopher Nolan's thriller about the possibility of shared dreaming, *Inception* (2010); Luc Besson's high-octane action film, *Lucy* (2014), which literalizes the process of becoming posthuman (or what Deleuzians might term "becoming-computer"); and Lars von Trier's lush, genre-defying end-of-the-world epic, *Melancholia* (2011). These exemplary sf films combine innovative techniques of narration with the kinds of speculative visual style and aesthetic surplus made possible by big-budget production and auteur-driven projects. With Marcuse's insistence that "form becomes content" (1978, 41) within the aesthetic dimension, I consider the formal, aesthetic, and generic properties of these recent films. Under the lens of a utopian hermeneutic, we can identify how these cinematic texts reveal complex formal strategies for grasping the complexities and contradictions of contemporary subjectivity: distributed beyond the bounds of the individual subject, straining heliotropically toward intersubjective and interspecies encounters.

## *INCEPTION*: VIEWER-AS-PLAYER AND THE SANDBOX MODE

With its interlacing narrative structure, nested layers of reality, and preoccupation with the relationship between dream and reality, Christopher Nolan's surreal sf thriller *Inception* is an obvious candidate for utopian analysis. Developed over a decade and eventually produced with an impressive budget (Warner Brothers invested $160 million, only $20 million less than *The Dark Knight* [2008]) (Mooney 2018, 87), *Inception* builds on Nolan's noir-ish *Memento* (2000), challenging viewers with its porous boundaries between dreams, memories, reality, and anticipation. The film's protagonist Tom Cobb (Leonardo DiCaprio) is a professional "dream stealer" who has developed a technique for extracting information from targets' subconscious. Cobb is hired by a Japanese businessman to undertake a high-risk operation that involves reversing this process and attempting to implant an idea in the mind of a business rival: convincing the heir to an energy conglomerate, Robert Fischer (Cillian Murphy), that his recently deceased father's business should be broken up. With its fantastical remediation of military-style special forces assignments and surveillance technology, this already outlandish narrative setup promises a slick tale of corporate espionage, updating Orwell's Big Brother for the data mining era. Cobb's team of talented criminals builds an elaborate virtual-reality labyrinth in which the kidnapped, drugged target Fischer can unwittingly participate in a shared dream with Cobb's team that proceeds through a number of levels to descend into Fischer's subconscious and there "incept" the idea that he should dismantle his father's industrial empire. However, the heist framework is quickly dispensed with and *Inception* proceeds more like a video game narrative, as

Cobb's experimental inception technology requires an increasing amount of improvisation from the team to satisfy unforeseen ludological rules. This formal aspect of the film lends itself to a utopian analysis, encouraging viewers to consider how Nolan aestheticizes noncinematic narrative structures in his film's remediation of the video game. In allowing viewers a player experience through speculative participation in the film's broader story world, *Inception* thus offers the utopian experience of an active ludic participation within the game world—an experience that critically reflects upon utopia's colonial origins in a manner akin to the decolonization of games by feminist game studies, which similarly insists on using utopia as a feminist *method* rather than a masculinist *telos* (Harvey 2019, 219–20).

Like the elaborate world-building in much sf, this kind of video game subcreation requires scrupulous design,[4] a feature that is in direct tension with the film's supposed subject matter of dreaming. As Darren Mooney notes, "When *Inception* was released, one of the big criticisms was the fact that it wasn't really *about* dreams. Its depiction of its characters' dreamscapes was rigidly rational and meticulously structured, failing to accurately capture the surreal or intangible aspects of a dream state" (2018, 88). Rather than the phantasmagoric landscapes we find in Andrei Tarkovsky's cinematic oeuvre, which enable viewers to experience the aesthetic and affective properties of Freudian *Traumwerk* (dreamwork), Nolan's psychological thriller is less interested in the dream ipso facto than the dreamlike possibilities of cinematic storytelling, as it remediates video game structures and the different types of play for which they allow. *Inception*'s plot literalizes the metaphor of cinematic narrative as a maze, offering its viewers the situational and limited ground-level perspective that the characters within the maze possess: frequently making mistakes, subsequently realizing their lack of knowledge, and struggling toward the maze's conclusion (Mooney 2018, 89). This often painful progress rehearses the narrative and ludic structures found in video games. While reviewers and scholars discussing *Inception* have tended to focus on the film's various nested dream levels, its generic violence, and complicated temporality, they have, as Warren Buckland notes, overlooked the significance of the film's engagement with video game structures. Drawing on the relevance of software studies for film analysis, Buckland writes, "*Inception* is not simply the result or an aggregate of narrative and video game structures; collectively, these structures combine to create a distinctive hybrid that cannot be reduced to its simple components" (2015, 197).

This remediation is a significant aspect of the film's overall aesthetic and points toward the utopian qualities of Marcuse's "aesthetic dimension" introduced earlier. As Buckland demonstrates, *Inception* combines the rules of narrative cinema (setup, complicating action, development, and so on) with specific video game rules, including the necessity of repeat viewings, in-game tutorials, an "emphasis on strategy and tactics," and what game designers refer to as the sandbox mode (Buckland 2015, 195). We might build on Buckland's reading of the centrally embedded layer of Cobb's subconscious known as Limbo, in which he fashions a utopian city with his wife, Mal (Marion Cotillard), to consider how the utopian function of the sandbox mode enables a different mode of participatory viewing and a creative praxis of unconstrained imaginative world-building.

We first learn about Limbo during a scene in which Cobb responds to repeated questioning by Ariadne, the architecture student he has hired to design the labyrinth located in the dream landscapes within which his team operates. Cobb reveals that he and his wife (now deceased) explored the concept of building "a dream within a dream":

> We were working together; we were exploring the concept of a dream within a dream. I kept pushing things. I wanted to go deeper and deeper; I wanted to go . . . further. I just didn't understand the concept that hours could turn into years down there. That we could get trapped so deep that when we wound up on the shore of our own subconscious we lost sight of what was real. We created; we built the world for ourselves. We did that for years. We built our own world.

There are two ways that a utopian reading might parse Limbo. The first is to approach Limbo as a chronotope within the nested dream worlds of *Inception* and consider what kind of *sociological utopia* Cobb and Mal build together. This vantage, however, has limited merit. The second is to approach Limbo as another level within the remediated video game structure of *Inception* that offers its own ludic possibilities for the "players" (the film's viewers) and their "avatars" (Cobb and Mal), that is, as a *sandbox utopia*. This second reading reveals the usefulness of a utopian method for formal analysis.

Let us proceed with the first approach and consider Limbo as an embedded sociological utopia within *Inception*'s dreamscapes (see figure 20.1). What Cobb describes to Ariadne (Eliot Page) is an escapist lovers' paradise. This has some similarities with, as well as important differences from, Fredric Jameson's theorization of the utopian enclave in *Archaeologies of the Future* (2005). Like Jameson's abstraction of utopian production, Cobb and Mal's simulated Limbo is a "pocket of stasis within the ferment and rushing

**FIGURE 20.1:** Cobb (Leonardo DiCaprio) reshapes the city in Limbo (*Inception*, 2010). Warner Bros.

forces of social change," which functions as a "mental space in which the whole system can be imagined as radically different" (Jameson 2005, 16). This time of "social change" is evidenced by the film's high-concept depiction of a data mining technology penetrating into our deepest unconscious dreams and memories, sold by black-ops criminal outfits to wealthy capitalists in an ever-more-sinister version of neoliberal globalization. This reading of the sociological utopia within the game world fails somewhat when we consider that *Inception* has denuded the utopian enclave of its sociality. Rather than the limited social world of even the Renaissance court from which Jameson takes inspiration, what remains in Limbo is the nuclear family as the smallest possible social unit (Cobb, Mal, and their two young children). At no point during their fifty years in Limbo does Cobb describe the couple encountering, or even generating, any other people.

As a sociological utopia, Limbo is "inhabited" by Cobb, Mal, and their two children as a kind of digitally remodeled frontier family adrift in what appears to be an empty cityscape, but which we know from the film's underlying dream logic ought to be populated by innumerable people from their past (who would remain alive, even in their dreams). Limbo is less of a *utopian* enclave than a *nostalgic* enclave, which we might understand better if we treat it as a Robinsonade. As Frank and Fritzie Manuel remind us, Daniel Defoe's *Robinson Crusoe* (1719) launched a new utopian form that depopulated More's island-kingdom of Utopia, leaving in its place an "aggressive bourgeois Prometheus" and his capitalist will to survive and dominate (Manuel and Manuel 1979, 433). In visual terms, the Robinsonade is conveyed through a sequence of recognizable scenes: washing up on the shore of the island (More's *Utopia*, Defoe's *Robinson Crusoe*, Shakespeare's *The Tempest* [c. 1610]), strolling hand in hand through the lovers' city like the romantically entwined tour guides and traveler-protagonists of utopian literature (Edward Bellamy's *Looking Backward* [1888], Leo Tolstoy's *Aelita, or The Decline of Mars* [1923]), and Limbo's warped temporality in which fifty years is compressed into the length of an afternoon nap (recalling the euchronian motif of waking up in the future, as in William Morris's *News from Nowhere* [1890] and Bellamy's *Looking Backward*). Where Defoe's *Homo economicus* recycles the original Morean utopia, ripping it out of its courtly Renaissance context and flinging it into the delusional colonial cartography of *terra nullius*, Cobb and Mal's nostalgic retreat remediates both of these foundational narratives within an elongated temporality of subconscious regression.

If Limbo fails as an embedded *sociological utopia*, it does succeed as a *sandbox utopia*, that is, as a different level within the film's remediated video game structure that enables viewers to become players. The distinction here is one of perspective and moves us from narrative content to narrative form. As a sandbox utopia, Limbo offers a speculative glimpse into an entire, and enticing, backstory for the viewer-as-player's "avatars," Cobb and family. Sandbox mode is a freeform, creative style of game playing. In contrast to a *constrained* mode within video games, "*Sandbox* mode lets the player do whatever she wants but usually doesn't offer the same rewards as the constrained mode—and may not offer any rewards at all. In this mode, the game resembles a tool more than a conventional video game" (Adams 2014, 175). Examples of games that allow for exploring sandbox mode (also referred to as sandbox or open-world games) include *The Sims, Sim*

*City*, massive multiplayer online role-playing games (MMORPGs) that utilize sandbox and progression gaming, and first-person shooter games such as *Grand Theft Auto*, *Red Dead Redemption*, and *Assassin's Creed* (Hjorth 2011, 41). As Hartmut Koenitz suggests, sandboxes offer players a playful experience of utopian engineering: "we can build what we can describe as utopias—considerable improvements over the current situation—and test them before they can fail in the real world" (2019, 254). Read as a sandbox, Limbo encourages its viewer-turned-player to ask: "What if" you could redesign the spatial and temporal coordinates around you? What if, like Ariadne, you could collapse Euclidean geometry and imagine a weird topographical realm where Parisian streets fold onto themselves?

This formal utopian reading reveals that Limbo's failure as a *sociological utopia* does not preclude its utopian potential as a *sandbox utopia*. Indeed, the film's labored symbolism encourages viewers to interpret Cobb's wife Mal, the film's malevolent antagonist, as an intertextual avatar figure rather than a fully realized character: not only is Mal literally "bad" or "evil," as her name suggests, but she also becomes Bertha Mason, the original "madwoman in the attic" haunting Charlotte Brontë's *Jane Eyre* (1847), when she is incarcerated in the basement of Cobb's subconscious. This situation complicates older cinematic conventions of realism and the bourgeois subjectivity that embodies such verisimilitude. As Fredric Jameson writes in *The Antinomies of Realism*, *Inception* repudiates such narrative coordinates: "as with much of modern philosophy, it evades the subject/object division altogether" (2015 299–300). Watching *Inception*, we become video game players exploring a sequence of interconnected virtual realities whose scenery is more intricate and interesting than the stock characters who afford us an avatar's point of view. Through its remediated video game structure, the film's own narrative logic of shared dreaming is thus put to the test, or subjected to sandbox mode, and linear narrative progression of the kind we associate with realism becomes increasingly untenable. Rather, the film encourages its viewers to reconceive of themselves as players, not spectators, in what Malte Hagener suggests is a utopian immersive spectatorship of "total cinema" (121).

# MORE-THAN-HUMAN: *LUCY* AND THE PREGNANT POSTHUMAN

Scarlett Johansson's trilogy of sf films—*Her* (Jonze 2013), *Under the Skin* (Glazer 2013), and *Lucy* (Besson 2014)—similarly examines what happens when narrative dispenses with the subject-object division, underscoring the importance of feminist politics to the production and reception of contemporary sf cinema. Each film, as Janice Loreck notes, portrays "different versions of the perfect woman: a virtual girlfriend in *Her*, a super-intelligent being in *Lucy*, and a fatal seductress in *Under the Skin*," and Johansson "deploys her sex symbol status" to bring these complex sf characters to life on screen

(2019, 175). I want to focus on Luc Besson's *Lucy*, which combines high-concept, high-production sf with glossy, fast-paced action, and a pulpy sense of *jouissance* that draws on sf fandom's love of intertextual and generic allusions. The film is also, in its content as well as its form, a meditation upon the question of the human subject.

Filmed in Taipei, Paris, and New York, Besson's film features the eponymous Lucy, a twenty-five-year-old student who is unwittingly catapulted into a drug-trafficking gang developing a dangerous mind-enhancing drug called CBH4. Upon its release, the film received mixed reviews. The more positive ones emphasized the film's "goofy, high-concept speculative thriller" character and "gratuitously globe-trotting pulp-trash extravaganza" (Chang 2014), while less generous critics such as Christopher Orr called *Lucy* a "mind-bendingly miscalculated sci-fi vehicle for Scarlett Johansson," concluding that the film is "so beyond-the-pale sloppy, so disastrous in both conceit and execution, that it simply defies conventional analysis" (Orr 2014). Early reviewer responses remind us of the centrality of liberal humanist conceptions of discrete, sovereign subjectivity, and its primary aesthetic terrain of cinematic realism. Indeed, the concept of a woman transmogrifying into a distributed computational consciousness, it seems, can only be appreciated at the level of sf pulp. Seen through our utopian hermeneutic, Lucy becomes a good example of *filmic utopianism*. Examining the formal properties of the film will reveal its challenging and experimental approach to the philosophical question of the human subject's place within the broader ecosphere.

Our utopian analysis might begin with the film's primary concern with what it means to be human as considered from the non-anthropocentric timescale of Darwinian evolution. *Lucy* begins with a viscous tableau of cells dividing, their microscopic scale rendered immense as they proliferate within an amorphous, extracellular gunk that recalls the ooze of sf body horror. This image cuts to a live-action scene of the early hominid Australopithecus (nicknamed Lucy) drinking from a lake as Johansson's voiceover narrates the question: "Life was given to us a billion years ago. What have we done with it?" This is a film that unambiguously plays with scale, vertiginously jumping about in space and time, as the opening sequence demonstrates by alluding to Stanley Kubrick's "Dawn of Man" sequence from *2001: A Space Odyssey* (1968). In the film's first scene, Lucy's involvement with a criminal gang extirpates her humanity and immediately reduces her to the status of an animal. Waiting at a hotel reception desk, her palpable fear is underscored by an accelerating heartbeat on the diegetic soundtrack, while cross-cutting emphasizes her vulnerability by juxtaposing this scene with wildlife footage of a cheetah stalking an antelope (Lucy's leopard-print biker jacket accentuating the predator-prey analogy), circling and closing in for the kill as Lucy is kidnapped. From this point on, Lucy fights for her survival and revenge. As her fate is decided by the psychopathic gang boss Mr. Jang (Min-sik Choi), parallel editing cuts between this violent scene and a public lecture on neuroscience being delivered by Professor Norman (Morgan Freeman), who argues that humans only employ 10 percent of their brain capacity and speculates about human abilities if the remaining brain capacity were liberated.

The anthropocentric focus of Norman's hypothesis—the assumption that humans are "at the top of the animal chain"—is undercut by Lucy's cognitive transformation during the film, which renders her progressively *less* human and *more* computational. This transformation reflects upon the explicitly gendered sf trope of imagining woman-as-machine, which we might date back to the stilted erotic gyrations of the robot Maria in Fritz Lang's *Metropolis* (1927), a character that, like Lucy, becomes an object of desire, fascination, fear, and revulsion. The genre's misogynistic treatment of its machinic femmes fatales is alluded to in the scene that catalyzes Lucy's transformation: a brutal beating involving kicks to her stomach (where the drugs have been surgically inserted), as she resists an attempted rape by her captor. In generic terms, this raises *Lucy* from sf action thriller to rape-revenge narrative and confirms the film's feminist concerns. As the packet of drugs ruptures inside her, the cinematic point of view dramatically dissolves boundaries of skin, flesh, inside and outside, veering like a roller coaster along Lucy's arterial network. In this special effects sequence, the contrast between Lucy's internal organs (meaty red) and the drug's trajectory (electric blue) recalls her earlier *Matrix*-style choice between the red pill and the blue pill; meanwhile, the visualization of her bodily interior alludes to Joe Dante's ground-breaking effects in *Innerspace* (1987), a story of experimental miniaturization in which a test pilot is injected into the bloodstream of a supermarket manager. This intertextuality thus mixes iconic moments from sf cinema into a referential collage, producing a cut-up aesthetic that frames the present dissolution of corporeal boundaries between epidermis and dermis, tissue and blood vessels, arterial channels and organs—between, that is, interiority and exteriority—which brings hemic verisimilitude to a speculative narrative about cognitive enhancement. Here, body horror meets what will soon become the Singularity, when Lucy reaches 100 percent of her brain capacity and transmogrifies into computational consciousness, dispensing with the need for her body altogether.

As this early scene of ingestion demonstrates, *Lucy* modulates not only between different intertextual and generic registers, but also between two distinct registers of philosophical posthumanism: cyborgian, transhumanist posthumanism and its more recent cousin, environmental posthumanism. As Lorrie Palmer observes, the film sutures extradiegetic perspectives (Norman's lectures, the wildlife documentary footage) and digital effects (as demonstrated in the visualization of Lucy's "inner space") to demonstrate that "the feminine is technological, scopic, and organic" (2019, 447). The film's feminist preoccupation with Lucy's abuse and subsequent retribution, signified most clearly when we read the film as a *rape revenge*, reminds us of the entanglement between feminism and environmental posthumanism, which intersects with a crucial point in the film's utopianism. As the beating scene suggests, with its brutal sense of the vulnerability of a woman's stomach and Johannsen's performance of futile resistance in the fetal position, Lucy's posthumanism is enabled by the inescapably feminist issue of pregnancy, which feminist phenomenologists have recently discussed as a key site of porousness and intersubjectivity, or plural subjectivity. The provenance of the synthetic drug that affords Lucy her preternatural cerebral abilities is important here. As the film explains, and Besson attests in interviews, CBH4 is a fictional name for a real molecule produced by pregnant women in their first trimester to accelerate the fetus's skeletal growth (Ebiri 2014). If Lucy is subject to a violent process of *becoming-computer* through the film, she is equally asked to endure the metamorphic process of becoming a *pregnant woman*. As Rodante van der Waal suggests, the pregnant woman is the "daughter of Donna Haraway's cyborg and Rosi Braidotti's posthuman.... She is a singular subject, but inside her subjectivity there is another subject growing, one that nobody can see yet. She is in a singular plural state" (Van der Waal 2018, 368).

The scene in which Lucy ascends into this *singular-plural* state of pregnancy, or of computational consciousness, is worth analyzing in some detail and reminds us of the film's preoccupation with the question of scale. As she describes to the wide-eyed scientists in Professor Norman's hastily assembled Paris laboratory,

> Humans consider themselves unique, so they've rooted their whole theory of existence on their uniqueness. We've codified our existence to bring it down to human size, to make it comprehensible; we've created a scale so we can forget its unfathomable scale.

Lucy demonstrates the challenges to this kind of human exceptionalism in a display of her rapidly growing posthuman abilities. As she rotates her hand it transmutes first into two hands conjoined at the wrist in mirror image and then flickers into the gnarly elongated fingers with grotesque yellow nails that we recognize from a variety of filmic precursors, including F. W. Murnau's 1922 expressionist horror classic *Nosferatu*; Guillermo del Toro's grotesque Pale Man in his 2006 fantasy, *Pan's Labyrinth*; and the iconic hydraulically operated puppetry of John Landis's 1981 horror comedy, *American Werewolf in London*.

The intertextual body horror of Lucy's elongating fingers conveys what Laura Marks (2002) calls "haptic visuality," wherein an estranging use of close-ups and out-of-focus or grainy shots requires viewers to use their eyes "like an organ of touch." "Haptic images," she writes, "invite the viewer to dissolve his or her subjectivity in the close bodily contact with the image. The oscillation between the two creates an erotic relationship, a shifting between distance and closeness" (13). Here, content meets form and the grotesque corporealization of Lucy's distributed subjectivity matches the dissolution of the viewer's own subjectivity through the film's use of estranging close-up, which forces a haptic, intimate encounter with Lucy's inhuman materiality (see figure 20.2). Indeed, Lucy's monstrous fingers become the launching point for her transformation, projecting themselves across the room in a sinister materialization of data that takes an undeniably organic form, with body horror shifting into eco-horror and the alien monstrosity of the vegetal.

If we consider the use of sound effects in this scene, *Lucy*'s vegetal subtext comes to the fore, enabling an environmentally posthumanist reading of the film's playful approach to generic discontinuity. As her monstrous fingers transmute into tentacles, the sound design layers an ambient digitized tinkling (that we can associate with the sound effects of "digital rain" in the Wachowskis' *The Matrix*) (Whittington 2007, 224), as well as the cracking and heaving of accelerated vegetal growth. As the film's sound designer Shannon Mills describes, creating the acoustic ambience for *Lucy* involved

FIGURE 20.2: Lucy (Scarlett Johansson) examines the haptic image of her grotesque fingers (*Lucy*, 2014). Universal Pictures.

generating new sound effects that mixed the computational and the digital with the organic. The inner space scene discussed earlier, for example, inspired the sound design team to record household items such as liquid foam, as well as animals: "We recorded some geese that were used when [the camera moves through] cells and veins" (Giardina 2014). This experimentation in sf sound design demonstrates how attention to the aesthetic dimension of a film like *Lucy* can extend our understanding of the ways in which sf films move between cyborgian and environmental registers, confirming *Lucy*'s experimental posthumanism as a vision of distributed subjectivity beyond the human. Johansson's defamiliarizing performance style masterfully conveys this vision, gradually paring back Lucy's gestural range to the staccato reptilian scale of a lizard inclining its head with deeply nonhuman insight and, ultimately, becomes properly *postsubjective* in Lucy's shedding of her human corporeal form at the end of the film.

What Lucy earlier referred to as the "unfathomable scale" of planetary existence finally becomes legible to her when she reaches 100 percent of brain capacity. Seated in a black office chair, Lucy is propelled through space to a series of dramatic locations—the Eiffel Tower, the chalk cliffs of Étretat in Normandy, and Times Square in Manhattan, where she discovers her ability to manipulate time, pausing the rush of people and swiping them backward into accelerated analepsis with a flick of her hand, as easily as swiping left on a smartphone. This demonstration affords the film its most utopian gesture. Lucy's manipulation of historical time propels her into Manhattan's preindustrial past where she encounters indigenous tribesmen, before swiping even further backward through the ice age into the volcanic formation of the landscape, finally stopping at the pterodactyls and dinosaurs of the Mesozoic Era. Like Steven Spielberg's *Jurassic Park* (1993) and *The Lost World* (1997), this post-anthropocentric perspective is made possible by the "new aesthetic realism" that big-budget, auteur-driven Hollywood sf can achieve (Buckland 2004, 26). While its literalization of posthuman, distributed subjectivity in the figure of Lucy-as-computer is undoubtedly spectacular, this closing sequence demonstrates that it is the properly epic scale of planetary evolution that reveals Lucy's utopian, more-than-human capabilities.

# Beyond the Torments of Subjectivity: Becoming Animal in *Melancholia*

A final example of an sf film that lends itself to a Blochian utopian reading is Lars von Trier's epic treatment of depression and the end of the world, *Melancholia* (2011), which effectively illustrates how cinematic form can extend an image of distributed subjectivity into a fully posthumanist, post-anthropocentric frame of reference. While the film's subject matter might sound distinctly anti-utopian, Adrian J. Ivakhiv notes that, "despite the bleak nihilism of its seeming message, it is quite possible to leave the theatre, as critic J. Hoberman did, feeling 'light, rejuvenated and unconscionably happy'" (2013, 321). Similarly, Mikkel Krause Frantzen (2019) observes that although *Melancholia* presents viewers with an "act of uncorrupted cinematic killjoy," the film also germinates a radical utopian impulse in its apocalyptic destruction of human life. The cosmological implications of von Trier's apocalyptic vision thus inspire an oddly utopian kind of "eschatological hope":

> The end result is not cosmic pessimism nor is it a cynical and capitalist version of depressive realism. The realism in question is one of a different kind, a realism based on what could be rather than on what is. . . . The "intention" of *Melancholia* is not to confront the characters or the audience with the pure void, the ultimate horror vacui, or the world-without-us, but to erect a principle of hope which only becomes relevant when there is none; a hope that emerges at the horizon of pure hopelessness, or even at the prospect of the end of the world. (Frantzen 2019, 192–93)

Another identifiably utopian quality of the film is its odd sense of untimeliness, or what Bloch would call its *Ungleichzeitigkeit* (noncontemporaneity or nonsimultaneity). As he writes in *Heritage of Our Times* (*Erbschaft dieser Zeit*) (1935), "Impulses and reserves from pre-capitalist times and superstructures are then at work . . . which a sinking class revives or causes to be revived in its consciousness" (Bloch 1991, 106). Although *Melancholia*'s noncontemporaneity cannot be said to be precapitalist, the film does pair contemporary experience with a much older sensibility. Thus Mark Cauchi suggests, "One of the reasons that *Melancholia* seems so untimely is that it takes up aesthetic and philosophical themes and issues—including melancholy itself—which quite literally belong to another era and which largely ceased to be concerns after the end of Modernism" (2018, 107).

Von Trier underscores this untimeliness, or noncontemporaneity, through his references to German Romanticism (as indicated through Wagner's music in the film) as well as iterated visual allusions to the Baroque paintings of Dutch master Johannes

Vermeer. Like Vermeer's *Milkmaid* (1657), the two sisters Justine and Claire are repeatedly positioned in front of an ornate leaded window, with natural light illuminating a dark wood-paneled room that exudes historic presence (the film was shot largely on location at Tjolöholm Castle in Sweden, a famous example of Arts and Crafts architecture that enacts utopian noncontemporaneity in evoking medieval and Gothic architectural styles). In formal terms, this *Ungleichzeittigkeit* is literalized in the film's nonlinear narrative rganization. Like the planet Melancholia's trajectory, this organization is elliptical: "The end," as T. S. Eliot writes in *Little Gidding*, "is where we start from" ([1943] 1971, 49). The film begins *after the end*, with a compressed sequence of highly stylized slow-motion images that will feature throughout, presented as an apocalyptic-phantasmagoric prelude. Let us first consider the utopian aesthetics of this sequence, before analyzing how it establishes an aestheticized mode of posthumanism that allows for the disintegration of subjectivity later in the film.

The prologue opens with an unnerving introduction to Justine (Kirsten Dunst) in an uncompromising close-up: opening her eyes in slow-motion to reveal a blank, dead stare straight into the camera. Justine's accusatory gaze interpellates the viewer into an uncomfortable position of complicity in whatever is unfolding. Dead birds fall around her, as the lush orchestral score of Wagner's *Tristan und Isolde* matches their descent with the music's emotive rising and swelling. Justine's wet hair is the color of sand; her face partly in shadow echoes this coloration. The sky behind her is a similarly desaturated flesh color with a suggestive hint of gaseous nebula, and the undulating birds are the same muted sandy brown (see Figure 20.3). This use of an almost monochromatic color palette gives forceful aesthetic expression to the idea that Justine's subjectivity is in a process of disintegration which, by extension, suggests the collapse of human exceptionalism at a moment of impending planetary catastrophe. Here is a subject, we are being told, whose human uniqueness has been dethroned within a flattened landscape of fungible parts; woman, hair, birds, and the planet's atmosphere are simultaneously dying, as well

**FIGURE 20.3:** The disintegration of Justine's subjectivity within the broader ecosphere (*Melancholia*, 2011). Nordisk Film/Magnolia Pictures.

as being animated by far greater forces. After several seconds, the shot cuts to the austere symmetry of a formal garden dominated by a preternaturally large sundial in the foreground (unambiguously announcing temporality as a central theme), followed by a full-screen shot of Pieter Bruegel the Elder's iconic painting *The Hunters in the Snow* (1565), which depicts the severe winter of 1565. Art historians note that Bruegel's painting enacts a doubled temporality, which situates the viewer both *inside* the present time of the pictorial frame and its narrative action, as well as heterodiegetically *outside* of it within some extratemporal sphere of aesthetic contemplation, which has been referred to as Bruegel's ecological impulse (Bonn 2006, 21–22). The doubled temporality of Bruegel's painting suggests *Melancholia*'s own structural contradiction as a narrative of apocalypse that begins after the apocalyptic event has taken place and makes iterated reference to the phantasmagorical prelude throughout the film. This extradiegetic apocalyptic time frame—narrated from the timeless perspective of some temporal location that stands outside of or beyond human chronology altogether—reminds us of utopia's formal function as an impossible figure of totality, a character that supports reading the film's utopian impulse at the formal level of the aesthetic, rather than the sociological level of embedded utopian enclaves within the sf narrative world.

The opening sequence plays with the borders of life and death, liveliness and deathliness, body-as-matter, and a dialectic between the static photographic or painterly image and cinematic motion. It is worth pausing to reflect upon von Trier's use of a nondiegetic musical score accompanying the prelude and the utopian implications of cinematic music in general and Wagnerian Romanticism in particular. As Johan Siebers explains, Bloch's interest in the film music of the silent era uncovers its utopian function as an accompaniment to the visual that is also a surrogate for the remaining senses (smell, taste, and touch). This function connects the cinematic image to lived reality:

> Music creates the whole sensorium within the art form, without which the moving image cannot become expressive. In the silent movie, *the material world becomes music*, and so we experience the utopian tone of reality itself in film. The fact that film puts all weight on the optical and renders the rest of the sensorium peripheral heightens this utopian potential even more, because the utopian, the absolute, is given to us only in indirect experience. (Siebers 2014, 54)

In *Melancholia*, the nondiegetic use of Wagner's overture to *Tristan und Isolde* signifies Justine's relationship to the approaching planet. We can begin to understand the utopian significance of the Wagnerian score by returning to Bloch's reading of the *Ring* cycle. As he writes in "The Philosophy of Music," Wagner's *Ring* invokes a "strangely unsubjective animal lyricism" through the orchestra's symphonic cadences. Inspired by Schopenhauer's immersive metaphysics, Wagner's characters are not presented as "*dramatis personae* advancing into the space of an encounter with each other and with their own profound destiny," but rather as "blossoms on a tree, indeed even just bobbing ships unresistingly obeying their subhuman ocean's sufferings, labors, love and yearning for redemption" (Bloch 2000, 154–55). We can trace a proto-posthumanist impulse in Bloch's analysis, which draws on the antihumanist perspective of Schopenhauer's cosmic pessimism (Mathäs 2020, 200). Echoing this reading, *Melancholia* privileges Wagner's Schopenhauer-inspired "epochal pessimism" (Korstvedt 2010, 161) in *Tristan und Isolde*, pairing the music with key scenes in which Justine moves from what we might call, in Kantian terms, the relative nothingness (*nihil privativum*) of catatonic depression within an elite circle of almost parodically nasty capitalists to the absolute nothingness (*nihil negativum*) of approaching cataclysm. This movement is at once a radical utopian gesture and also an aestheticization of antihumanism.

Let us look at two key scenes in which this disintegration of anthropocentric subjectivity into *unsubjective animal experience* plays out: first, what I refer to as Justine's nude bathing scene in the blue glow of Melancholia, and second, Claire's desperate attempt to escape impending apocalypse in a golf cart. The nude bathing scene occurs late in the evening, as Justine is drawn toward the planet's blue light and we see her walking somnambulantly across the lawn followed at a distance by Claire (Charlotte Gainsbourg), who looks both fascinated and horrified by what is about to unfold. Under cover of trees, Claire's uncomfortable voyeurism is rewarded with a sudden crescendo of *Tristan und Isolde*, as we see a wide shot of Justine reclining naked on an outcrop of rock at the riverbank. With arms outstretched in a pose of extreme relaxation, but also extreme contrivance (not to mention discomfort, lying naked among the scratchy undergrowth and craggy bedrock), Justine's pale body is subject to a weird illumination, presented in contrast to the dark forest that surrounds her. Under the aquamarine fluorescence of Melancholia, she gazes upward and slowly caresses herself as Claire watches in disgust from beneath her canopy of trees. For Justine, Melancholia's advance produces an erotic encounter with approaching death and inspires her to transform herself *into a work of art*; posing like the reclining nudes of classical painting, she is both painter and subject, relocating art's perennial celebration of the unattired human form from a soft chaise-longue to this woodland setting. This odd image thus combines a libidinal investment in the very forces that will shortly annihilate her, as Justine welcomes the weird weather, eerie nocturnal glow of the approaching planet, and vacillating atmospheric conditions that make it difficult to breathe (suggestive, here, of autoerotic asphyxiation, but more akin to a panic attack when suffered by Claire), precisely because they threaten her precarious selfhood and offer the unsubjective promise of animal temporality. This is a temporal experience that involves luxuriating in the present moment with seemingly no thought for the catastrophic future just around the corner, although, ironically, this kind of occupation of the *animal present* is only possible through an anticipatory consciousness that has already looked annihilation squarely in the face (a utopian engagement with death-as-alterity that Levinas's 1976 lectures on Bloch similarly uncover).[5]

In her oddly staged eroticism, Justine embodies Marcuse's utopian "Great Refusal," as the nude forest bathing scene becomes an unambiguously aesthetic gesture. It is prefigured in an earlier scene when Justine leaves the wedding reception for the sanctity of the library and, after an angry confrontation with Claire and her husband, Michael (Alexander Skarsgård), rearranges a display of modern paintings. Tossing to the floor images by Kandinsky and Malevich, she defiantly replaces these modernist experimenters with works including Pieter Bruegel's *The Hunters in the Snow* (1565) and *The Land of Cockaigne* (1567), John Everett Millais's pre-Raphaelite paintings *Ophelia* (1852) and *The Woodsman's Daughter* (1851), Caravaggio's *David with the Head of Goliath*, and Carl Frederik Hill's *Crying Deer* (1883). As Todd McGowan notes, Justine's aesthetic choices in this scene reveal "a romantic belief that modernity's toppling of traditional authority doesn't portend the elimination of all authority. . . . We can still discover a substantial authority in nature, in community or even in death" (2017, 194–95). However, we should note that "death" here represents not simply Justine's subjective annihilation, nor even the planetary destruction rapidly approaching, but the final liberation from a repressive social world; as Fredric Jameson famously quipped, "It is easier to imagine the end of the world than the end of capitalism" (2003, 76). Through her disastrous wedding reception in part 1 of the film, we have seen how Justine endures workaholism, depression-shaming, and parental abandonment, all played out in the extravagant luxury setting of her brother-in-law's palatial chateau. If Justine welcomes the planetary disaster, the film makes clear that she savors the imminent destruction of these capitalist-bourgeois ties, as her sarcastic rejection of the entreaty that they mark the end of the world by sipping a glass of wine on the terrace unambiguously demonstrates.

As the second scene demonstrates, it is harder for Claire to relinquish this repressive social world, as her response to impending apocalypse conveys a different kind of *unsubjective animal experience*. Desperate to undertake any kind of action, no matter how futile, Claire drags her son Leo into a golf cart and accelerates across the chateau's golf course. The mournful Wagnerian motif returns as she scoots from green to green in the electric buggy, while the cables overhead crackle with alarming voltage. The scene mixes the sublime with the ridiculous: slumped forward, Claire's almost comically careful driving speaks of her need for security and a sense of purpose. Finally, she abandons the cart and carries Leo, both of them panting heavily. The weird weather releases a torrential hailstorm as Claire founders onto the golfing green, cradling Leo's head before sinking into a posture of abject despair. Charlotte Gainsbourg's performance in this scene brings to the fore Claire's previous attributes of upper-class nonchalance (sporting the unkempt hair, oversized woolen jumpers, loose-fitting slacks, and Hunter wellington boots of a patrician lady of the manor) that easily collapses as she undertakes the basic animalistic labor of seeking safety and shelter. As Claire limps across the golf course, Leo keeps slipping from her grip, the boy seemingly infantilized by his mother's panic into the pose of a much younger child, taciturn with fear. If Justine's earlier response to Melancholia's approach in the nude bathing scene demonstrated the strangely unsubjective animal *lyricism* that Bloch notes in Wagner's musical drama, Claire's animalistic labor during the hailstorm suggests a strangely unsubjective animal *desperation*, peeling back the character's layers of urbanity and prim good manners to reveal the futile maternalistic gestures of an animal that knows its end is drawing close. As Deleuze and Guattari write, "Who has not known the violence of these animal sequences, which uproot one from humanity, if only for an instant, making one scrape at one's bread like a rodent or giving one the yellow eyes of a feline?" (2013, 280). The unsubjectivity of this scene is emphasized through a technique of distancing. Whereas in previous scenes the viewer has been given access to Claire's psychological interiority through prolonged close-ups of her face (particularly in scenes that take place on the chateau's terrace), here this access is negated and we only see Claire at a distance, adding to the inscrutability of her animal desperation.

Read comparatively, these two scenes offer viewers contrasting models of distributed, disintegrating subjectivity in which the women arrive at the unsubjective ontological experience of *"becoming-animal."* In dismantling their subject positions as anthropo-centrically human, *Melancholia* suggests that Claire and Justine, in very different ways, approach what Deleuze and Guattari call the "point that the human being encounters the animal . . . in the interstices of [the] disrupted self" (2013, 280). Where earlier we identified the repudiation of older models of cinematic subjectivity that maintained the subject-object division in Nolan's *Inception*, and the posthuman temporal scale of Lucy's computational consciousness in Besson's *Lucy*, in *Melancholia* this radical pulling apart of an older understanding of discrete human subjectivity hints at a grim sort of utopian anticipation of apocalyptic extermination. As Todd McGowan notes, "By destroying her, nature spares Justine from the torments of subjectivity" (2017, 193).

As this chapter demonstrates, a utopian reading of these sf films' sophisticated treat-ment of human subjectivity and its disintegration should enrich our understanding of them. Attending to their complex formal, generic, and stylistic strategies—that is, the films' *aesthetic dimension*—enables us to parse the utopian possibilities of enacting con-tent through form. Through their use of innovative narrative structures, remediation of other media, experimental sound design, arresting effects, and defamiliarizing per-formance styles, *Inception*, *Lucy*, and *Melancholia* challenge liberal humanist notions of discrete subjectivity. And in so doing, these films demolish anthropocentric frameworks of understanding, insisting on an increasingly porous conception of the human sub-ject, as well as multispecies and postanthropocentric readings of agency, possibility, and experience.

## NOTES

1. While this field is too large to list all relevant scholarly works here, for key interventions into debates about sf literature and utopia, see Suvin (1979) and Moylan (1986).
2. These arguments are summarized in Edwards (2019).
3. However, I should note that occasionally scholars working within the fields of utopian studies and/or sf studies do publish research that considers utopian (and dystopian) films as a genre (see Pordzik 2015).
4. J. R. R. Tolkien introduced the term "subcreation" to describe how the fantasy author establishes a new secondary world, taking God's original creation as its foundation (Wolf 2012, 6).
5. For a brief discussion of Levinas's 1976 lectures on Ernst Bloch's theorization of utopia and death, see Edwards 2019 (80–85).

## Works Cited

Adams, Ernest. *Fundamentals of Game Design*. 3rd ed. San Francisco: New Riders, 2014.

Bloch, Ernst. *Heritage of Our Times*. Trans. Neville Plaice and Stephen Plaice. Cambridge: Polity Press, 1991.

Bloch, Ernst. "Significant Change in Cinematic Fables." In *Literary Essays*. Trans. Andrew Joron, 59-63. Stanford, CA: Stanford UP, 1998.

Bloch, Ernst. "The Philosophy of Music." In *The Spirit of Utopia*. Trans. Anthony A. Nassar, 34–164. Stanford, CA: Stanford UP, 2000.

Bloch, Ernst. *The Principle of Hope*. Volume 1 [1959]. Trans. Neville Plaice, Stephen Plaice, and Paul Knight. Cambridge, MA: MIT Press, 1986.

Bonn, Robert L. *Painting Life: The Art of Pieter Bruegel, the Elder*. New York: Chaucer Press, 2006.

Bronte, Charlotte. *Jane Eyre*. 1847. New York: Random House, 2006.

Buckland, Warren. "Between Science Fact and Science Fiction: Spielberg's Digital Dinosaurs, Possible Worlds and the New Aesthetic Realism." In *Liquid Metal: The Science Fiction Film Reader*, edited by Sean Redmond, 24–34. Chichester: Wallflower, 2004.

Buckland, Warren. "*Inception*'s Video Game Logic." In *The Cinema of Christopher Nolan: Imagining the Impossible*, edited by Jacqueline Furby and Stuart Joy, 189–200. Chichester: Wallflower, 2015.

Cauchi, Mark. "The Death of God and the Genesis of Worldhood in von Trier's *Melancholia*." In *Immanent Frames: Postsecular Cinema between Malick and von Trier*, edited by John Caruana and Mark Cauchipp, 105–28. Albany, NY: SUNY Press, 2018.

Chang, Justin. "Film Review: *Lucy*." *Variety*, July 23, 2014. https://variety.com/2014/film/revi ews/film-review-lucy-1201267405/.

Deleuze, Gilles, and Félix Guattari. "Becoming-Intense, Becoming-Animal, Becoming Imperceptible." In *A Thousand Plateaus: Capitalism and Schizophrenia*. Trans. Brian Massumi, 271–360. London: Bloomsbury, 2013.

Dyer, Richard. *Only Entertainment*. 2nd ed. London: Routledge, 2002.

Ebiri, Bilge. "Luc Besson on *Lucy* and Knowing the Limits of the Human Brain." *Vulture* 25 (2014). https://www.vulture.com/2014/07/luc-besson-director-lucy-chat.html.

Edwards, Caroline. *Utopia and the Contemporary British Novel*. Cambridge: Cambridge UP, 2019.

Eliot, T. S. *Four Quartets*. New York: Harcourt Books, [1943] 1971.

Fitting, Peter. "Utopia, Dystopia and Science Fiction." In *The Cambridge Companion to Utopian Literature*, edited by Gregory Claeys, 135–53. Cambridge: Cambridge UP, 2010.

Frantzen, Mikkel Krause. *Going Nowhere, Slow: The Aesthetics and Politics of Depression*. London: Zero Books, 2019.

Giardina, Carolyn. "*Lucy*: How Sound Design Helped Unlock the Brain." *Hollywood Reporter*, July 28, 2014. https://www.hollywoodreporter.com/behind-screen/lucy-how-sound-des ign-helped-721484.

Hagener, Malte. *Moving Forward, Looking Back: The European Avant-Garde and the Invention of Film Culture, 1919–1939*. Amsterdam: Amsterdam UP, 2007.

Harvey, Alison. "Feminist Interventions for Better Futures of Digital Games." In *Playing Utopia: Futures in Digital Games*, edited by Benjamin Beil, Gundolf S. Freyermuth, and Hanns Christian Schmidt, 211–34. Bielefeld: Transcript Verlag, 2019.

Hjorth, Larissa. *Games and Gaming: An Introduction to New Media*. Oxford: Berg, 2011.

Ivakhiv, Adrian J. *Ecologies of the Moving Image: Cinema, Affect, Nature*. Waterloo, CA: Wilfrid Laurier UP, 2013.

Jameson, Fredric. *Archaeologies of the Future: The Desire Called Utopia and Other Science Fictions*. London: Verso, 2005.

Jameson, Fredric. *The Antinomies of Realism*. London: Verso, 2015.

Jameson, Fredric. "Future City." *New Left Review* 21 (2003): 65–79.

Koenitz, Hartmut. "Playful Utopias: Sandboxes for the Future." In *Playing Utopia: Futures in Digital Games*, edited by Benjamin Beil, Gundolf S. Freyermuth, and Hans Christian Schmidt, 253–66. Bielefeld: Transcript Verlag, 2019.

Korstvedt, Benjamin M. *Listening for Utopia in Ernst Bloch's Musical Philosophy*. Cambridge: Cambridge UP, 2010.

Loreck, Janice. "Man, Meat and *Bêtes-Machines*: Scarlett Johansson in *Under the Skin*." In *Screening Scarlett Johansson: Gender, Genre, Stardom*, edited by Janice Loreck, Whitney Monaghan, and Kirsten Stevens, 165–82. Basingstoke: Palgrave Macmillan, 2019.

Manuel, Frank E., and Fritzie P. Manuel. *Utopian Thought in the Western World*. Cambridge, MA: Harvard UP, 1979.

Marcuse, Herbert. *The Aesthetic Dimension: Toward a Critique of Marxist Aesthetics*. Trans. and revised from the original German by Herbert Marcuse and Erica Sherover. Boston: Beacon Press, 1978.

Marks, Laura. *Touch: Sensuous Theory and Multisensory Media*. Minneapolis: U of Minnesota P, 2002.

Mathäs, Alexander. *Beyond Posthumanism: The German Humanist Tradition and the Future of the Humanities*. New York: Berghahn, 2020.

McGowan, Todd. "Not Melancholic Enough: Triumph of the Feminine in *Melancholia*." In *Lars von Trier's Women*, edited by Rex Butler and David Denny, 181–200. London: Bloomsbury, 2017.

Mooney, Darren. *Christopher Nolan: A Critical Study of the Films*. Jefferson, NC: McFarland, 2018.

Moylan, Tom. *Demand the Impossible: Science Fiction and the Utopian Imagination*. New York: Methuen, 1986.

Moylan, Tom. *Scraps of the Untainted Sky: Science Fiction, Utopia, Dystopia*. Boulder, CO: Westview, 2000.

Orr, Christopher. "*Lucy*: The Dumbest Movie Ever Made about Brain Capacity." *The Atlantic*, July 25, 2014. https://www.theatlantic.com/entertainment/archive/2014/07/life-is-futile-so-heres-what-to-do-with-it-according-to-lucy-a-spoilereview/375006/.

Palmer, Lorrie. "A Digital Nature: *Lucy* Takes Technology for a Ride." In *A Companion to the Action Film*, edited by James Kendrick, 439–55. Oxford: Wiley-Blackwell, 2019.

Pordzik, Ralph. "Biopleasures: Posthumanism and the Technological Imaginary in Utopian and Dystopian Film." In *The Palgrave Handbook of Posthumanism in Film*, edited by Michael Hauskeller, Thomas D. Philbeck, and Curtis D. Carbonell, 259–68. Basingstoke: Palgrave Macmillan, 2015.

Shelton, Robert. "The Utopian Film Genre: Putting Shadows on the Silver Screen." *Utopian Studies* 4, no. 2 (1993): 18–25.

Siebers, Johan. "The Utopian Function of Film Music." In *Marx at the Movies: Revisiting History, Theory and Practice*, edited by Ewa Mazierska and Lars Kristensen, 46–61. Basingstoke: Palgrave Macmillan, 2014.

Spiegel, Simon. "Some Thoughts on Utopian Film." *Science Fiction Film and Television* 10, no. 1 (2017): 53–79.

Suvin, Darko. *Metamorphoses of Science Fiction: On the Poetics and History of a Literary Genre.* New Haven, CT: Yale UP, 1979.

van der Waal, Rodante. "The Pregnant Posthuman." In *Posthuman Glossary*, edited by Rosi Braidotti and Maria Hlavajova, 368–70. London: Bloomsbury, 2018.

Whittington, William. *Sound Design and Science Fiction.* Austin: U of Texas P, 2007.

Wolf, Mark J. P. *Building Imaginary Worlds: The Theory and History of Subcreation.* New York: Routledge, 2012.

# FILMOGRAPHY

The following is a select list of films corresponding to each of the various "New Science Fiction Cinemas" categories (that is, those of Section II). The listing for each area is not meant to be complete, nor is it exclusive. As the individual chapters frequently note, other films, both earlier and later than the ones listed, could easily be added to these categories, while some films could well be slotted into other, closely allied groups. This listing is mainly intended to be *representative* of these significant types and to provide readers with some of the categories' more significant titles (including some not discussed in the respective essays) for their further consideration and study.

## Afrofuturist Films

*Afronauts* (2014) Dir. Nuotama Bodomo. Prod. Isabella Wing-Davey. Cast: Diandra Forrest, Yolonda Ross, Hoji Fortuna. Sundance Institute.

*After Earth* (2013) Dir. M. Night Shyamalan. Prod. Caleeb Pinkett, Jada Pinkett Smith, Will Smith, James Lassiter, Shyamalan. Cast: Jaden Smith, Will Smith, Zoe Kravitz. Sony Pictures.

*Black Panther* (2018) Dir. Ryan Coogler. Prod. Kevin Feige. Cast: Chadwick Boseman, Michael B. Jordan, Lupita Nyong'o. Marvel/Walt Disney Pictures.

*Born in Flames* (1983) Dir. Lizzie Borden. Prod. Lizzie Borden. Cast: Honey, Adele Bertei, Kathryn Bigelos. First Run Features.

*The Brother from Another Planet* (1984) Dir. John Sayles. Prod. Peggy Rajski, Maggie Renzi. Cast: Joe Morton, Darryl Edwards, Steve James, David Strathairn. Cinecom Pictures.

*Crumbs* (2015) Dir. Miguel Llanso. Prod. Miguel Llanso, Daniel Taye Workou. Cast: Daniel Tadesse, Quino Pinero, Selam Tesfayie. BiraBiro Films/Lanzadera Films.

*Destination: Planet Negro* (2013) Dir. Kevin Wilmott. Prod. Wilmott, Grant Fitch, J. S. Hampton. Cast: Tosin Morohunfola, Danielle Cooper, Kevin Willmott. Candy Factory Films.

*Pumzi* (2009) Dir. Wanuri Kahiu. Prod. Simon Hansen, Amira Quinlan, Hannah Slezacek, Steven Markovitz. Cast: Kudzani Moswela, Chantelle Burger. Inspired Minority/Focus Features.

*Ratnik* (2020) Dir. Dimeji Ajobola. Prod. Dimeji Ajobola. Cast: Osas Ighodaro, Bolanie Ninalowo, Adunni Ade. Flipsyde Studios.

*Robots of Brixton* (2011) Dir. Kibwe Tavares. Prod. Kibwe Tavares. Cast (Voice Actors): Kibwe Tavares, Yung Swizz'Agg. Factory Fifteen.

*See You Yesterday* (2019) Dir. Stefon Bristol. Prod. Spike Lee. Cast: Eden Duncan-Smith, Dante Crichlow, Marsha Stephanie Blake. 40 Acres & A Mule.

*Space Is the Place* (1974) Dir. John Coney. Prod. Jim Newman. Cast: Sun Ra, Raymond Johnson, Seth Hill. North American Star System.

*Strange Frame* (2012) Dir. G. B. Hajim. Prod. G. B. Hajim. Cast: Tim Curry, Claudia Black, Ron Glass, Juliet Landau. Screaming Wink/Wolfe Releasing.

318  FILMOGRAPHY

*They Charge for the Sun* (2016) Dir. Terence Nance. Prod. Guilia Caruso, Kady Kamakate, Ana Souza. Cast: Rylee Nykhol, Jontille Gerard, Crystal Cotton. Film Independent/ MVMT.

*Virtuosity* (1995) Dir. Brett Leonard. Prod. Gary Lucchesi. Cast: Denzel Washington, Russell Crowe, Kelly Lynch, Stephen Spinella. Paramount Pictures.

*A Wrinkle in Time* (2018). Dir. Ava DuVerney. Prod. Jim Whitaker, Catherine Hand. Cast: Oprah Winfrey, Reese Witherspoon, Storm Reid, Deric McCabe.


## Biopunk Films

*Antiviral* (2012) Dir. Brandon Cronenberg. Prod. Niv Fichman. Cast: Caleb Landry Jones, Sarah Gadon, Malcolm McDowell. Alliance Films.

*Blade Runner* (1982) Dir. Ridley Scott. Prod. Michael Deeley. Cast: Harrison Ford, Rutger Hauer, Sean Young, Edward James Olmos. Ladd Company/Warner Bros.

*Blade Runner 2049* (2017) Dir. Denis Villeneuve. Prod. Bud Yorkin, Cynthia Sikes Yorkin, et al. Cast: Ryan Gosling, Harrison Ford, Robin Wright, Ana de Armas. Columbia Pictures/ Warner Bros.

*Code 46* (2003) Dir. Michael Winterbottom. Prod. Andrew Eaton. Cast: Tim Robbins, Samantha Morton, Jeanne Balibar. BBC/Revolution Films.

*eXistenZ* (1999) Dir. David Cronenberg. Prod. Cronenberg, Andras Hamori, Robert Lantos. Cast: Jude Law, Jennifer Jason Leigh, Ian Holm. Miramax.

*Gattaca* (1997) Dir. Andrew Niccol. Prod. Danny DeVito, Michael Shamberg, Stacey Sher, Gail Lyon. Cast: Ethan Hawke, Uma Thurman, Alan Arkin, Jude Law. Columbia Pictures.

*The Girl with All the Gifts* (2016) Dir. Colm McCarthy. Prod. Camille Gatin, Angus Lamont. Cast: Gemma Arterton, Paddy Considine, Glenn Close, Sennia Nanua. BFI/Warner Bros.

*Lucy* (2014) Dir. Luc Besson Prod. Virginie Besson-Silla. Cast: Scarlett Johansson, Morgan Freeman, Choi Min-sik, Amr Waked. EuropaCorp.

*Repo Men* (2010) Dir. Miquel Sapochnik. Prod. Scott Stuber. Cast: Jude Law, Forest Whitaker, Liev Schreiber, Alice Braga. Universal Pictures.

*Splice* (2009) Dir. Vincenzo Natali. Prod. Steve Hoban. Cast: Adrien Brody, Sarah Polley, Delphine Chaneac. Dark Castle/Warner Bros.

*Tetsuo: The Iron Man* (1989) Dir. Shinya Tsukamoto. Prod. Tsukamoto. Cast: Tomorowo Taguchi, Kei Fujiwara, Shinya Tsukamoto. Kaijyu Theatres.


## Cli-Fi Films

*Aniara* (2018) Dir. Pella Kagerman and Hugo Lilja. Prod. Annika Rogell. Cast: Emelie Garbers, Anneli Martini, Bianca Cruzeiro. Film Capital Stockholm/Magnolia Pictures.

*Children of Men* (2006) Dir. Alfonso Cuaron. Prod. Hilary Shor. Cast: Clive Owen, Julianne Moore, Michael Caine, Chiwetel Ejiofor. Strike Entertainment/Universal Pictures.

*The Colony* (2013) Dir. Jeff Renfroe. Prod. Pierre Even, Matthew Cervi, Paul Barkin, Marie-Claude Poulin. Cast: Kevin Zegers, Bill Paxton, Charlotte Sullivan. RLJ Entertainment/ Entertainment One.

*The Day After Tomorrow* (2004) Dir. Roland Emmerich. Prod. Emmerich, Mark Gordon. Cast: Dennis Quaid, Jake Gyllenhaal, Iam Holm, Sela Ward. Centropolis/20th Century-Fox.

*Elysium* (2013) Dir. Neill Blomkamp. Prod. Blomkamp, Bill Block, Simon Kinberg. Cast: Matt Damon, Jodie Foster, Sharlto Copley, Alice Braga. Tristar Pictures/Sony Pictures.

FILMOGRAPHY 319

*Geostorm* (2017) Dir. Dean Devlin. Prod. Devlin, David Ellison, Dana Goldberg. Cast: Gerard Butler, Jim Sturgess, Abbie Cornish. Warner Bros.

*Hardware* (1990) Dir. Richard Stanley. Prod. JoAnne Sellar, Paul Truybitts. Cast: Dylan McDermott, Stacey Travis, John Lynch, Iggy Pop. British Satellite/Millimeter Films.

*Interstellar* (2014) Dir. Christopher Nolan. Prod. Emma Thomas, Nolan, Lynda Obst. Cast: Matthew McConaughey, Anne Hathaway, Jessica Chastain, Bill Irwin. Legendary Pictures/ Paramount Pictures.

*The Road* (2009) Dir. John Hillcoat Prod. Nick Wechsler, Steve Schwartz, Paula Mae Schwartz. Cast: Viggo Mortensen, Kodi Smit-McPhee, Robert Duvall, Charlize Theron. Dimension Films.

*Silent Running* (1972) Dir. Douglas Trumbull. Prod. Trumbull, Michael Gruskoff, Harty Hornstein. Cast: Bruce Dern, Cliff Potts, Ron Rifkin. Universal Pictures.

*Snowpiercer* (2013) Dir. Bong Joon-ho. Prod. Jeong Tae-sung, Steven Nam, Park Chan-wook, Lee Tae-hun. Cast: Chris Evans, Song Kang-ho, Tilda Swinton, Jamie Bell. Weinstein Company.

*Sunshine* (2007) Dir. Danny Boyle. Prod. Andrew Macdonald. Cast: Cillian Murphy, Rose Byrne, Chris Evans. Fox Searchlight Pictures.

*Take Shelter* (2011) Dir. Jeff Nichols. Prod. Sophia Lin, Tyler Davidson. Cast: Michael Shannon, Jessica Chastain, Shea Whigham, Katy Mixon. Hydraulx Entertainment/Sony Pictures.

*2012* (2009) Dir. Roland Emmerich. Prod. Harald Kloser, Mark Gordon, Larry J. Franco. Cast: John Cusack, Chiwetel Ejiofor, Amanda Peet, Oliver Platt. Centropolis/ Columbia Pictures.

*WALL-E* (2008) Dir. Andrew Stanton. Prod. Jim Morris. Cast (Voice Actors): Ben Burtt, Elissa Knight, Jeff Garlin, Fred Willard. Pixar/Walt Disney Pictures.

*Waterworld* (1995) Dir. Kevin Reynolds. Prod. Kevin Costner, John Davis, Charles Gordon, Lawrence Gordon. Cast: Kevin Costner, Dennis Hopper, Jeanne Tripplehorn. Universal Pictures.

## Ethnogothic Films

*Blacula* (1972) Dir. William Crain. Prod. Samuel Z. Arkoff, Joseph T. Naar. Cast: William Marshall, Vonetta McGee, Denise Nicholas. American International Pictures.

*Brown Girl Begins* (2017) Dir. Sharon Lewis. Prod. Vince Buda, Sharon Lewis. Cast: Moua Traore, Nigel Shawn Williams, Shakura S'Aida, Rachael Crawford. Urbansoul.

*Get Out* (2017) Dir. Jordan Peele. Prod. Sean McKittrick, Jason Blum, Edward H. Hamm Jr., Jordan Peele. Cast: Daniel Kaluuya, Allison Williams, Bradley Whitford. Universal Pictures.

*The People under the Stairs* (1991) Dir. Wes Craven. Prod. Stuart M. Besser, Marianne Maddalena. Cast: Brandon Adams, Everett McGill, Wendy Robie. Universal Pictures.

*Les Saignantes* (aka *The Bloodettes*, 2005) Dir. Jean-Pierre Bekolo. Prod. Bekolo, Andre Bennett, Lisa Crosato. Cast: Dorylia Calmel, Adele Ado. Bekolo SARL.

*Sankofa* (1993) Dir. Haile Gerima. Prod. Gerima. Cast: Kofi Ghanaba, Oyafunmike Ogunlano, Alexandra Duah. Mypheduh Films.

*Sugar Hill* (1974) Dir. Paul Maslansky. Prod. Elliot Schick. Cast: Marki Bey, Robert Quarry, Don Pedro Colley. American International Pictures.

*Da Sweet Blood of Jesus* (2014) Dir. Spike Lee. Prod. Lee, Chiz Schultz. Cast: Stephen Tyrone Williams, Rami Malek, Zaraah Abrahams, Felicia Pearson. 40 Acres and a Mule/Gravitas.

320 FILMOGRAPHY

*Thriller* (2018) Dir. Dallas Jackson. Prod. Adam Hendricks, John H. Lang, Greg Gilreath, Jackson. Cast: Jessica Allain, Tequan Richmond, Chelsea Rendon. Blumhouse Productions/Netflix.

*Us* (2019) Dir. Jordan Peele. Prod. Peele, Jason Blum, Ian Cooper, Sean McKittrick. Cast: Lupita Nyong'o, Winston Duke, Elisabeth Moss, Tim Heidecker. Universal Pictures.

## Femspec Films

*Alien* (1979) Dir. Ridley Scott. Prod. Gordon Carroll, David Giler, Walter Hill. Cast: Sigourney Weaver, John Hurt, Tom Skerritt. 20th Century-Fox.

*Born in Flames* (1983) Dir. Lizzie Borden. Prod. Lizzie Borden. Cast: Honey, Adele Bertei, Flo Kennedy. First Run Features.

*Captain Marvel* (2019) Dir. Anna Boden, Ryan Fleck. Prod. Kevin Feige. Cast: Brie Larson, Samuel L. Jackson, Gemma Chan, Annette Bening. Marvel/Walt Disney Pictures.

*Ex Machina* (2015) Dir. Alex Garland. Prod. Andrew Macdonald, Allon Reich. Cast: Alicia Vikander, Oscar Isaac, Domhnall Gleeson. Universal Pictures.

*Gravity* (2013) Dir. Alfonso Cuaron. Prod. Cuaron, David Heyman. Cast: Sandra Bullock, George Clooney. Warner Bros.

*Lucy* (2014) Dir. Luc Besson. Prod. Virginie Besson-Silla. Cast: Scarlett Johansson, Morgan Freeman, Choi Min-sik. EuropaCorp.

*Lucy in the Sky* (2019) Dir. Noah Hawley. Prod. Reese Witherspoon, Bruna Papandrea. Cast: Natalie Portman, Jon Hamm, Dan Stevens. Pacific Standard/Fox Searchlight.

*The Machine* (2013) Dir. Caradog W. James. Prod. John Giwa-Amu. Cast: Caity Lotz, Sam Hazeldine, Toby Stephens. Red and Black/Content Media.

*Mad Max: Fury Road* (2015) Dir. George Miller. Prod. Doug Mitchell, George Miller, P. J. Voeten. Cast: Charlize Theron, Tom Hardy, Hugh Keays-Byrne. Roadshow/Warner Bros.

*Tank Girl* (1995) Dir. Rachel Talalay. Prod. Richard B. Lewis, Pen Densham, John Watson. Cast: Lori Petty, Ice-T, Naomi Watts, Malcolm McDowell. Trilogy/United Artists.

*Under the Skin* (2013) Dir. Jonathan Glazer. Prod. James Wilson, Nick Wechsler. Cast: Scarlett Johansson, Jeremy McWilliams, Adam Pearson. BFI/Film 4/StudioCanal.

*Wonder Woman 1984* (2020) Dir. Patty Jenkins. Prod. Jenkins, Deborah Snyder, Zack Snyder, Charles Roven. Cast: Gal Gadot, Chris Pine, Kristen Wiig, Pedro Pascal. Warner Bros.

## Heterotopic/Utopian Films

*The Abyss* (1989) Dir. James Cameron. Prod. Gale Anne Hurd. Cast: Ed Harris, Mary Elizabeth Mastrantonio, Michael Biehn. 20th Century-Fox.

*Arrival* (2016) Dir. Denis Villeneuve. Prod. Shawn Levy, Dan Levine, Aaron Ryder, David Linde. Cast: Amy Adams, Jeremy Renner, Forest Whitaker, Michael Stuhlbarg. FilmNation/Paramount.

*Avatar* (2009) Dir. James Cameron. Prod. Cameron, Jon Landau. Cast: Sam Worthington, Zoe Saldana, Stephen Lang, Sigourney Weaver. 20th Century-Fox.

*Big Hero 6* (2014) Dir. Don Hall, Chris Williams. Prod. Roy Conli. Cast (Voice Actors): Scott Adsit, Ryan Potter, T. J. Miller, Damon Wayans Jr. Walt Disney Pictures.

*Dark City* (1998) Dir. Alex Proyas. Prod. Andrew Mason, Proyas. Cast: Rufus Sewell, Keifer Sutherland, Jennifer Connelly. New Line Cinema.

FILMOGRAPHY 321

*District 9* (2009) Dir. Neill Blomkamp. Prod. Peter Jackson, Carolynne Cunningham. Cast: Sharlto Copley, Jason Cope, David James. Sony Pictures.

*Guardians of the Galaxy, Vol. 2* (2017) Dir. James Gunn. Prod. Kevin Feige. Cast: Zoe Saldana, Chris Pratt, Dave Bautista, Karen Gillan. Marvel/Walt Disney Pictures.

*Interstellar* (2014) Dir. Christopher Nolan. Prod. Emma Thomas, Nolan, Lynda Obst. Cast: Matthew McConaughey, Anne Hathaway, Jessica Chastain, Bill Irwin. Legendary Pictures/Paramount Pictures.

*The Lobster* (2015) Dir. Yorgos Lanthimos. Prod. Ceci Dempsey, Ed Guiney, Lee Magiday. Cast: Colin Farrell, Rachel Weisz, Jessica Barden, Olivia Colman. Element Pictures.

*The Matrix* (1999) Dir. The Wachowskis. Prod. Joel Silver. Cast: Keanu Reeves, Laurence Fishburne, Carrie-Anne Moss, Hugo Weaving. Warner Bros.

*Oblivion* (2013) Dir. Joseph Kosinski. Prod. Peter Chernin, Dylan Clark, Barry Levine. Cast: Tom Cruise, Morgan Freeman, Olga Kurylenko. Universal Pictures.

*Pleasantville* (1998) Dir. Gary Ross. Prod. Ross, Jon Kilik, Robert J. Degus, Steven Soderbergh. Cast: Tobey Maguire, Jeff Daniels, Joan Allen, Reese Witherspoon. New Line Cinema.

*Stalker* (1979) Dir. Andrei Tarkovsky. Prod. Aleksandra Demidova. Cast: Alexander Kaidanovsky, Anatoly Solonitsyn, Alisa Freindlich. Mosfilm/Goskino.

*Sucker Punch* (2011) Dir. Zack Snyder. Prod. Snyder, Deborah Snyder. Cast: Emily Browning, Abbie Cornish, Jena Malone, Carla Gugino. Warner Bros.

## Kaiju Films

*Cloverfield* (2008) Dir. Matt Reeves. Prod. J. J. Abrams, Bryan Burk. Cast: Lizzy Caplan, Jessica Lucas, T. J. Miller, Mike Vogel. Paramount Pictures.

*Colossal* (2016) Dir. Nacho Vigalondo. Prod. Nahikari Ipina, Russell Levine, Nicolas Charter, et al. Cast: Anne Hathaway, Jason Sudeikis, Dan Stevens. Voltage Pictures/Neon.

*Godzilla* (2014) Dir. Gareth Edwards. Prod. Thomas Tull, Jon Jashni, et al. Cast: Aaron Taylor-Johnson, Juliette Binoche, Ken Watanabe, Elizabeth Olsen. Warner Bros.

*Godzilla: King of the Monsters* (2019) Dir. Michael Dougherty. Prod. Mary Parent, Thomas Tull, Jon Jashni, et al. Cast: Kyle Chandler, Vera Farmiga, Millie Bobby Brown, Bradley Whitford. Warner Bros.

*The Meg* (2018) Dir. Jon Turteltaub. Prod. Lorenzo di Bonaventura, Colin Wilson, Belle Avery. Cast: Jason Stathan, Li Bingbing, Rainn Wilson, Ruby Rose. Warner Bros.

*Monsters* (2010) Dir. Gareth Edwards. Prod. Allan Niblo, James Richardson. Cast: Scoot McNairy, Whitney Able. Vertigo Films.

*Pacific Rim* (2013) Dir. Guillermo del Toro. Prod. del Toro, Thomas Tull, Jon Jashni, Mary Parent. Cast: Charlie Hunnam, Idris Elba, Rinko Kikuchi, Charlie Day. Warner Bros.

*Pacific Rim: Uprising* (2018) Dir. Steven S. DeKnight. Prod. Mary Parent, Cale Boyter, Guillermo del Toro, Thomas Tull, Jon Jashni. Cast: John Boyega, Scott Eastwood, Jing Tian, Cailee Spaeny, Rinko Kikuchi. Universal Pictures.

*Rampage* (2018) Dir. Brad Peyton. Prod. Peyton, Beau Flynn, John Rickard, Hiram Garcia. Cast: Dwayne Johnson, Naomie Harris, Malin Akerman, Jake Lacy. New Line/Warner Bros.

*Shin Godzilla* (2016) Dir. Hideaki Anno, Shinji Higuchi. Prod. Minami Ichikawa, Taichi Ueda, Yoshihiro Sato, et al. Cast: Hiroki Hasegawa, Yutaka Takenouchi, Satomi Ishihara. Toho Pictures.

322   FILMOGRAPHY

*Transformers* (2007) Dir. Michael Bay. Prod. Lorenzo di Bonaventura, Tom DeSanto, Don Murphy, Ian Bryce. Cast: Shia LaBeouf, Tyrese Gibson, Josh Duhamel, Megan Fox. Dreamworks/Paramount Pictures.

*Transformers: The Last Knight* (2017) Dir. Michael Bay. Prod. Lorenzo di Bonaventura, Tom DeSanto, Don Murphy, Ian Bryce. Cast: Mark Wahlberg, Josh Duhamel, Stanley Tucci. Paramount Pictures.

## Magical Realism Films

*Beasts of the Southern Wild* (2012) Dir. Benh Zeitlin. Prod. Dan Janvy, Josh Penn, Michael Gottwald. Cast: Quvenzhane Walls, Dwight Henry. Fox Searchlight.

*Being John Malkovich* (1999) Dir. Spike Jonze. Prod. Michael Stipe, Sandy Stern, et al. Cast: John Cusack, Cameron Diaz, John Malkovich. Gramercy Pictures/USA Films.

*Birdman* (2014) Dir. Alejandro G. Inarritu. Prod. Inarritu, John Lesher, Arnon Milchan, James W. Skotchdopole. Cast: Michael Keaton, Zach Galifianakis, Edward Norton, Emma Stone, Naomi Watts. Fox Searchlight.

*Defending Your Life* (1991) Dir. Albert Brooks. Prod. Robert Grand, Michael Grillo, Herb Nanas. Cast: Albert Brooks, Meryl Streep, Rip Torn, Lee Grant. Warner Bros.

*Edward Scissorhands* (1990) Dir. Tim Burton. Prod. Burton, Denise Di Novi. Cast: Johnny Depp, Winona Ryder, Dianne Wiest, Vincent Price. 20th Century-Fox.

*The Fall* (2006) Dir. Tarsem Singh. Prod. T. Singh, Ajit Singh, Tommy Turtle. Cast: Lee Pace, Catinca Untaru, Justine Waddell. Radical Media/Roadside Attractions.

*Hello Rain* (2018) Dir. C. J. Obasi. Prod. Oge Obasi. Keira Hewatch, Tunde Aladese, Ogee Nelson. Fiery Film Co./Igodo Films/Matanya Films.

*Kiki's Delivery Service* (1989) Dir. Hayao Miyazaki. Prod. Miyazaki. Cast (Voice Actors): Minami Takayama, Rei Sakuma, Kappel Yamaguchi. Studio Ghibli/Toei Company.

*Pan's Labyrinth* (2006) Dir. Gullermo del Toro. Prod. del Toro, Bertha Navarro, Alfonso Cuaron, Frida Torresblanco. Cast: Sergi Lopez, Maribel Verdu, Ivana Baquero, Doug Jones. Telecino/Warner Bros.

*Scott Pilgrim vs. the World* (2010) Dir. Edgar Wright. Prod. Wright, Eric Glitter, Nira Park, Marc Platt. Cast: Michael Cera, Mary Elizabeth Winstead, Kieran Culkin, Chris Evans. Universal Pictures.

*The Shape of Water* (2017) Dir. Guillermo del Toro. Prod. del Toro, J. Miles Dale. Cast: Sally Hawkins, Michael Shannon, Richard Jenkins, Doug Jones. Fox Searchlight.

## Steampunk Films

*The Adventures of Baron Munchausen* (1988) Dir. Terry Gilliam. Prod. Thomas Schuhly. Cast: John Neville, Eric Idle, Sarah Polley. Allied Filmmakers.

*Alice in Wonderland* (2010) Dir. Tim Burton. Prod. Richard Zanuck. Cast: Johnny Depp, Mia Wasikowska, Helena Bonham Carter, Anne Hathaway. Walt Disney Pictures.

*City of Lost Children* (1995) Dir. Marc Caro, Jean-Pierre Jeunet. Prod. Felicie Dutertre. Cast: Ron Perlman, Daniel Emilfork, Judith Vittet, Dominique Pinon. Canal+/Union General Cinematographique.

*The Fabulous World of Jules Verne* (aka *Invention for Destruction*, 1958) Dir. Karel Zeman. Prod. Zdenek Novak. Cast: Lubor Tokos, Arost Navratil, Miloslav Holub. Ceskoslovensky Statni Film/Warner Bros.

FILMOGRAPHY 323

*Howl's Moving Castle* (2004) Dir. Hayao Miyazaki. Prod. Toshio Suzuki. Cast (Voice Actors): Chieko Baisho, Takuya Kimura, Asihiro Miwa. Studio Ghibli/Toho.

*The League of Extraordinary Gentlemen* (2003) Dir. Stephen Norrington. Prod. Trevor Albert, Rick Benattar, Mark Gordon, et al. Cast: Sean Connery, Shane West, Peta Wilson, Stuart Townsend. 20th Century-Fox.

*Mortal Engines* (2018) Dir. Christian Rivers. Prod. Zane Weiner, Amanda Walker, Peter Jackson, et al. Cast: Hera Hilmar, Hugo Weaving, Robert Sheehan, Ronan Raftery. Universal Pictures.

*9* (2009) Dir. Shane Acker. Prod. Jim Lemley, Tim Burton, Timur Bekmambetov, Dana Ginsburg. Cast (Voice Actors): Elijah Wood, John C. Reilly, Jennifer Connelly, Crispin Glover. Focus Features.

*Sherlock Holmes* (2009) Dir. Guy Ritchie. Prod. Joel Silver et al. Cast: Robert Downey Jr., Jude Law, Mark Strong, Rachel McAdams. Warner Bros.

*Steamboy* (2004) Dir. Katsuhiro Otomo. Prod. Shinji Komori, Hideyuki Tomioka. Cast (Voice Actors): Anne Suzuki, Manami Konishi, Katsuo Nakamura, Masane Tsukayama. Sunrise/ Toho.

*Treasure Planet* (2002) Dir. Ron Clements, John Musker. Prod. Clements, Musker, Roy Conli. Cast (Voice Actors): Joseph Gordon-Levitt, Brian Murray, Emma Thompson, Martin Short. Walt Disney Pictures.

*Van Helsing* (2004) Dir. Stephen Sommers. Prod. Sommers, Bob Ducsay. Cast: Hugh Jackman, Kate Beckinsale, Richard Roxburgh. Universal Pictures.

*Wild Wild West* (1999) Dir. Barry Sonnenfeld. Prod. Sonnenfeld, Jon Peters. Cast: Will Smith, Kevin Kline, Kenneth Branagh, Salma Hayek. Warner Bros.

## Superhero Films

*Ant Man* (2015) Dir. Peyton Reed. Prod. Kevin Feige. Cast: Paul Rudd, Evangeline Lilly, Michael Pena, Michael Douglas. Marvel/Walt Disney Pictures.

*Avengers: Age of Ultron* (2015) Dir. Joss Whedon. Prod. Kevin Feige. Cast: Robert Downey Jr., Chris Hemsworth, Mark Ruffalo, Scarlett Johansson, Chris Evans. Marvel/Walt Disney Pictures.

*Avengers: Endgame* (2019) Dir. Anthony and Joe Russo. Prod. Kevin Feige. Cast: Robert Downey Jr., Chris Hemsworth, Mark Ruffalo, Scarlett Johansson, Chris Evans. Marvel/Walt Disney Pictures.

*Captain America: The First Avenger* (2011) Dir. Joe Johnston. Prod. Kevin Feige. Cast: Chris Evans, Tommy Lee Jones, Hugo Weaving, Hayley Atwell. Marvel/Walt Disney Pictures.

*Captain Marvel* (2019) Dir. Anna Boden, Ryan Fleck. Prod. Kevin Feige. Cast: Brie Larson, Samuel L. Jackson, Gemma Chan, Annette Bening. Marvel/Walt Disney Pictures.

*Glass* (2019) Dir. M. Night Shyamalan. Prod. Shyamalan, Jason Blum, Marc Bienstock, Ashwin Rajan. Cast: Bruce Willis, James McAvoy, Samuel L. Jackson, Anya Taylor-Joy. Blinding Edge Pictures/Universal Pictures.

*Guardians of the Galaxy* (2014) Dir. James Gunn. Prod. Kevin Feige. Cast: Chris Pratt, Zoe Saldana, Dave Bautista, Michael Rooker. Marvel/Walt Disney Pictures.

*Hellboy* (2004) Dir. Guillermo del Toro. Prod. Lawrence Gordon, Mike Richardson, Lloyd Levin. Cast: Ron Perlman, Selma Blair, Jeffrey Tambor, John Hurt. Sony Pictures/Columbia Pictures.

## 324    FILMOGRAPHY

*The Incredibles* (2004) Dir. Brad Bird. Prod. John Walker. Cast (Voice Actors): Craig T. Nelson, Holly Hunter, Sarah Vowell, Spencer Fox. Pixar/Walt Disney Pictures.

*Shazam!* (2019) Dir. David F. Sandberg. Prod. Peter Safran. Cast: Zachary Levi, Mark Strong, Asher Angel, Djimon Hounsou. New Line/Warner Bros.

*Spider-Man: Homecoming* (2017) Dir. John Watts. Prod. Kevin Feige, Amy Pascal. Cast: Tom Holland, Michael Keaton, Jon Favreau, Robert Downey Jr. Gwyneth Paltrow. Marvel/Columbia/Sony Pictures.

*Watchmen* (2009) Dir. Zack Snyder. Prod. Lawrence Gordon, Lloyd Levin, Deborah Snyder. Cast: Malin Akerman, Billy Crudup, Carla Gugino, Jackie Earle Haley. Paramount Pictures/Warner Bros.

*Wonder Woman* (2017) Dir. Patty Jenkins. Prod. Charles Roven, Deborah Snyder, Zack Snyder, Richard Suckle. Cast: Gal Gadot, Chris Pine, Robin Wright, Danny Huston. DC Films/Warner Bros.

*Wonder Woman 1984* (2020) Dir. Patty Jenkins. Prod. Jenkins, Deborah Snyder, Zack Snyder, Charles Roven. Cast: Gal Gadot, Chris Pine, Kristen Wiig, Pedro Pascal. Warner Bros.

# Index

*For the benefit of digital users, indexed terms that span two pages (e.g., 52–53) may, on occasion, appear on only one of those pages.*

Figures are indicated by *f* following the page number

Abels, Michael, 83–84
*Abigail* (film), 156–58
Abraham, John, 53
Abrams, J. J., 191, 224–25
Academy Award, 92, 108–9, 119–20, 135–36
Ackerman, Forrest, 67n.3
Acland, Charles, 233
*Adventurer: The Curse of the Midas Box, The* (film), 156
*Adventures of Baron Munchausen, The* (film), 142, 151, 153
*Adventures of Pluto Nash, The* (film), 25
*Aelita, or The Decline of Mars* (novel), 301
*Aeon Flux* (film), 92
Afrofuturism, 14, 82–83
  abjection in, 32
  actors in, 25
  definitions of, 25, 28
  diaspora, 23, 36
  dystopian elements of, 23, 27–28, 30–31, 32
  and horror, 23–24
  iconography, 23, 35–36
  politico-aesthetic character of, 36
  production types of, 29
  revolutionary mode, 25, 28, 29
  strong mode, 24–25, 27–28, 29–30, 31, 32
  thin mode, 24, 27, 29
  utopian elements, 33
*Afronauts* (film), 24–25
*After Earth* (film), 24, 26, 26f, 58
Agamben, Giorgia, 193, 194, 199
*Age of Stupid, The* (film), 62–63
*A.I.: Artificial Intelligence* (film), 31, 55–56, 62–63, 272–73

Alaimo, Stacy, 250
*Alien* (film), 90, 91f, 124, 270–71, 278, 283–84
  androids in, 282–83, 286–87
  performance in, 288–89
*Alien: Covenant* (film), 278, 290–94
*Alien Resurrection* (film), 270–71, 276, 278, 286f
  alternate version of, 294n.1
  gendered empathy in, 285–87, 286f
  and male heterosexual desire, 283–85
  performance in, 287–88
  repositioning human heterosexuality, 293–94
*Alien 3* (film), 270–71
*Alita: Battle Angel* (film), 16
Alpert, Robert, 292–93
Altman, Rick, 155
Amado, Jorge, 132
*Amazing Stories* (magazine), 165
American International Pictures, 117–18
*American Werewolf in London* (film), 305
Anderson, Chris, 227
Anderson, Michael, 295
Andrejevic, Mark, 230–31
*Andromeda Strain, The* (film), 38–39
*Aniara* (film), 62–63
*Annihilation* (film), 191, 241
Anthropocene, 17–18, 191, 266
  and biopolitics, 196
  and biopunk, 46–47
  and climate change, 55–56, 63–64, 237–38
  and cultural disorientation, 185, 189–90
  as deep ecology, 181–82, 183
  defined, 183–85, 192n.1
  and environmentalism, 181–82

## 326    INDEX

*Antiviral* (film), 39–40, 45–47
apocalypse, 53–54, 310–12
*Apocalypse Now* (film), 57–58
Arau, Alfonso, 132–33, 134
Arendt, Hannah, 194–95
Arnheim, Rudolf, 186–87
Arnold, Jack, 265, 276
*Around the World in 80 Days* (film), 148
*Arrival* (film), 17–18, 274f
   as adaptation, 108–9
   as clif-fi narrative, 63–67
   heterotopic space in, 101–2, 109–10
   language in, 274–75
   and posthumanism, 273, 275
   speciesism, 273–75
   temporality in, 66
Asma, Stephen T., 73
Aspin, Les, 96
*Astounding Science Fiction* (magazine), 165
Asturias, Ángel, 130–31
Atkinson, Michael, 122
*Atlantis: The Lost Empire* (film), 156
*Attach of the 50 Foot Woman* (film), 89–90
Attebery, Brian, 165
Atwood, Margaret, 209–10
avant-garde film, 39–40, 61, 192n.2, 242–43
*Avatar* (film), 14–15, 224–25, 227–28, 241
*Avengers, The* (film), 92
*Avengers: Age of Ultron* (film), 55–56, 168,
   173–74
*Avengers: Endgame* (film), 191
*Avengers: Infinity War* (film), 55–56, 191

*Back to the Future* (film), 230
Baker, Brian, 43–44
Barad, Karen, 238–41
*Barbarella* (film), 208–9
Bardini, Thierry, 44–45
Barr, Jason, 113, 116, 124
Barreto, Bruno, 132
*Barton Fink* (film), 206–7
*Batman and Robin* (film), 122
*Batman Begins* (film), 122
Baudrillard, Jean, 220–21, 250
Bay, Michael, 120
Beachler, Hannah, 35

*Beast from 20,000 Fathoms, The* (film), 115–16
*Beasts of the Southern Wild* (film), 67n.7, 133
*Being John Malkovich* (film), 206–7
Bellamy, Edward, 301
Bennett, Gertrude Barrows, 88
Bennett, Jane, 238–39, 240–41, 242, 246, 253
Benshoff, Harry M., 79
Berry, Halle, 92
Besson, Luc, 265–66, 297–98, 302–3, 304–5, 312
*Bicentennial Man* (film), 31
Binoche, Juliette, 125–26
bioethics, 16–17
   *See also* biopolitics
biopolitics, 193, 203–4
   and the alien, 196
   and the Anthropocene, 196
   and colonialism, 193–94
   economics as crucial element in, 195–96
   and strategies of governance, 199, 200–3
biopunk, 14
   and the anthropocene, 46–47
   and capitalism, 44–45, 49
   as cultural formation, 39–40, 50n.2
   and cyberpunk, 38, 149
   defined, 39–40, 49–50n.1
   genetic determinism, 39–42, 44
   genetic engineering, 39, 43
   human genome 39, 44–45
   posthumanism, 38–40, 43, 48–49
   and zombie narratives, 39–40, 46–47
*Birdman* (film), 131, 133
*Black Moon* (film), 74–75
*Black Panther* (film), 24–25, 27, 35f, 175
   as adventure narrative, 33–34
   costuming in, 34–36
   and colonialism, 32
   production design, 35
   screenplay, 33–34
   success of, 30, 32, 36
   as utopian narrative, 32, 34–35, 36
   *See also* Marvel Cinematic Universe
*Black Widow* (film), 175
*Blacula* (film), 76–78, 81, 85–86
*Blade Runner* (film), 11f, 16, 31
   and climate change, 55–56
   cult status, 208–9
   as cyberpunk film, 38

## INDEX

*Blade Runner 2049* (film), 83, 191
blaxploitation films, 30, 76–77, 78
Blaylock, James, 141, 142
*Blindness* (film), 53–54
Bloch, Ernst, 296, 297–98, 307, 311–12, 312n.5
Blomkamp, Neill, 60, 107
Bloom, Dan, 52, 67n.3
Bogost, Ian, 249
Bohr, Niels, 238–39
Bond, James, III, 79
Bong, Joon-ho, 119, 260–61
*Book of Eli, The* (film), 53–54, 209–10
Boorman, John, 295
Borden, Lizzie, 28
Bordwell, David, 132, 232
Borensztein, Sebastian, 134
*Born in Flames* (film), 25, 28
Boseman, Chadwick, 33
Bould, Mark, 6–7, 8
Bowers, Maggie Ann, 131
Braidotti, Rosi, 268–70, 271–72, 273, 274–75, 304–5
*Brazil* (film), 151, 206, 210–13, 212f, 295
Breton, André, 130–31
Brody, Richard, 81
Brontë, Charlotte, 302
Brooks, Kinitra D., 83
*Brother from Another Planet, The* (film), 25
*Brown Girl Begins* (film), 10, 24–25, 30, 82–83
*Brown Girl in the Ring* (novel), 30–31, 82–83
Bruegel, Pieter, 308–9
Bryant, Levi, 5–6, 17–18
Buckland, Warren, 299
Bukatman, Scott, 269–70
Bullock, Sandra, 58–59
*Bumblebee* (film), 125–26
Burgmann, J. R., 53
Burke, Liam, 124–25
Burroughs, Edgar Rice, 165
Burton, Tim, 151–52, 153, 155
   *Alice in Wonderland* (film), 151–52
   *Alice through the Looking Glass* (film), 151–52
Butler, Judith, 278–79
   and gendered identity, 280–81
Butler, Octavia, 247n.5
Byars, Jackie, 98–99
*Byt* (film), 252–53, 254

Caldwell, John Thornton, 223–24
Cameron, James, 14–15, 90, 133–34, 224–25, 227–28, 241
Campbell, John W., Jr., 103
   "Who Goes There?" 103–4
Cannes Film Festival, 216
Capra, Frank, 295
*Captain Marvel* (film), 94–95, 97, 98f
   feminist ideology in, 98, 99
   narrative construction, 95
   as origin story, 95–99
Carpenter, John, 103, 206–7, 208–9, 214
Carruth, Shane, 224–25, 242–45, 246
Casetti, Francesco, 223–24, 229
*Castle of Otranto, The* (novel), 73
Cather, Willa, 75
Cauchi, Mark, 307
Cavell, Stanley, 131
Chaplin, Charlie, 131
   *Modern Times* (film), 157
*Charly* (film), 270
Cheshire, Godfrey, 222–23
Chiang, Ted, 108–9, 273
*Children of Men* (film), 53–54
*Chinese Tale, A* (film), 131
*City of Lost Children, The*, 148, 150, 152, 155, 157
Cixous, Hélène, 237
Clarens, Carlos, 7
cli-fi, 14–15
   apocalyptic environmentalism, 53–54, 60–62
   climate destabilization, 17–18
   forms of climate change, 56
   as literary form, 52–53, 67n.3
   narrative options, 62–63
   and real science, 53
   special effects in, 54–55
   strategic realism in, 56–58
   as subgenre, 53, 67n.1
   techno-utopian character of, 58–60
Clinton, George, 27–28
*Clockwork Orange, A* (film), 208–9
*Close Encounters of the Third Kind* (film), 118
*Cloud Atlas* (film), 17–18
*Cloverfield* (film), 15, 120–21, 124, 125–26
*Cloverfield Paradox, The* (film), 55–56

328 INDEX

*Code 46* (film), 39–40, 42–43
  capitalist exploitation in, 44–45
  and Oedipus myth, 43
Coen Brothers, 206–7
Cohen, Jeffrey Jerome, 46–47
Cold War, 9–10, 110–11, 135–36, 137–38
colonialism, 32, 64, 103, 193–94
*Color Out of Space, The* (film), 67n.9
*Colossal* (film), 126
*Colossus of New York, The* (film), 270
Columbia Pictures, 115
comic books, 162–63, 164–65, 173
  genres in, 166
  influence of Stan Lee and Jack Kirby, 165–66
Comic-Con, 122–23
Coney, John, 28
Conrad, Joseph
  *Heart of Darkness* (novel), 35
Coogan, Peter, 163–64, 166–67
Coogler, Ryan, 33–34
Cooper, Melinda, 195–96
*Coraline* (film), 152–53
Corman, Roger, 118
Couldry, Nick, 221–22
COVID-19 15, 228, 256–57
  impact on film attendance, 230
Cox, Alex, 206
Crain, Caleb, 242–43
Cranston, Bryan, 125–26
*Crawl* (film), 67n.7
*Creature from the Black Lagoon* (film), 134–35,
  137, 138f, 276
*Creature Walks among Us, The* (film), 276
Creekmur, Corey, 162
Crichton, Michael, 118, 208–9
Cronenberg, Brandon, 45
Cronenberg, David, 208–9
crowdsourcing, 223–24, 231, 233
Crowther, Bosley, 116
Cruise, Tom, 252
Crutzen, Paul J., 183–85
Csicsery-Ronay, Istvan, Jr., 163, 167–68, 169,
  170, 174–76
  and the sublime, 171–72, 187
  the Technologiade, 172–73
*Cube* (film), 261–63
cult film, 18–19

as black comedy, 210, 212
cultural markers of, 207, 208–9
and dystopian science fiction, 206, 209–11
excess in, 207–8
as midnight movie, 207
as social critique, 214, 217, 218
as subgenre of science fiction, 206–7, 208, 209
*Curious Case of Benjamin Button, The* (film),
  224–25
cyberpunk. *See* biopunk
cyborg, 16, 149, 153–54, 162, 173–74, 269–70,
  304–6
  as posthuman species, 268, 269, 275
  racialized, 31
  as superhero, 164, 168, 170, 172

*Damnation Alley* (film), 209–10
Dante, Joe, 303–4
Darabont, Frank, 251–52
*Dark City* (film), 150
*Dark Knight, The* (film), 298–99
Dash, Julie
  *Daughters of the Dust* (novel), 30–31
*Da Sweet Blood of Jesus* (film), 81–82
Davis, Elizabeth, 84
Day, Charlie, 123–24
*Day after Tomorrow, The* (film), 54–56, 54f
  impact on climate awareness, 55
  narrative structure of, 55
  strategic realism in, 57
  as thriller, 62–63
*Day the Earth Caught Fire, The* (film), 55–56
*Day the Earth Stood Still, The* (film), 55–56,
  89–90, 106
  aliens in, 105, 108–10, 137–38
  contact zone in, 106
  and heterotopic space, 101–2
DC Universe, 15, 162–63
*Deadly Mantis, The* (film), 115
*Def by Temptation* (film), 79–81, 80f, 85–86
Defoe, Daniel, 172–73, 301
Delany, Samuel R., 15–16
Deleuze, Gilles, 238–39, 247n.9, 253–54, 297–98,
  311–12
del Toro, Guillermo, 15, 121, 133–36, 305
  importance of fantasy, 138–39

## INDEX

influence of *Creature from the Black Lagoon*
(film), 134–35, 137
and narrative construction, 134
*Deluge* (film), 271–72
*Demolition Man* (film), 25
Denis, Claire, 16–17, 196–97, 200, 204n.3,
204n.6
Derrida, Jacques, 66, 246n.2
Dery, Mark, 28
*Destination: Planet Negro* (film), 24–25, 27–28,
29–30
Devlin, Dean, 17–18, 118
dieselpunk. *See* steampunk
digital cinema, 220, 221–22
democratizing effects, 222, 224–25, 230–32,
233
digital projection systems, 222–24, 227–28
distribution of, 226–27, 228, 232
motion capture, 224–25
reediting of, 225–26, 230
and special effects, 224–25, 231–32
theory of, 18–19
Disney, 230
and adaptation, 147
and blockbuster films, 27–28
streaming service, 225, 226–27
theme parks, 147–48
*District 9* (film), 16–17, 53, 101–2, 108f
aliens in, 109–10
as heterotopia, 106–8
*Dona Flor and Her Two Husbands* (film), 132, 133
*Donnie Darko* (film), 18–19, 206–7
*Doomsday* (film), 57
*Downsizing* (film), 58
Doyle, Arthur Conan, 113–14
*Dr. Cyclops* (film), 19n.1
*Dracula* (film), 73
*Dracula* (novel), 73
*Dune* (novel), 173
Dussel, Enrique, 183
DuVerney, Ava, 27
Dyer, Richard, 296–97
dystopia. *See* utopia

*E. T. the Extraterrestrial* (film), 225–26
*Earth vs. the Flying Saucers* (film), 89–90

Ebert, Roger, 126n.1, 132–33
Eco, Umberto, 218n.2
ecosophy. *See* anthropocene
Edwards, Gareth, 122–23, 124–25
Eliot, T. S., 307–8
*Elysium* (film), 59–60, 60f
Emmerich, Roland, 54–55, 118, 121–22
*Empire of Corpses, The* (film), 156
environmentalism, 3, 56, 181
and apocalypse, 60–61
geostories, 241–42
*Eraserhead* (film), 206–7
Erb, Cynthia, 114
Esposito, Roberto, 193, 203–4
Esquivel, Laura, 132–33
ethnogothic, 14
and African diaspora, 71–72, 81
contemporary forms of, 81–85
defined, 71–72
folklore in, 83
Gothic antecedents, 73–75
as Gothic cinema, 72–73
as heterotopia, 71–72
and vampirism, 76–77, 78, 79, 81
*Ex Machina* (film), 16, 93–94, 276,
283–84
expressionism, 129–30, 151–52, 305

*Fabulous World of Jules Verne, The. See
Invention for Destruction* (film)
fandom, 13–14, 53, 115, 116–18, 222, 225, 228–29,
232
collective film experience, 228–29, 232,
233
conventions, 122–23, 141
cosplay, 141, 148
and crowdsourcing, 223–24, 231, 233
fan fiction, 141
fan films, 222, 223–24, 230
harmful behavior of, 230–31
recognition by media companies, 124–25
and social media, 220–21, 230–31
websites, 146, 232
*See also* cult film
Fassbender, Michael, 278, 289, 291–92
Fay, Jennifer, 53–54

## 330 INDEX

feminism, 98, 99, 304–5
  cyberfeminism, 19n.2, 304–5
  and the film industry, 92–93
  feminist materialism, 16–17, 236, 237, 238–41, 243–44
  fourth wave, 96–97, 98
  influence of modern physics, 238–40
  and masculinity, 237–38
  second wave, 89
  third wave, 91, 96–97, 98
femspec, 88, 90, 91f, 94–95, 96, 97–98
  defined, 89
  feminist future, 14, 88
  and feminist ideology, 93, 96–97, 98, 99
  *Femspec* (journal), 88
  gender construction, 88, 92
  and gender stereotypes, 89–90, 91–92
  and voyeurism, 89
  *See also* feminism
Ferrando, Francesca, 266–67, 268
film noir, 11f, 94, 129, 135–36, 208–9
*Finitude* (novel), 52
*Firefly* (tv series), 18–19
first contact narratives, 101–2, 103, 192n.1
Fisher, Elizabeth, 236–37, 238, 240–41
Fitting, Peter, 296–97
*Five Children and It* (novel), 52–53
*Flash Gordon* (film), 208–9
Fleck, Ryan, 94–95
Foucault, Michel, 71–72, 75–76, 280
  and biopolitics, 193, 194
  critique of the self, 280
  and heterotopias, 101–2, 103, 105
  on racism, 204n.1
  and subjugated knowledge, 266–67
*Frankenstein* (novel), 280, 292–93
Frantzen, Mikkel Krause, 307
Friedberg, Anne, 13, 15–16

*Ganja & Hess* (film), 76, 78–79, 81
Garland, Alex, 241
*Gattaca* (film), 39–40, 41f
  genetic determinism in, 40–42
gender identity, 3
  cultural production of, 16–17, 93
genetic manipulation, 3

genre, 3–6, 53, 141, 207–8
  and communities of practice, 9
  conventions of, 19, 162, 163
  as cultural formation, 39–40, 50n.2, 50n.4
  definitions of, 4–5, 129
  fractured forms of, 13
  hybrid forms, 146, 162–63, 167, 175, 207, 208–9, 242
  improvisation in, 11–12
  and Marxism, 16
  narrative patterns of, 10
  semantic and syntactic dimensions of, 155
  slippery nature, 8–9, 12–13
*Geostorm* (film), 17–18, 55–56
Gernsback, Hugo, 53, 165–66
  and naming science fiction, 7–8, 165
*Get Out* (film), 71, 72, 83–86, 85f
Geyrhalter, Nikolaus, 62
Ghosh, Amitav, 53
*Giant Claw, The* (film), 15, 115
Gilliam, Terry, 151, 153, 155, 206, 218n.4, 295
Gioia, Ted, 132
*Girl with All the Gifts, The* (film), 39–40, 46–49, 48f
Gitelman, Lisa, 223–24
Glass, Rodge, 53
Glassy Mark, 38–39
*Godzilla/Gojira* (film), 15, 38–39, 115–16
  American remakes of, 119, 121–26, 123f
  marketing of, 116
  mythic character of, 116
  re-editing of, 116
*Godzilla: Final Wars* (film), 119
*Godzilla: King of the Monsters* (film), 125–26
*Godzilla 2000: Millennium* (film), 119
*Godzilla vs. Kong* (film), 125–26
*Godzilla vs. Hedorah* (film), 125
*Golden Compass, The* (film), 156
Goodman, Martin, 166
Goss, Brian, 43
Grant, Rebecca, 96
*Gravity* (film), 58–59, 172
Gray, W. Russel, 212–13
*Guardians of the Galaxy* (film), 92
Guattari, Félix, 183, 238–39, 247n.9, 253–54, 311–12
Gunn, Bill, 78
Gunning, Tom, 147

Hagener, Malte, 302
Hairston, Andrea, 119–20
Halberstam, Jack, 73–74
Halperin, Victor, 73–74
*Hancock* (film), 25
Hansen, Mark, 46
*Happening, The* (film), 53–54
Haraway, Donna, 63–64, 238–40, 242–43, 247n.8
   and cyberfeminism, 16, 269–70, 304–5
   and feminist materialism, 240–42, 246, 305–6
   posthumanism, 268–69, 273, 304–5
   and technoscience, 237–38
*Hardware* (film), 55–56
Harman, Graham, 249–51
Harryhausen, Ray, 114–15
Hartlaub, Gustav, 130
Hassler-Forest, Dan, 233
Hawkins, Sally, 125–26
Heinlein, Robert
   and *Starship Troopers* (novel), 26
Helford, Elyce Rae, 89, 92
*Hellboy* (film), 133–34, 137
*Hello, Rain* (film), 24–25
Hemingway, Ernest, 75
Herbert, Frank, 173
heterotopia, 14–15, 71–72
   and Afrofuturism, 24
   and asymetrical relations, 102, 103
   as contested space, 101–2
   defined, 75–76, 110
   and hyperobjects, 110–11
   and the Other, 102, 105, 107–8, 109–10
*Hidden Fortress, The* (film), 260
Higgins, David M., 88
*Highlander II* (film), 55–56
*High Life* (film), 16–17, 196–97, 203f
   biopolitical framing of, 197, 200, 201–4
   ecology in, 197, 198f
   economic dimension of, 200–1
   and technology issues, 198–200
Holding, Sarah, 52–53
*Homo sapiens* (film), 61
Honda, Ishiro, 115–17, 122
Hopkinson, Nalo, 82–83
*Host, The* (film), 119

*House of Dracula* (film), 38–39
*Hugo* (film), 156, 157–58
Hugo Award, 108–9
*Hunt, The* (film), 228

*I Am Legend* (film), 25, 57, 271–72
*I Married a Monster from Outer Space* (film), 89–90
*I, Robot* (film), 24, 25
*I Walked with a Zombie* (film), 74–75
icons of science fiction, 3–4, 7, 10, 17–18, 27–28, 32, 36, 38, 64, 72, 89, 102, 196, 225, 236, 303–4
*Inception* (film), 297–300, 300f, 312
   as sandbox utopia, 301–2
   as sociological utopia, 300–1
   and video game rules, 298–99, 301–2
*Inconvenient Truth, An* (film), 181
*Incredible Shrinking Man, The* (film), 265–66, 272
independent cinema, 27–28, 29, 55, 115, 122, 126, 224–25, 227
Industrial Light and Magic, 118, 120
*Inhospitable World* (film), 53–54
*Innerspace* (film), 303–4
*Interstellar* (film), 58–59, 63
*Into Eternity* (film), 53–54
*Invisible Man, The* (film), 228
*Invasion of Astro-Monster* (film), 116–17
*Invention for Destruction* (film), 142–44, 143f, 147
   and extrapolative steampunk, 145–46
   and transdimensional kinemation, 151
Irigaray, Luce, 237
*Iron Man* film series, 173–74, 174f, 175–76
   fictive neologies in, 168–69
   fictive novum in, 169–70
   hybridity in, 167
   imaginary science in, 170
   shadow mage figure, 174
   and the sublime, 171–72, 171f
   as Technologiade, 172–73, 174–75
*Iron Sky* (film), 18–19, 222, 224–25, 232f
   and crowdsourcing, 230, 231, 233
   as media franchise, 231
   satire in, 233

*Iron Sky: The Coming Race* (film), 222, 231
*Island of Dr. Moreau, The* (film), 39
*Island of Lost Souls* (film), 38–39
*It Came from Beneath the Sea* (film), 89–90, 115
*It Came from Outer Space* (film), 89–90, 137–38
Ivakhiv, Adrian J., 307

Jackson, Peter, 119–20
James, Edward, 3–4
  ambiguities of science fiction, 4
  defining science fiction, 3–5, 11–12
  and the market place, 8–9, 10
Jameson, Fredric, 58, 66–67, 131, 300–1, 302, 310–11
*Jane Eyre* (novel), 302
*Janelle Monaé: Dirty Computer* (film), 24–25, 27–28, 31–32
  as music video, 31
Jarvis, Michael, 84
*Jaws* (film), 124
Jeffords, Susan, 90–91
Jeffrey, Scott, 170
Jenkins, Henry, 163, 166, 220–21, 229, 230–31
Jeter, K. W., 141
Johansson, Scarlett, 302–3
Johnson, Rian, 232
Jones, Gerald, 165–66
Jonze, Spike, 206–7
Joon-ho, Bong, 119
*Jungle Fever* (film), 82
*Jurassic Park* (film), 39, 118, 119, 306–7
*Jurassic World* (film), 18–19

kaiju, 15, 55–56, 57
  atomic imagery in, 115–16, 122–23
  blockbuster approach, 118–19, 123–24, 126
  campy elements of, 117–18, 122, 123–24
  conventions of, 116–17, 125
  franchising, 119, 120, 121, 125–26
  kabuki influence on, 124
  marketing of, 116
  and the MonsterVerse, 125–26
  origins, 113–15

reinvigoration of, 120–21
  as transnational type, 113
Kakoudaki, Despina, 279–80, 282, 290–91
Kaneko, Shusuke, 119
Kaplan, E. Ann, 53–54, 66
Keaton, Buster, 53–54, 131
Keller, Evelyn Fox, 237–38
Kelly, Richard, 206–7
Kern, Stephen, 186
Kernan, Lisa, 7–8
Kile, Meredith B., 94–95
*King Kong* (film), 113–15
  remake of, 119–20
*King Kong vs. Godzilla* (film), 116–18, 117*f*
Kirby, Jack, 33–34, 165–66
Koenitz, Hartmut, 301–2
Korda, Alexander, 295
Kubrick, Stanley, 23, 185–86, 197, 208–9, 236–37, 241, 246, 272–73, 303

Lacan, Jacques, 262, 280
Landis, John, 305
Landon, Brooks, 110, 175–76
Landy, Marcia, 272–73
Lang, Fritz, 27–28, 89, 270, 295, 303
Lanthimos, Yorgos, 206
*Last Angel of History, The* (film), 24–25
*Last Man on Earth, The* (film), 271–72
Latham, Rob, 3–4
Latour, Bruno, 241
Lavender, Isiah, III, 24
Lavery, David, 186–87
*Lawrence of Arabia* (film), 288–89
*League of Extraordinary Gentlemen, The* (film), 149–50, 155, 158
Lee, Spike, 81–82
Lee, Stan, 33–34
*Leech Woman, The* (film), 89–90
Legendary Pictures, 121–22, 124–26
Le Guin, Ursula K., 110–11, 111n.2, 236–37, 240–41, 246
Levine, Joseph E., 116
Lewis, C. S., 52–53
Lewis, Sharon, 82–83
*Like Water for Chocolate* (film), 131, 132–33, 134
Lim, Bliss Cua, 144

*Limitless* (film), 270
Lindee, Susan, 40
*Lion, the Witch and the Wardrobe, The* (novel), 52–53
*Lobster, The* (film), 10, 206, 210, 215–17
*Logan's Run* (film), 295
*Looking Backward* (novel), 301
*Looper* (film), 232
*Lost Horizon* (novel), 295
*Lost River* (film), 67n.7
*Lost World, The* (film), 113–14
*Lost World: Jurassic Park, The* (film), 118, 306–7
Lotz, Amanda, 226–27
Lovecraft, H. P., 63–64, 67n.9
Lucas, George, 7–8, 27–28, 225
*Lucy* (film), 17–18, 265–66, 267f, 302–5, 306f
   and body horror, 305
   posthumanism in, 305–7, 312
   sound design in, 305–6
Lugosi, Bela, 73–74
Lumiere brothers, 131, 157–58
Lynch, David, 206–7

MacDonald, Hamish, 52
*Machine, The* (film), 16
*Mad Max: Fury Road* (film), 16–17, 61, 210
*Mad Max 2* (film), 55–56
magical realism, 15
   and cli-fi, 52–53
   defined, 130–31, 133
   as genre, 130–31, 136–37
   origins of, 129–31, 132
   and science fiction, 137–39
   and surrealism, 130–31
   *See also* del Toro, Guillermo
*Manufactured Landscapes* (film), 53–54
Marcuse, Herbert, 296, 297–98, 299, 310–11
Marks, Laura, 305
Márquez, Gabriel García, 132, 133–34
*Martian, The* (film), 58–59
Marvel Cinematic Universe, 12, 15, 32–33, 162–63, 164, 167, 168
   femspec characters in, 92, 94–95, 99
   and the fictive novum, 169–70
   and Marvel comics, 166–67
Maslin, Janet, 132–33

*Master of the World, The* (film), 148
Mastow, Jonathan, 254
*Matrix, The* (film), 14–15, 18–19, 25, 55–56, 220–22, 221f, 230, 233, 305–6
*Matrix Reloaded, The* (film), 220–21
Mauricette, Carolyn, 83
Mbembe, Achille, 193, 195
McCann, Hannah, 278
McCarthy, Colm, 46–47
McGarvey, Seamus, 122–23
McGowan, Todd, 310–11
McHale, Brian, 38
McLuhan, Marshall, 274–75
Meillassoux, Quentin, 249–50
*Melancholia* (film), 297, 307–9, 308f
   and art history, 308–9, 310–11
   music in, 308–10
   subjectivity in, 307–9, 310, 312
Méliès, Georges, 89, 131, 142–43, 144, 156, 157–58
   *Hallucinations of Baron Munchausen, The* (film), 142
   and the trick film, 142–43, 145, 147–48, 151, 224
   *Trip to the Moon* (film), 89, 147, 224
Melville, Herman, 75
*Memento* (film), 298–99
*Men in Black* (film), 25
*Men in Black: International* (film), 25
Mendelson, Scott, 125–26
Menzies, William Cameron, 295
*Metropolis* (film), 17–18, 27–28, 31, 94, 270, 283–84, 295, 303
*Midnight Special* (film), 10
Miéville, China, 63–64, 67n.10
*Mighty Joe Young* (film), 114–15
Miller, George, 210
Milner, Andrew, 53
*Mimic* (film), 133–34, 137
*Minority Report* (film), 23, 175–76, 252
*Mist, The* (film), 251–52
Mittel, Jason, 6, 8–9
Miyazaki, Hayao, 151, 153–54, 155
   *Castle in the Sky* (film), 153, 154, 156
   compared to Tim Burton, 153
   *Howl's Moving Castle* (film), 154, 156
   *Nadia: The Secret of Blue Water* (film), 153
   *Nausicaa of the Valley of the Wind* (film), 153
   use of depth, 153–54

334 INDEX

*Mo' Better Blues* (film), 82
Monaghan, Whitney, 278
*Monsters* (film), 57–58, 122
*Monty Python and the Holy Grail* (film), 206–7
*Monty Python's Flying Circus* (television series), 211–12
Mooney, Darren, 299
Moore, Jason W., 185
Morrison, Toni, 75
*Mortal Engines* (film), 14–15
Morton, Timothy, 65–66, 110–11
Mosco, Vincent, 221–22
Mulvey, Laura, 89–90
Murnau, F. W., 305
Murphy, Richard, 130

Næss, Arne, 181–82
Nama, Adilifu, 25
Natali, Vincenzo, 261
Neale, Steven, 5–6
  naming science fiction, 7–8
Nelkin, Dorothy, 40
Nesbit, E., 52–53
New Wave science fiction, 145
New World Pictures, 118
*Night of the Living Dead* (film), 73–74
*9* (film), 152, 156
*1984* (novel), 209–10
Nolan, Christopher, 297–99
*Nosferatu* (film), 73, 305
*Nuoc 2030* (film), 62–63

object-oriented ontology, 249–50, 263
  and gravity, 255–57
  Kantian influence on, 249–51
  mereology branch of, 260–61
  and phenomenology, 250–51
  and pluralism, 250
  and realism, 250–52
*Oblivion* (film), 58
O'Brien, Willis, 113–15
*Omega Man, The* (film), 38–39, 271–72
*On the Beach* (film and novel), 55–56, 67n.4
*One Hundred Years of Solitude* (novel), 132
*One Million B.C.* (film), 114–15

*Orphanage, The* (film), 73
Orr, Christopher, 303
Ortega y Gasset, José, 130
Orwell, George, 53, 209–11, 298–99
*Others, The* (film), 73
Otomo, Katsuhiro, 14–15, 149
*Ouanga* (film), 74–75
"Ozymandias" (poem), 292

*Pacific Rim* (film), 15, 57, 121, 123–24, 133–34, 137
Palmer, Lorrie, 304–5
*Pan's Labyrinth* (film), 131, 133–34, 305
*Paradise Lost* (poem), 293
*Parasite* (film), 67n.7
Parker, Helen, 38–39
*Park Film* (film), 192n.2
Pearce, Philippa, 52–53
Peele, Jordan, 71, 83–86
Penrose Process, 196–97, 202, 204n.3
Pierson, Michele, 9–10
Pignarre, Philippe, 156
*Plan 9 from Outer Space* (film), 18–19, 206–7
*Planet of the Apes* (film), 271–72
*Plastic Bag* (film), 191
Poe, Edgar Allan, 75, 165
Poll, Ryan, 84–85
posthumanism, 16–17, 46, 94, 297–98, 304–6
  and biopolitics, 194–95, 196
  and biopunk, 38–40, 49
  as challenge to the human, 169–70, 269–70, 274–75
  and desire, 45
  in film history, 270–72
  and human extinction, 48–49, 271–72
  as hybrid species, 46, 49–50n.1, 270–71, 273–74
  political dimensions of, 266–67
  and postmodernism, 265–66
  and subjectivity, 45–47, 269, 305–8
  and technology, 43–44
  as transcendence, 197, 272, 275
postmodernism, 91, 93, 131, 265–66
Powers, Tim, 141
Pratt, Mary Louise, 101–2, 103, 110
*Primer* (film), 224–25

## INDEX

*Prometheus* (film), 278
  androids in, 288, 289, 290–91
  CGI spectacle in, 290
  denaturalized gender roles, 289
  performance in, 288–90, 288*f*, 291
Proyas, Alex, 24, 25, 150
Puar, Jsabir, 193, 194, 195
pulp magazines, 162–63, 165–66
  characters in, 166–67
*Pumzi* (film), 191

queer theory, 16–17, 278
  deconstructing gender roles, 293–94
  depictions of deviance, 291–92
  and the heteronormative community, 293

Ra, Sun, 27–30, 32
Rajan, Kaushik Sunder, 195–96
Rampton, Martha, 96–97
*Ready Player One* (film), 233
*Rebecca* (film), 73
*Reefer Madness* (film), 206–7
Reeves, Matt, 120–21
*Reign of Fire* (film), 55–56
Rennie, Michael, 106
*Repo Man* (film), 18–19, 206, 208–9, 210,
  213–16, 213*f*, 218, 218n.6
Republic Pictures, 29–30
*Resident Evil* (film), 92
*Resident Evil: The Final Chapter*
  (film), 55–56
*Return of Godzilla, The* (film), 118
  reissue as *Godzilla 1985* (film), 118
*Revenge of the Zombies* (film), 74–75
Reynolds, Richard, 164–65
Rieder, John, 8–9, 13–14
  and genre communities, 9, 16
Riley, Boots, 206
Rios, Albert, 133
RKO Pictures, 113–15
*Road, The* (film), 53–54, 55–56
Robinson, Kim Stanley, 183
*Robinson Crusoe* (novel), 172–73, 301
*Robot Monster* (film), 208–9
*Robots of Brixton* (film), 25

Robu, Cornell, 187
*Rocky Horror Picture Show, The* (film), 207
Roh, Franz, 129–30
Rombes, Nicholas, 225
*Room, The* (film), 206–7
*Rosemary's Baby* (film), 73
Rosen, Phillip, 224–25
Rushdie, Salmon, 133
Ruyer, Raymond, 155

Sargent, Pamela, 88
Schwartzenegger, Arnold, 257–58
science fiction journals, 13–14
Science Fiction Research Association, 12–13
Scorsese, Martin, 157–58
Scott, Ridley, 38, 208–9
Scott-Travis, Shane, 207
Scranton, Roy, 66
*Serenity* (film), 18–19
Sessions, George, 182
*Seventh Seal, The* (film), 28–29
*Shape of Water, The* (film), 15, 131, 133, 134–35,
  136*f*
  Academy Award, 135–36
  critical reaction to, 135–36
  links to *Creature from the Black Lagoon*
    (film), 134–35, 137
  science fictional character of, 137–38
  tradition of magical realism, 136–39
*Sharknado* (film), 67n.6
Shaviro, Steven, 44–45
Shelley, Mary, 88, 165
Shelley, Percy Bysshe, 292
Shelton, Robert, 295
*Sherlock Holmes* (film), 158
*Sherlock Holmes: Game of Shadows* (film), 158
*Shin Godzilla* (film), 126
*Shining, The* (film), 230
Shute, Neville, 67n.4
Shyamalan, M. Night, 26, 124
Siebers, Johan, 297, 309
Siegel, Jerry, 164–66
*Signs* (film), 124
*Silent Running* (film), 197
*Sky Captain and the World of Tomorrow* (film),
  156

336 INDEX

Smith, Will, 25
*Snowpiercer* (film), 14–15, 18*f*, 55–56, 61, 191, 260–61
  *See also* cli-fi
Sobchack, Vivian, 3–4
*Solo: A Star Wars Story* (film), 225
Solomon, Brian, 116–17
*Son of Frankenstein* (film), 38–39
*Son of Ingagi* (film), 81
*Son of Kong* (film), 114–15
Sontag, Susan, 3, 9–10
*Sorry to Bother You* (film), 206, 210, 217–18
*Soylent Green* (film), 55–56
*Space Children, The* (film), 137–38
*Space is the Place* (film), 25, 28–29, 32
space opera, 52, 172–73, 191
special effects, 12, 14–15, 18–19, 54–55, 89, 150–51, 183, 230
  and budgets, 113–15, 122–23, 158, 222, 224–25, 231–32
  computer generated, 118, 119–20, 220, 222, 224–25, 290
  in *kaiju* films, 115–17, 124
  stop-motion, 113–15
  in superhero films, 172, 175
  and time lapse photography, 185–87, 191
  in "trick" films, 142–43, 145, 224
  *See also* digital cinema
*Species* (film), 39
speculative realism. *See* object-oriented ontology
Spiegel, Simon, 295–96
Spielberg, Stephen, 23, 118, 120, 133–34, 175–76, 225, 252, 272, 306–7
*Split Second* (film), 55–56
*Stalker* (film), 67n.9
Stanfill, Mel, 124–25
*Star Trek* (television series), 52
*Star Wars* (film), 52, 191, 230, 231, 260
*Star Wars Episode 1: The Phantom Menace* (film), 222–23
*Star Wars: A New Hope* (film), 225
*Star Wars: The Rise of Skywalker* (film), 224–25, 230
*Steamboat Bill, Jr.* (film), 53–54
*Steamboy* (film), 14–15, 149–50, 150*f*

steampunk, 14–15, 211–12
  and the action film, 158
  adventure of energy in, 154–60, 159*f*
  defined, 141
  and dieselpunk, 156
  internal polarization of, 146
  kino-aesthetic mode of, 143–44, 145–46
  origins, 141–42
  as science fiction, 141–42
  steam worlds of, 146–51, 150*f*
  subgenres of, 156
  thematic component of, 155–56, 158–59
  transdimensional kinemation in, 151–54, 155
  Victorian context, 141, 142, 148, 151–52, 156–57, 158
  world mode of, 143–44, 145–46
Stengers, Isabelle, 156, 241
*Stepford Wives, The* (film), 84
Stewart, Garrett, 18–19
*Still Life* (film), 53–54
Stone, James, 120–21
Strathairn, David, 125–26
Striphas, Ted, 226–27
*Summer of Sam* (film), 82
Sundance Film Festival, 224–25, 242
superhero narrative, 15
  as cultural dialogue, 163
  as hybrid form, 162–63, 165, 175
  influence of Jerry Siegel, 164–65
  special effects technologies in, 175
*Supernova* (film), 25
*Surrogates* (film), 254, 255*f*
Suvin, Darko, 4–5, 8, 145, 167, 169
  and cognitive estrangement, 3, 133–34, 208, 251–52, 254
  on extrapolative science fiction, 144–45
  and steampunk cinema, 145
Švankmajer, Jan, 252–53, 254
Szeman, Imre, 58, 67n.8

*Taiji* film series, 158–60, 159*f*
*Take Shelter* (film), 53–54
Tanaka, Tomoyuki, 115–16
Tarkovsky, Andrei, 299
Telotte, J. P., 130, 208, 224

*Terminator, The* (film), 90–91
*Terminator 2: Judgment Day* (film), 90–92
Terranova, Fabrizio, 247n.6, 247n.8
*Terror of Mechagodzilla* (film), 116–17
Terry, Clifford, 132–33
*Them!* (film), 38–39, 89–90
*They Live* (film), 208–9, 214
*Thing from Another World, The* (film), 104*f*, 105, 135, 208–9
  advertising for, 7
  alien in, 109–10
  and heterotopic space, 101–2, 104–5, 109
  violence in, 106
*Thing, The* (film), 103
*Things to Come* (film), 5*f*, 17–18, 295
  *See also* Wells, H. G.
Thompson, Kristin, 132
*THX 1138* (film), 27–28
*Time after Time* (film), 149
*Time Machine, The* (film), 150, 187–91, 188*f*, 190*f*, 271–72
*Tobor the Great* (film), 270
Toho Studios, 115–16, 117–18, 119, 121–22, 126
Tolkien, J. R. R., 312n.4
Tolstoy, Leo, 301
*Total Recall* (film), 257–61, 259*f*
*Transformers* (film), 120–21, 123–24
Travers, Peter, 120
Trexler, Adam, 53
*Tristan und Isolde* (opera), 308–10
Tristar Pictures, 118
*Trolls World Tour* (film), 228
Tsuburaya, Eiji, 115–17
Tudor, Andrew, 4–6
Tull, Thomas, 121–23, 124–25
Turing test, 93
Turner, Terry, 116
Turturro, John, 123–24
*20 Million Miles to Earth* (film), 89–90
*20,000 Leagues under the Sea* (film), 148
  *See also* Verne, Jules
*2001: A Space Odyssey* (film), 23, 185–86, 197, 199–200, 204n.5, 224–25, 236–37, 241, 246, 303
  posthumanism in, 272–73
  *See also* Kubrick, S.

*Ultraman* (television series), 116–17
*Ultraviolet* (film), 92
Universal Pictures, 115, 218n.4, 228
*Upstream Color* (film), 16–17, 242–43, 245–46
  braiding of time in, 244–45
  cinemicroscopy, 243–44, 244*f*
  as feminist materialist science fiction, 242
  formal composition in, 243–44, 245*f*
utopia, 101–2, 233, 245–46, 262–63
  classic concepts of, 296–97, 301
  and digital cinema, 233
  dystopia, 58, 101–2, 209–11, 212, 215–16, 218
  as fantasy, 76
  as fictional genre, 296, 312n.3
  sandbox mode of, 301–3
  sociological mode of, 300–1
  and subjectivity, 307–9, 310, 312
  and Thomas More, 295
  untimely quality of, 307–8, 310
utopianism, 3, 17–18, 303
  as analytical method, 296, 300, 303
  entertainment effect, 296–97
  as sandbox, 299, 300, 302

*V for Vendetta* (novel), 209–10
VanderMeer, Jeff, 63–64
Verhoeven, Paul, 257
Verne, Jules, 141, 142, 144, 145, 165
  adaptations of, 147–48
  *From the Earth to the Moon* (novel), 144–45
  *20,000 Leagues under the Sea* (novel), 147–48, 153
video games, 4
*Vidocq* (film), 150
Vigalondo, Nacho, 126
Villeneuve, Denis, 108–9, 273
Vint, Sherryl, 6–7, 8, 208, 209
von Trier, Lars, 206–7, 297–98, 307–8, 309
Vuorensola, Timo, 222

Wachowskis, 220, 305–6
Wagner, Richard, 290–91, 307–10, 311–12
*Walking Dead, The* (film), 73–74
*WALL-E* (film), 58, 191

# 338 INDEX

Walpole, Horace, 73
*War of the Worlds, The* (film), 89–90, 124
Warner Bros. Pictures, 115, 125–26, 298–99
Warner, Marina, 74–75
Warren, Bill, 103
*Wasp Woman, The* (film), 89–90
Watanabe, Ken, 121–22, 125–26
Weaver, Sigourney, 90, 91*f*, 283–84, 286*f*
*Welcome to the Terrordome* (film), 25
Wells, H. G., 141, 142, 145, 149, 295
    adaptations of, 190–91, 228
    influence of, 165
    *Invisible Man, The* (novel), 144–45, 147
    *Island of Doctor Moreau, The* (novel), 38–39
    *Things to Come* (film), 5*f*, 17–18, 295
    *Time Machine, The* (novel), 190–91
Wells, Simon, 187–88
Wester, Maisha, 72–73
*Westworld* (film), 208–9
*When Worlds Collide* (film), 271–72
Whissel, Kristen, 233
*White Zombie* (film), 73–75
*Wild Blue Yonder, The* (film), 58
*Wild Wild West* (film), 149–50, 155, 158
Williams, Robin, 153
Williamson, Jack, 165–66
Wilmington, Michael, 132–33
Winfrey, Oprah, 27
Winterbottom, Michael, 42–43

Wiseau, Tommy, 206–7
Wolfe, Gary K., 110
Womack, Ytasha, 72–73
Wood, Aylish, 281–83
Wood, Ed, Jr., 206–7
*World, the Flesh and the Devil, The* (film), 25
*World War Z* (film), 57
*Wrinkle in Time, A* (film), 24
    novel version, 27
Wynter, Sylvia, 193, 194–95

*X-Men* (film), 92

Yaszek, Lisa, 14
Young, Helen, 24

*Zardoz* (film), 295
Zeman, Karel, 142, 144–46, 147–48
    and the adventure of energy, 146
    dimensionality in the films of, 142–43, 144, 151–52
    Méliès's influence on, 142–44, 145
    and the past, 144, 148, 151, 154
    Verne's influence on, 142–43, 144, 145, 153
Zhangke, Jia, 53–54
Žižek, Slavoj, 183